Management of Shared Fish Stocks

Edited by

A.I.L. Payne, C.M. O'Brien and S.I. Rogers
Centre for Environment, Fisheries and Aquaculture Science (CEFAS)
Lowestoft Laboratory, Pakefield Road, Lowestoft, Suffolk

Blackwell Publishing

CEFAS
The Centre for Environment, Fisheries & Aquaculture Science

Blackwell Publishing Ltd,
Editorial Offices:
Blackwell Publishing Ltd, 9600 Garsington Road, Oxford OX4 2DQ, UK
Tel: +44 (0)1865 776868
Iowa State Press, a Blackwell Publishing Company, 2121 State Avenue, Ames, Iowa 50014-8300, USA
Tel: +1 515 292 0140
Blackwell Publishing Asia, 550 Swanston Street, Carlton, Victoria 3053, Australia
Tel: +61 (0)3 8359 1011

First published 2004 by Blackwell Publishing Ltd

Library of Congress Cataloging-in-Publication Data
Management of shared fish stocks / edited by A.I.L. Payne, C.M. O'Brien, and S.I. Rogers.
p.cm.
Includes bibliographical references (p.).
ISBN 1-4051-0617-4 (alk. Paper)
1. Fish populations—Congresses. 2. Fishery management—Congresses. I. Payne, A.I.L. (Andrew I. L.), 1946- II. O'Brien, C. M. (Carl M.) III. Rogers, S. I. (Stuart I.)

QL618.3.M365 2004
333.95'6—dc22

2003063025

ISBN 978-1-4051-0617-7

A catalogue record for this title is available from the British Library

For further information on Blackwell Publishing, visit our website:
www.blackwellpublishing.com

Contents

Foreword

During 2002, the Centre for Environment, Fisheries and Aquaculture Science (CEFAS) celebrated its centenary of fisheries research at Lowestoft. As one of several events celebrating that centenary, CEFAS hosted a forward-looking international symposium entitled "International Approaches to Management of Shared Stocks – Problems and Future Directions" in July 2002. Personal invitations resulted in the attendance of some 80 scientists, policy-makers and managers from more than 20 countries, covering many of the world's main fishing areas and a variety of resources. Presentations were both verbal and poster, and four keynote speakers (Doug Butterworth, South Africa; Kevern Cochrane, FAO, Italy; Geoff Kirkwood, UK; and John Pope, UK) led presentations and discussion on four interwoven themes:

- The consequences and management of unregulated/unreported catches
- Competition
- External drivers and resource behaviour
- Ecosystems and migration.

The 20 papers that form the bulk of this volume are the peer-reviewed result of some of the presentations, including the four keynotes, and the order of publication is the same as the four themes listed above; the interwoven nature of the themes is clear from the content of the papers. The two discussion papers that follow the 20 scientific papers were not peer-reviewed, but the content was collated by rapporteurs and the co-chairs (the keynote speakers) from the discussions.

CEFAS management, the various sponsors, the editors, the event organisers, the CEFAS Publications and Graphics Team, the indexer, the authors, the reviewers, the rapporteurs, the participants, the four keynote speakers and Blackwell are all acknowledged for their valued support for and input into what we hope will become a useful part of the reference literature on this crucial fisheries management topic. We hope that you, the reader, will find as much of value from this volume as we did in bringing it to publication.

Andrew I.L. Payne (Symposium Chair)
Sarah Rollo (Symposium Organiser)
July 2003

Back row: John Hoolihan, Michael Armstrong, Nick Bailey, Chris Darby, Hamed Al-Oufi, Moses Maurihungirire, Colin Bannister, John Pope. Doug Butterworth, Stephen Wentworth, Tore Jacobsen, Victor Restrepo, Jordi Lleonart.

3rd row: Evald Ojaveer, Graham Pilling, Georgs Kornilovs, Carl O'Brien, John Reynolds, Mahfuzul Haque, Chris Morrice, Al-Sayyid Ibrahim Al-Busaidi, Geoff Tingley, Robert Aps, Brian Robinson, Tomasz Linkowski, Mike Pawson, Ted Potter, John Casey, Julian Metcalfe.

2nd row: Indroyono Soesilo, Clive Fox, Kevern Cochrane, Douglas Beveridge, Ana Baloi, Peter Greig-Smith, Joe Horwood, Sidney Holt, Stratis Gavaris, Tony Smith, Geoffrey Kirkwood, Tony Beeching, Sagir Ahmed.

Front row: Charlotte Mogensen, Akaki Komakhidze, Sarah Rollo, Evgeny Romanov, Laurence Kell, Andrew Payne, Matthew Dunn, John Barton, Eugene Sabourenkov, David Agnew, Ewen Bell, Mary Brown, Georgi Daskalov.

Photograph: © Victoria Harrison

List of participants

Agnew, Dr David
RRAG
Imperial College
Royal School of Mines
Prince Consort Road
London SW7 2BP UK
Tel: +44 207 5949273
E-mail: d.agnew@ic.ac.uk

Ahmed, Dr Sagir
Department of Zoology
University of Dhaka
Dhaka - 1000
Bangladesh
Tel: +880 2 9555233
Fax: +880 2 8615583
E-mail: sagir@du.bangla.net

Al-Busaidi, Mr Al-Sayyid Ibrahim
Ministry of Agriculture & Fisheries - Oman
P.O.Box 852
121 Seeb - Oman
Tel: +968 9323352
Fax: +968 605634
E-mail: ibrooo@hotmail.com

Al-Oufi, Dr Hamed
Assistant Professor & Assistant Dean for PGS&R
College of Agriculture & Marine Sciences
Sultan Qaboos University
P.O. Box 34 Al-Khod 123
Sultanate of Oman
Tel: +968 515235 / 515257
Fax: + 968 513418
E-mail: hamdoufi@squ.edu.om

Aps, Dr Robert
Estonian Marine Institute
University of Tartu
18 B Viljandi Road
11216 Tallinn
Estonia
Tel: +372 6281574
Fax: +372 6281563
E-mail: robert.aps@ness.sea.ee

Armstrong, Dr Michael
Department of Agriculture & Rural Development
Agricultural & Environmental Sciences Division
Newforge Lane
Belfast BT9 5PX UK
Tel: +44 28 90 255507
Fax: +44 28 90 255004
E-mail: mike.armstrong@dardni.gov.uk

Bailey, Mr Nick
Fisheries Research Services
P.O. Box 101
Victoria Road
Aberdeen AB9 8DB UK
Tel: +44 1224 295398
Fax: +44 1224 295511
E-mail: baileyn@marlab.ac.uk

Baldry, Ms Suzy
CEFAS
Lowestoft Laboratory
Pakefield Road
Lowestoft
Suffolk NR33 0HT UK
Tel: +44 1502 524446
Fax: +44 1502 524546
E-mail: s.a.baldry@cefas.co.uk

Baloi, Ms Ana Paula
Fisheries Research Institute - Moçambique
Av. Mao Tse Tung
P.O. Box 4603
389 - Maputo - Moçambique
Tel: +2581 490406
Fax: +2581 492112
E-mail: apbaloi@moziip.org

Bannister, Dr Colin
CEFAS
Lowestoft Laboratory
Pakefield Road
Lowestoft
Suffolk NR33 0HT UK
Tel: +44 1502 524360
Fax: +44 1502 524511
E-mail: r.c.a.bannister@cefas.co.uk

Barton, Mr John
Falkland Islands Government
Fisheries Department
P.O. Box 598
Stanley
Falkland Islands
Tel: +(500) 27260
Fax: +(500) 27265
E-mail: jbarton@fisheries.gov.fk / director@fisheries.gov.fk

Beeching, Mr Tony
Department of Marine & Wildlife Resources, American Samoa
c/o 174 Grove Road
Rayleigh
Essex SS6 8UA UK
Tel: +684 633 4456
Fax: +684 633 5944
E-mail: beeching@blueskynet.as

Bell, Dr Ewen
CEFAS
Lowestoft Laboratory
Pakefield Road
Lowestoft
Suffolk NR33 0HT UK
Tel: +44 1502 524238
Fax: +44 1502 524511
E-mail: e.d.bell@cefas.co.uk

Beveridge, Mr Douglas
National Federation of Fishermen's Organisations
Marsden Road
Fish Docks
Grimsby UK
Tel: +44 (0) 1472 352141
Fax: +44 (0) 1472 242486
E-mail: dbeveridge@nffo.org.uk

Brown, Mrs Mary
CEFAS
Lowestoft Laboratory
Pakefield Road
Lowestoft
Suffolk NR33 0HT UK
Tel: +44 1502 524227
Fax: +44 1502 524511
E-mail: m.j.brown@cefas.co.uk

Butterworth, Professor Doug
Department of Mathematics & Applied Mathematics
University of Cape Town
Rondebosch 7701
South Africa
Tel: +27 21 650 2343
Fax: +27 21 650 2334
E-mail: dll@maths.uct.ac.za

Casey, Dr John
CEFAS
Lowestoft Laboratory
Pakefield Road
Lowestoft
Suffolk NR33 0HT UK
Tel: +44 1502 524251
Fax: +44 1502 524511
E-mail: j.casey@cefas.co.uk

Cochrane, Dr Kevern
Senior Fishery Resources Officer
Fisheries Department
Food & Agriculture Organisation of the United Nations (FAO)
Viale delle Terme di Caracalla
00100 Rome
Italy
Tel: +39 06 570 56109
Fax: + 39 06 570 53020
E-mail: kevern.cochrane@fao.org

Darby, Dr Chris
CEFAS
Lowestoft Laboratory
Pakefield Road
Lowestoft
Suffolk NR33 0HT UK
Tel: +44 1502 524329
Fax: +44 1502 524511
E-mail: c.d.darby@cefas.co.uk

Daskalov, Dr Georgi M.
CEFAS
Lowestoft Laboratory
Pakefield Road
Lowestoft
Suffolk NR33 0HT UK
Tel: +44 1502 524584
Fax: +44 1502 524511
E-mail: g.m.daskalov@cefas.co.uk

Dunn, Dr Matthew
CEFAS
Lowestoft Laboratory
Pakefield Road
Lowestoft
Suffolk NR33 0HT UK
Tel: +44 1502 524363
Fax: +44 1502 524511
E-mail: m.r.dunn@cefas.co.uk

Fox, Dr Clive
CEFAS
Lowestoft Laboratory
Pakefield Road
Lowestoft
Suffolk NR33 0HT UK
Tel: +44 1502 524258
Fax: +44 1502 524511
E-mail: c.j.fox@cefas.co.uk

Gavaris, Dr Stratis
Fisheries & Oceans, Canada
531 Brandy Cove Road
St Andrews NB E5B 2L9
Canada
Tel: +506 529 5912
Fax: +506 529 5862
E-mail: GavarisS@mar.dfo-mpo.gc.ca

Gillatt, Mrs Deborah
CEFAS
Lowestoft Laboratory
Pakefield Road
Lowestoft
Suffolk NR33 0HT UK
Tel: +44 1502 524448
Fax: +44 1502 524569
E-mail: d.a.gillatt@cefas.co.uk

Greig-Smith, Dr Peter
CEFAS
Lowestoft Laboratory
Pakefield Road
Lowestoft
Suffolk NR33 0HT UK
Tel: +44 1502 524217
Fax: +44 1502 524515
E-mail: p.w.greig-smith@cefas.co.uk

Haque, Dr Mahfuzul
Department of Fisheries Management
Bangladesh Agricultural University
Mymensingh
Bangladesh
Tel: +880 91 55485
Fax: +880 91 55810
E-mail: mahfuz@royalten.net / salehamk@bdcom.com

Holt, Dr Sidney
4 Upper House Farm
Crickhowell NP8 1BZ UK
Tel: +44 1873 812388
Fax: +44 1873 812389
E-mail: sidneyholt@cs.com

Hoolihan, Mr John
Environmental Research & Wildlife Development Agency
P.O. Box 45553
Abu Dhabi
United Arab Emirates
Tel: +971 2 6817171
Fax: +971 2 6810008
E-mail: jhoolihan@erwda.gov.ae

Horwood, Dr Joe
CEFAS
Lowestoft Laboratory
Pakefield Road
Lowestoft
Suffolk NR33 0HT UK
Tel: +44 1502 524248
Fax: +44 1502 524515
E-mail: j.w.horwood@cefas.co.uk

Jakobsen, Mr Tore
Institute of Marine Research
P.O. Box 1870 Nordnes
N-5817 Bergen
Norway
Tel: +47 55 23 86 36
Fax: +47 55 23 86 87
E-mail: torej@imr.no

Jennings, Dr Simon
CEFAS
Lowestoft Laboratory
Pakefield Road
Lowestoft
Suffolk NR33 0HT UK
Tel: +44 1502 524363
Fax: +44 1502 513865
E-mail: s.jennings@cefas.co.uk

Kell, Dr Laurence
CEFAS
Lowestoft Laboratory
Pakefield Road
Lowestoft
Suffolk NR33 0HT UK
Tel: +44 1502 524257
Fax: +44 1502 524511
E-mail: l.kell@cefas.co.uk

Kirkwood, Dr Geoffrey
RRAG
Imperial College
Royal School of Mines
Prince Consort Road
London SW7 2BP UK
Tel: +44 207 5949272
Fax: +44 207 5895319
E-mail: g.kirkwood@ic.ac.uk

Komakhidze, Dr Akaki
Georgian Marine Ecology & Fisheries Research Institute
51 Rustaveli Str
Batumi 6010
Georgia
Tel: +995 222 74640
Fax: +995 222 74642 (43)
E-mail: mefri@batumi.net

Kornilovs, Dr Georgs
Latvian Fisheries Research Institute
Daugavgrivas Str
8 Riga
LV-1008
Latvia
Tel: +370 7613775
Fax: +370 7616946
E-mail: georgs_k@latfri.lv

Linkowski, Dr Tomasz
Sea Fisheries Institute (MIR)
Kollataja 1
PL-81-332 Gdynia
Poland
Tel: +48 58 620 2825
Fax: +48 58 620 2831
E-mail: tlink@mir.gdynia.pl

Lleonart, Dr Jordi
Institut de Ciències del Mar (CSIC)
Passeig Marítim 37-49
08003 Barcelona
Spain
Tel: +34 93 230 95 48
Fax: +34 93 230 95 55
E-mail: lleonart@icm.csic.es

Maurihungirire, Dr Moses
(Sponsored by NORAD)
Ministry of Fisheries & Marine Resources
P.O. Box 912
Swakopmund
Namibia
Tel: +264 64 410 1000
Fax: +264 64 404 385
E-mail: mmaurihungirire@mfmr.gov.na

Metcalfe, Dr Julian
CEFAS
Lowestoft Laboratory
Pakefield Road
Lowestoft
Suffolk NR33 0HT UK
Tel: +44 1502 524352
Fax: +44 1502 524511
E-mail: j.d.metcalfe@cefas.co.uk

Mogensen, Ms Charlotte
Joint Nature Conservation Committee
Monkstone House
City Road
Peterborough PE1 1JY UK
Tel: +44 (0) 1733 866832
Fax: +44 (0) 1733 555948
E-mail:
Charlotte.Mongensen@jncc.gov.uk

Morrice, Mr Chris
c/o FCO (Bangladesh)
King Charles Street
London
SW1A 2AH UK
E-mail: chris.sufer@fmsbd.org

O'Brien, Dr Carl
CEFAS
Lowestoft Laboratory
Pakefield Road
Lowestoft
Suffolk NR33 0HT UK
Tel: +44 1502 524256
Fax: +44 1502 524511
E-mail: c.m.obrien@cefas.co.uk

Ojaveer, Professor Evald
Estonian Marine Institute
18 B Viljandi Road
11216 Tallinn
Estonia
Tel: +372 6281 568
Fax: +372 6281 563
E-mail: e.ojaveer@ness.sea.ee

Pawson, Dr Mike
CEFAS
Lowestoft Laboratory
Pakefield Road
Lowestoft
Suffolk NR33 0HT UK
Tel: +44 1502 524436
Fax: +44 1502 524511
E-mail: m.g.pawson@cefas.co.uk

Payne, Dr Andrew
CEFAS
Lowestoft Laboratory
Pakefield Road
Lowestoft
Suffolk NR33 0HT UK
Tel: +44 1502 524344
Fax: +44 1502 524511
E-mail: a.i.l.payne@cefas.co.uk

Perks, Miss Charlotte
CEFAS
Lowestoft Laboratory
Pakefield Road
Lowestoft
Suffolk NR33 0HT UK
Tel: +44 1502 524578
Fax: +44 1502 524569
E-mail: c.v.perks@cefas.co.uk

Perry, Ms Allison
(Sponsored by NRC Europe Ltd)
School of Biological Sciences
University of East Anglia
Norwich NR4 7TJ UK
Tel: +44 (0) 1603 473956
Fax: +44 (0) 1603 592250
E-mail: a.perry@uea.ac.uk

Pilling, Dr Graham
CEFAS
Lowestoft Laboratory
Pakefield Road
Lowestoft
Suffolk NR33 0HT UK
Tel: +44 1502 527730
Fax: +44 1502 513865
E-mail: g.m.pilling@cefas.co.uk

Pope, Professor John
NRC (Europe) Ltd
The Old Rectory
Staithe Road
Burgh St Peter
Norfolk NR34 0BT UK
Tel: +44 (0) 1502 677377
Fax: +44 (0) 1502 677377
E-mail: PopeJG@aol.com

Potter, Mr Ted
CEFAS
Lowestoft Laboratory
Pakefield Road
Lowestoft
Suffolk NR33 0HT UK
Tel: +44 1502 524260
Fax: +44 1502 524511
E-mail: e.c.e.potter@cefas.co.uk

Restrepo, Dr Victor
International Commission for the
Conservation of Atlantic Tunas
ICCAT
Corazon de Maria 8-6
28002 Madrid
Spain
Tel: +34 91 416 5600
Fax: +34 91 415 2612
E-mail: victor.restrepo@iccat.es

Reynolds, Dr John
School of Biological Sciences
University of East Anglia
Norwich NR4 7TJ UK
Tel: +44 (0) 1603 593210
Fax: +44 (0) 1603 592250
E-mail: Reynolds@uea.ac.uk

Robinson, Mr Brian
CEFAS
Lowestoft Laboratory
Pakefield Road
Lowestoft
Suffolk NR33 0HT UK
Tel: +44 1502 524458
Fax: +44 1502 524569
E-mail: b.j.robinson@cefas.co.uk

Rogers, Dr Stuart
CEFAS
Lowestoft Laboratory
Pakefield Road
Lowestoft
Suffolk NR33 0HT UK
Tel: +44 1502 524267
Fax: +44 1502 524511
E-mail: s.i.rogers@cefas.co.uk

Rollo, Mrs Sarah
CEFAS
Lowestoft Laboratory
Pakefield Road
Lowestoft
Suffolk NR33 0HT UK
Tel: +44 1502 524430
Fax: +44 1502 524569
E-mail: s.f.rollo@cefas.co.uk

Romanov, Dr Evgeny
2 Sverdlov St
Kerch
98300
Crimea
Ukraine
Tel: +380 6561 23530
Fax: +380 6561 21572
E-mail: island@crimea.com

Sabourenkov, Dr Eugene
Commission for the Conservation of Antarctic Marine Living Resources (C-CAMLR)
P.O. Box 213
North Hobart
Tasmania 7002
Australia
Tel: +61 (03) 62 310 366
Fax: +61 (03) 62 349 965
E-mail: eugene@ccamlr.org

Smith, Dr Tony
CSIRO Marine Research
G.P.O. Box 1538
Hobart
Tasmania 7001
Australia
Tel: +61 3 62 325 372
Fax: +61 3 62 325 053
E-mail: tony.d.smith@csiro.au

Soesilo, Dr Indroyono
Agency for Marine & Fisheries Research
Ministry of Marine Affairs and Fisheries
JL Mt Haryono 52-53
Jakarta
Indonesia
Tel: + 62 21 791 80159
Fax: +62 21 791 80458
E-mail: soesilo@rad.net.id

Tingley, Dr Geoff
CEFAS
Lowestoft Laboratory
Pakefield Road
Lowestoft
Suffolk NR33 0HT UK
Tel: +44 1502 524345
Fax: +44 1502 524569
E-mail: g.a.tingley@cefas.co.uk

Wentworth, Mr Stephen
Department for Fisheries and Rural Affairs
Room 505
Nobel House
London
Tel: +44 0207 238 5796
Fax: + 44 020 7 238 5814
E-mail:
stephen.wentworth@defra.gsi.gov.uk

Deterring IUU Fishing

Geoffrey P. Kirkwood and David J. Agnew
Department of Environmental Science and Technology, Imperial College London, London SW7 2AZ, UK

ABSTRACT: Illegal, unreported and unregulated (IUU) fishing is a problem that has been around since the first attempts at fishery management. However, it has deservedly been given high prominence in recent years as more and more instruments designed to manage fisheries on the high seas have come into force. The International Plan of Action on IUU fishing (IPOA–IUU), developed by FAO within the framework of its Code of Conduct for Responsible Fisheries, is a major step forward. The aim of this paper is review progress in implementing some of the measures outlined in the IPOA and to discuss future prospects for eliminating IUU fishing. The paper begins by examining the incentives to fish illegally and relating the various measures in the IPOA to how they decrease particular incentive factors and increase disincentive factors. Three of the measures are then discussed in more detail. The first is the means available to a State to prevent illegal fishing in waters over which it has jurisdiction. The second relates to measures that can be taken against flag of convenience vessels. The third covers imposition of trade-related measures, increasingly being pursued by Regional Fisheries Management Organizations, especially ICCAT and CCAMLR. The paper concludes with suggestions for further actions to deter and prevent IUU fishing.

INTRODUCTION

As an activity, illegal, unreported and unregulated (IUU) fishing has been with us since fisheries management first started. As an acronym, however, it is much younger. First used informally during the early 1990s by the Commission for the Conservation of Antarctic Marine Living Resources (CCAMLR)[1] in relation to Southern Ocean fishing, it began life as "IU" (illegal and unreported). Formal use of the term IUU can be found in the report of the Commission's XVIth Meeting in 1997 and in a letter to the Food and Agriculture Organization of the United Nations (FAO) that same year, in which the nature and seriousness of these problems were described[2]. IUU fishing is now commonly understood to refer to fishing activities that are inconsistent with or in contravention of the management or conservation measures in force for a particular fishery.

---

1 *The Commission established under Article VII of the Convention on the Conservation of Antarctic Marine Living Resources (CAMLR), 1980. Reprinted in International Legal Materials, **19**: 827.*

2 *Executive Secretary, CCAMLR to FAO (REF: 4.2.1. (l), 18 December 1997), as cited in Lugten, G. 1999. A review of Measures taken by Regional Marine Fishery Bodies to address contemporary Fishery Issues. FAO Fisheries Circular, **940**. FAO, Rome: Footnote 130 at 35.*

There are a number of international instruments that contain provisions that are of relevance to the control of IUU fishing. These include the 1982 United Nations Law of the Sea Convention[3] (the 1982 Agreement), the 1993 FAO Compliance Agreement, the 1995 United Nations Straddling Stocks Agreement[4] (the 1995 Agreement), and the 1995 FAO Code of Conduct for Responsible Fisheries (see Edeson, 1966). However, none of these was set up to deal directly with IUU fishing.

Concern over the growth of IUU fishing worldwide increased rapidly during the late 1990s. In early 1999, the need to prevent, deter and eliminate IUU fishing was addressed by the FAO Committee on Fisheries (COFI; FAO, 1999) and shortly afterwards the FAO announced its intention to develop a global plan of action to deal effectively with all forms of IUU fishing. In early 2000, the Government of Chile in cooperation with the FAO convened an International Conference on Monitoring, Control and Fishing Surveillance[5]. This was followed by a Government of Australia/ FAO Expert Consultation in Sydney in May 2000[6], which started the process of elaboration of an International Plan of Action (IPOA) on IUU fishing. Following two further Technical Consultations, the IPOA was adopted by COFI in March 2001 (FAO, 2001). The IPOA is a voluntary agreement, and it has been elaborated within the overall framework of the FAO Code of Conduct for Responsible Fishing.

IUU fishing is defined in paragraph 3 of the IPOA as follows:

"*Illegal* fishing refers to activities:

(1) conducted by national or foreign vessels in waters under the jurisdiction of a State, without the permission of that State, or in contravention of its laws and regulations;
(2) conducted by vessels flying the flag of States that are parties to a relevant regional fisheries management organization but operate in contravention of the conservation and management measures adopted by that organization and by which the States are bound, or relevant provisions of the applicable international law; or
(3) in violation of national laws or international obligations, including those undertaken by co-operating States to a relevant regional fisheries management organization.

Unreported fishing refers to fishing activities:

(1) which have not been reported, or have been misreported, to the relevant national authority, in contravention of national laws and regulations; or
(2) undertaken in the area of competence of a relevant regional fisheries management organization which have not been reported or have been misreported, in contravention of the reporting procedures of that organization.

---

[3] *United Nations Convention on the Law of the Sea, Montego Bay, 10 December 1982.*

[4] *Agreement for the Implementation of the Provisions of the United Nations Convention on the Law of the Sea relating to the Conservation and Management of Straddling Fish Stocks and Highly Migratory Fish Stocks. New York, 4 December 1995.*

[5] *Report of the International Conference on Monitoring, Control and Fisheries Surveillance, Santiago, Chile, 25–27 January 2000.*

[6] *Report of the Expert Consultation on Illegal, Unreported and Unregulated Fishing, Organized by the Government of Australia in Cooperation with FAO, Sydney, Australia 15–19 May 2000. Paper AUS:IUU/ 2000/3, available at www.affa.gov.au/ecoiuuf.*

Unregulated fishing refers to fishing activities :

(1) in the area of application of a relevant regional fisheries management organization that are conducted by vessels without nationality, or by those flying the flag of a State not party to that organization, or by a fishing entity, in a manner that is not consistent with or contravenes the conservation and management measures of that organization; or

(2) in areas or for fish stocks in relation to which there are no applicable conservation or management measures and where such fishing activities are conducted in a manner inconsistent with State responsibilities for the conservation of living marine resources under international law."

Not all unregulated fishing is necessarily conducted in contravention of applicable international law. This is because many high seas waters and/or fisheries are still unregulated by regional fishery management organisations (RFMOs). Examples of these include the orange roughy/alfonsino fishery in the southern Indian Ocean, and the toothfish fishery on the northern Patagonian shelf edge. The IPOA specifically acknowledges this exception (paragraph 3.4), but we consider it as another dimension to the IUU problem. While there is no doubt that the orange roughy/alfonsino fishery is currently legitimately unregulated, it certainly should become regulated, and the negotiations for the South-West Indian Ocean Convention address this concern. With the entry into force (in December 2001) of the 1995 UN Straddling Stocks Agreement, it has been argued that there are no areas of high seas fishing that may be considered legitimately unregulated in terms of States obligations under that Agreement and Part VII of the1982 Agreement. However, this appears to be an area of international law about which there remain differences of opinion (see, for example, Freestone & Makuch, 1996), and we will leave further comment on this issue to those more qualified to make it.

The main body of the IPOA outlines a lengthy series of measures designed to prevent, deter and eliminate IUU fishing. These are grouped under the headings: all State responsibilities, Flag State responsibilities, Coastal State measures, Port State measures, internationally agreed market-related measures, and regional fisheries management organizations.

In this paper we first examine the factors influencing the incentives for IUU fishing and place the various measures proposed in the IPOA in the context of how they are likely to decrease the incentives and increase the disincentives. We then discuss in more detail enforcement measures, actions against open register or flag of convenience IUU fishing and trade-related measures, which are three areas in which progress in deterring IUU fishing seems to be being made at present. We conclude by briefly discussing some other possible measures that may be brought to bear. We do not attempt, in this paper, to quantify the extent of IUU fishing worldwide. Nor do we discuss in detail the specifics of particular control measures. In relation to IUU fishing for Patagonian toothfish, such an assessment is provided by Sabourenkov and Miller (2003) in a separate paper to this Symposium.

INCENTIVES AND DISINCENTIVES FOR IUU FISHING

It is already clear that there is no single measure that, if taken, would immediately eliminate IUU fishing. Consequently, before attempting to evaluate ways in which IUU fishing might be deterred, it is valuable first to identify the incentives for IUU fishing. The extent to which possible measures may succeed will almost certainly depend on how much they act to reduce the incentives and enhance the disincentives.

The reasons that vessels are engaged in IUU fishing are solely economic, if we ignore the possibility that some vessel owners and crew may simply prefer to fish illegally. This also immediately implies that vessel owners will prefer to engage their vessels legally in regulated fisheries rather than in IUU fishing, as long as the opportunity to do so exists and legal fishing is sufficiently profitable. However, for a substantial and increasing number of vessels, the conditions of this proviso are not met. As estimated by FAO (2000), almost 70% of the world's fisheries are either fully exploited, overexploited, or in various stages of recovery from overexploitation. Management responses to this have led in many cases to substantially reduced allowable catches, and at last action is also being taken to reduce the overcapacity that exists in most of the world's major fishing fleets. In the absence of heavily subsidized decommissioning schemes, and with ageing vessels being replaced in regulated fleets by (heavily subsidized) newer and more efficient vessels, it is inevitable that owners of vessels unable to maintain past levels of profits will look for other options.

In previous eras, pressures such as these led to vessels looking offshore for new fishing opportunities. For example, the establishment of Exclusive Economic Zones (EEZs) led to many distant water fleets being excluded from fisheries in waters of coastal state jurisdiction, and the response was the development of then-unregulated fisheries on the high seas. This legitimate avenue is now no longer open to many such vessels, because most of these resources are now regulated by RFMOs and many are also subject to substantial levels of exploitation. There are therefore now strong incentives to engage in unregulated fishing by transferring vessels to the fishing vessel registers of open register States, thereby becoming what is otherwise known as Flag of Convenience (FOC) vessels, or to engage in illegal fishing.

The overall economic incentives underlying the rise in illegal and unregulated fishing are therefore clear. To progress further, however, it is useful to consider the incentives (and disincentives) a little more closely. We do this by examining the factors affecting the simple profit and loss equation for IUU fishing:

$$\text{Profit from IUU fishing} = \text{Benefit from IUU fishing} - \text{Cost of IUU fishing}$$

The benefit, obviously, arises from the sale of the catch. If access to markets is unrestricted, the future benefits for IUU fishing look rosy. While the demand for marine fish products continues to rise steadily, overall supply has been at best static for a number of years and, given the state of the world's marine fish stocks, it is unlikely to increase much above current levels in the near future. Buoyant and increasing fish prices are therefore to be expected. The key to reducing the incentive for IUU fishing arising from the benefits available, equally obviously, is to restrict access to markets for IUU-caught fish.

In the case of illegal trading in over-quota catches taken in many regulated fisheries (so-called "black fish"), many of the fish are put into a bulk market, where they are relatively easily disguised as other fish products. Other factors aside, the prices obtained for these fish are probably sufficiently low that they alone would discourage entry into such fisheries solely as an IUU vessel. Almost all such fishing is probably done by otherwise legitimate fishers, though the boundary between what is effectively opportunistic IUU fishing and specialist IUU fishing may be blurred. The benefits available from engaging in IUU fishing are greater if the fish taken are of high value – for example sashimi-grade tunas or Patagonian toothfish. The problem with black fish can really only be tackled by better or different fisheries management and enforcement, such that the incentive for misreporting is removed. While this approach should also be used for high value IUU fishing, in this case probably the most effective approach is to curb access to the market through introduction of trade-related measures, as we discuss later.

The costs faced by IUU vessels are also familiar. These include those costs faced by all fishing vessels – the cost of the vessel (capital cost plus depreciation), running costs (for vessel and crew) and costs associated with steaming to and from fishing grounds and transhipment points. To these are added the costs specifically associated with IUU fishing activities, including the direct costs potentially resulting from being apprehended and fined for illegal fishing and the indirect costs of needing to find a State willing to have the IUU fishing vessel on its register.

Compared with the average vessel legitimately engaged in regulated fisheries, the costs of IUU vessels are relatively low. Most of the IUU fleet (and especially those vessels engaged in illegal fishing) are old vessels no longer capable of competing with the modern fleets operating in regulated fisheries. The number of such vessels available is increased and to some extent their purchase costs are further decreased by the continued practice of some countries to provide subsidies for building new and more efficient fishing vessels. Clearly, effective decommissioning schemes for vessels retiring from regulated fleets and a cessation of subsidies for building new vessels would assist in curbing the supply of potential IUU vessels.

Because of the generally poor conditions on IUU fishing vessels and the potential risks associated with being arrested for illegal fishing, it might be expected that crew costs would be higher on IUU vessels than others. While this may be true to some extent for skilled fishing masters, our industrial sources suggest ordinary crew costs if anything are lower. This results on the one hand from shrinking opportunities for both officers and crew in legitimate fisheries, and on the other hand from an apparently ready supply of low-cost labour from some developing countries. Another factor reducing costs for IUU fishing is that rarely are standard international safety practices followed; maintenance of safety and employment rights on IUU vessels is of extremely low priority. Finally, because the pay of the fishing crew is usually linked to the sale of the catch, a crew on an IUU vessel will always have an incentive to fish beyond the limits of safety to bring as much catch as possible aboard. One potentially powerful means of addressing these issues contained in the IPOA is for States to take measures to ensure that nationals subject to their jurisdiction do not support or engage in IUU fishing. Though the basis for such measures is well founded in the 1982 Convention (Article 94), this idea does not appear to have found its way into many international fisheries management agreements, and enforcement of such provisions by individual countries may prove legally rather difficult.

While the marginal cost of occasionally taking and processing over-quota fish in regulated fisheries is minimal (provided this is undetected), full-time IUU vessels face an additional burden arising from the fact that many of the fishing grounds effectively available for IUU fishing are remote. This imposes higher costs associated with steaming to and from the fishing grounds and also higher costs for transhipment and/or port visits. The remoteness of some IUU fishing grounds and the high steaming costs to get there also mean that it is unprofitable for IUU vessels to undertake only short fishing campaigns on those grounds. The necessity to spend sufficient time on these grounds (often enough time to fill the hold) in turn increases the chances of being detected, a point to which we will return in the discussion.

Clearly, any actions that further restrict opportunities for landing IUU-caught fish, or even just for entering ports for re-supply, will act as a strong disincentive for IUU fishing. The IPOA outlines a number of measures to be taken by Port States against IUU fishing vessels. These include strengthened inspection procedures and measures to restrict or prevent access to their ports by IUU vessels.

The last category of costs comprises those specifically associated with IUU fishing. The first of these relates to flags and authorizations to fish. Whereas some IUU vessels apparently operate without flags or authorizations to fish from any State, most operate under the flags of open register states, which exercise little or no control over the activities of vessels to which they have issued authorizations. The IPOA proposes a series of measures to be taken by Flag States to discourage IUU fishing, and RFMOs are also active in this area, as we discuss in a later section. For present purposes, it is sufficient to note that to the extent that these measures are successful, they will act as disincentives in a similar manner to the Port State measures mentioned above.

Finally, where IUU vessels are actually fishing illegally, they then risk apprehension and subsequent penalties if apprehended. Currently, these risks are only faced by vessels fishing illegally in Coastal State EEZs or on designated straddling stocks. Obviously, in order to be apprehended, a system for detection of illegal fishing must be in place, as must a system for subsequent arrest and prosecution upon detection. The often-high cost of this must be borne by the Coastal State, though any fines imposed may be offset against this cost. The potential cost borne by an illegal fishing vessel is equal to the risk of being detected and arrested, multiplied by the likely fine imposed for illegal fishing. In practice, what matters in terms of deterring illegal fishing is not the actual risk of being detected, but the risk as it is perceived by the illegal fisher. We outline briefly optimal policies for enforcement and deterrence of illegal fishing by a coastal state in the next section.

In this overview, we have reviewed the various factors affecting the benefits associated with IUU fishing and the costs associated with it. For each factor, we have briefly identified how possible measures to deter IUU fishing, such as those listed in the IPOA, can decrease the benefits or increase the costs. As noted at the start of this section, it is unlikely that any one measure alone can serve to eliminate IUU fishing. Rather, the most productive approach will be to use a battery of measures, each designed to act incrementally to tip the balance between benefits and costs to a state where IUU fishing is no longer viable. We now examine three of these measures in more detail.

ACTIVITIES BY COASTAL STATES: CONTROLLING IUU FISHING IN EEZS

One area about which there is no dispute over the legal powers for a State to enforce regulations against IUU fishing is within its 200 nautical mile EEZ. That IUU fishing remains a problem in some EEZs is usually a problem of ineffective enforcement. This may not be so much a problem of a lack of will, but a lack of effective ability to do so in the face of difficult problems. Especially from the perspective of a developing coastal or small island State, EEZs are large, the areas they protect are often remote, and the costs of patrolling can be high.

Depending on the state of development of its domestic fishing industry, a coastal State may wish to reserve all licensed fishing for its own vessels, or it may wish to earn revenue just by licensing foreign fishing vessels to fish in its EEZ, or it may seek some combination of the two. The types of incentives for a coastal State to manage sustainably and responsibly may vary with the biological characteristics of the fish stocks found in its EEZ, but regardless of these, strong incentives still remain. If the stock(s) are solely or largely confined to the EEZ, then the incentives are strong, because if the stock(s) become overexploited, either through mismanagement or illegal fishing, recovery of the stock and the related income flows and employment opportunities may take a long time. However, if the available stock(s) are seasonal visitors from a highly migratory species, and management actions by the State within its EEZ are unlikely to affect stock status very much either way, vital potential revenue will still be foregone if unlicensed vessels can fish with impunity for these species in the zone.

For initial simplicity, let us suppose that the coastal State is solely interested in licensing foreign fishing vessels. Under what circumstances is it worthwhile for a vessel to undertake IUU fishing within the EEZ, rather than legal licensed fishing? Of prime importance, the expected marginal return from fishing in the EEZ at that time must exceed the expected marginal return from fishing in alternative places. Obviously, the greater the difference between marginal returns from fishing inside or outside the EEZ, the greater is the incentive to fish illegally within the EEZ.

If there is an incentive for IUU fishing, then in the absence of any deterrent measures by the state, it will occur. What can the state do about this? First, of course, it can legislate to make it illegal to fish in its zone without a licence issued by the State, for which it will normally charge a fee. However, it must back this up by a system of monitoring and surveillance of fishing activities within the EEZ and imposition of penalties for vessels detected fishing illegally. Monitoring, surveillance and prosecutions cost money, sometimes substantial amounts, which must be offset against licence fees charged, assuming that alternative sources of revenue to fund these activities are not available.

Why should a fisher opt to purchase a licence for fishing legally inside the State's EEZ, rather than fishing illegally inside the EEZ, or fishing legally outside the State's EEZ? Clearly, it will be worthwhile for the fisher to purchase a licence if the expected benefit of licensed fishing in the EEZ is both greater than zero and in excess of the expected benefit of fishing illegally inside the EEZ (or fishing legally outside the EEZ). The expected marginal benefit of licensed fishing within the EEZ is equal to the difference between the expected benefit of fishing inside the EEZ and that of fishing outside the EEZ, less the cost of the licence fee. The expected benefit of fishing illegally inside the EEZ is equal to the difference in expected benefits of fishing inside and outside the EEZ (as for licensed fishing), less the expected cost of being detected and fined for illegal fishing. This latter cost is simply the product of the probability of being detected and the expected fine if detected, prosecuted and convicted.

Even from this very simple analysis, there are two obvious immediate conclusions. If the expected benefit of fishing inside the EEZ compared with outside the EEZ is low, then any licence fee charged that is likely to go anywhere near covering surveillance and prosecution costs may be greater than that expected benefit. If so, then no licences will be taken up, but equally there will be little incentive for illegal fishing. On the other hand, if the expected benefit of fishing within the EEZ is high relative to that available in alternative fishing grounds, then the potential for damage from illegal fishing is high, but potentially so also are the opportunities for the coastal State to earn revenue from licensing fishing, even allowing for surveillance and prosecution costs.

To proceed further, it is useful to define some terms and turn to some diagrams. Let MR be the expected marginal benefit for a fishing vessel to fish within the EEZ, L be the licence fee charged for fishing within the EEZ, and E(F) be the expected fine for a vessel detected and successfully prosecuted for illegal fishing in the EEZ. The expected fine is the product of the actual fine F and the probability of being detected, Q.

For the fisher, the decision rules are a function of the three variables (Figure 1). If $L \leq MR$ and $L < E(F)$, then the fisher will opt to fish in the EEZ with a licence. This region is the lower triangle in Figure 1. If $E(F) \leq MR$ and $E(F) < L$, then the fisher will opt to fish illegally without a licence in the EEZ. This region is the upper triangle in Figure 1. If $L = E(F) \leq MR$, then a risk-neutral fisher will be indifferent between fishing legally or illegally within the EEZ. Obviously, if $L > MR$ or $E(F) > MR$ then the fisher will not wish to fish within the EEZ at all.

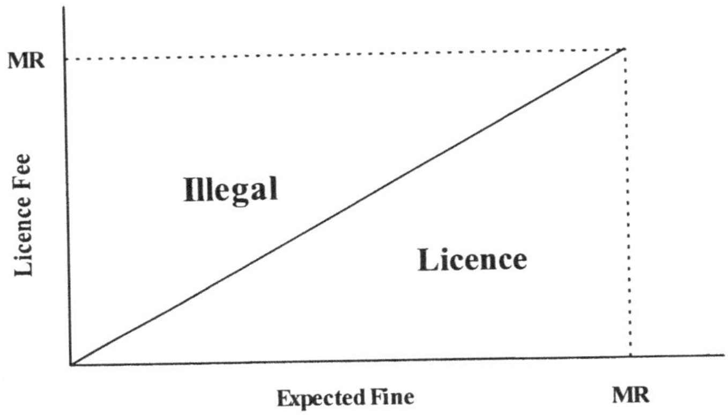

Fig. 1 Decision rules for fishing in an EEZ: fisher's perspective. MR is the marginal revenue associated with fishing inside the EEZ instead of alternative fishing areas outside the EEZ.

To examine the decision rules from the point of view of the coastal State, we also need to define S as the per-vessel surveillance (for simplicity we ignore prosecution costs). The decision rules for the State are illustrated in Figure 2. These are nearly the mirror image of those for the fisher. However, now the boundary between the regions for licensing or not licensing is not the line $L = E(F) < MR$, but rather the line $L = E(F) - S$. This is because the (possibly unscrupulous) State will make more net revenue from detecting and prosecuting illegal fishing if $L < E(F) - S$ than it would from the licence fee.

The two sets of decision rules are merged in Figure 3. Inspecting this figure, it is clear that there is only one area of agreement between the State and the fisher. This area lies between the two lines $L = E(F) < MR$ and $L = E(F) - S < MR$ and it coincides with fishers wanting licences and the State also wishing to issue licences.

From Figure 3, it is also clear that, for the State, the theoretically optimal licence fee would be fractionally below the marginal revenue, MR, and similarly the optimal expected fine would be just fractionally below the optimal licence fee.

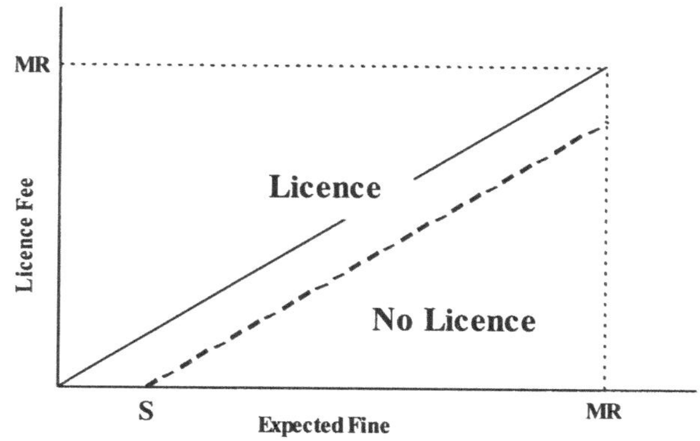

Fig. 2 Decision rules for fishing in an EEZ: State's perspective. MR is the marginal revenue associated with fishing inside the EEZ instead of alternative fishing areas outside the EEZ. S is the per-vessel cost of surveillance.

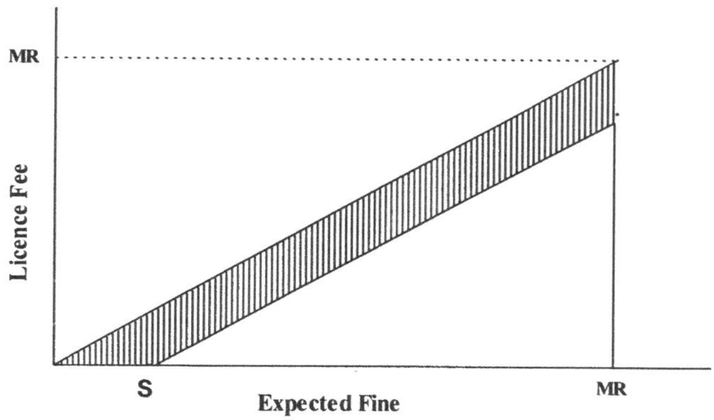

Fig. 3 Decision rules for fishing in an EEZ: combined fisher and State perspectives. MR is the marginal revenue associated with fishing inside the EEZ instead of alternative fishing areas outside the EEZ. S is the per-vessel cost of surveillance. The hatched area indicates the region where both fisher and State rules coincide.

In practice, of course, such a policy on the part of the State would not be particularly sensible, because the expected marginal benefit to the fisher for fishing in the EEZ would be very small, and it may therefore not be considered worth fishing in the EEZ at all. This would then leave the State with no revenue. Another important factor is that while the licence fee is fixed and known, being detected fishing illegally is a chance event. It is therefore quite likely that the fisher will be risk prone, rather than risk neutral. In such circumstances, a similar analysis shows that the optimal levels of licence fees and expected fines are lower than those in the risk-neutral case, the extent of the reduction depending on the degree of risk-proneness of the fisher.

To complete this simple analysis, it is necessary to deconstruct the expected fine into its two components, the actual fine and the probability of detection, with the latter varying with the amount spent on surveillance. It is immediately obvious that the higher the fine, the lower need be the detection probability and therefore surveillance expenditure. It seems reasonable to assume that the probability of detection is an increasing function of surveillance expenditure, and that the rate of increase in detection probability with surveillance expenditure decreases with increasing surveillance expenditure. When this is combined with the previous analysis, MRAG (1993) showed that the relationship between net revenue to the State from detecting and fining illegally fishing vessels as a function of surveillance costs has the form of the curve shown in Figure 4, and from that they derived optimal licence fees, maximum fines and surveillance expenditures as a function of the various parameters.

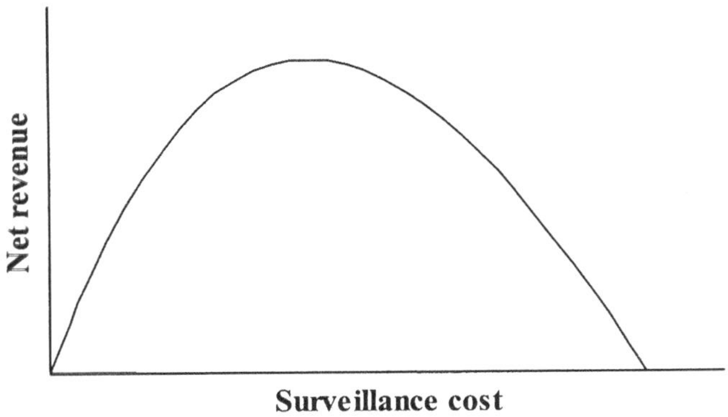

Fig. 4 Relationship between net revenue to State from fishing and surveillance expenditure.

The analysis described above assumed simply that the State wished to licence a single fleet of identical foreign fishing vessels to fish in its EEZ. Fortunately, as MRAG (1993) showed, this analysis is easily extended to cover the case where there are several different fleets of foreign vessels wishing to fish in the EEZ, and also the case where there is a conservation constraint on the maximum levels of fishing effort to be allowed, either because there is a domestic fleet to protect, or because of the potential for the stock to be overexploited. The analysis is more complicated and the results and optimal policies differ in detail, but the principles of the optimal policy remain the same. For a fishery in which the marginal benefit for fishing in the EEZ is high, the optimal policy is:

(1) Set the licence fee at as high a proportion of the potential marginal revenue as the fishers will bear.
(2) Set the fine at as high a value as can be levied (often equal to the value of the vessel, plus gear and catch.
(3) Allocate appropriate levels of surveillance expenditure commensurate with the levels of the licence fees and fines.

An attractive feature of this policy is that it accords directly with intuition, at least in hindsight, and it is easily explained. In at least one fundamental respect, however, it differs from conventional practice, in that the licence fees are set as a proportion of the potential marginal revenue from fishing in the EEZ, rather than as a proportion (e.g. 5%) of the gross value of catch taken within the zone. This latter commonly applied policy has the virtue of being simple to calculate and implement, but again in hindsight it is obviously inferior.

The policy also works in practice, having been applied to the extent possible to the licensing of foreign fishing vessels in the fishing zones around the Falkland Islands, South Georgia and the British Indian Ocean Territory. A similar approach, at least in terms of setting levels of fines for illegal fishing, has also been used successfully by France in its Antarctic Territories. The experience of application of this policy in British Indian Ocean Territory waters also illustrated another important point: it is the fisher's perception of the risk of being caught fishing illegally that matters, rather than the true risk. For some time after the policy was implemented, foreign fishing fleets that had previously fished freely in these waters were understandably unhappy at suddenly being required to take up licences and pay licence fees, so for a while they didn't. However, the subsequent arrest and heavy fines imposed on two vessels detected fishing illegally suddenly led to a rush of applications for licences, as the risk of unlicensed fishing was proved to be real, rather than theoretical.

The theoretical analyses (see also Hart, 1997)[7] and practical experience suggests that, where the incentive for illegal fishing is high, the policies described above should act as a powerful deterrent. Put another way, where the legal jurisdiction is such that IUU fishing is in clear violation of pertinent State laws or regulations and can be prosecuted as such, then the approach described above seems adequate to control IUU fishing.

There is, however, one potential caveat to this encouraging conclusion. The optimal policy with respect to fines for illegal fishing is that they should be set as high as possible, which usually means that they are set at the value of the vessel plus gear and catch held on board. If the vessel arrested is a large European or US purse-seiner, then the deterrent effect of such a fine is massive. However, as noted earlier, some specialist IUU fleets tend to consist of old relatively inexpensive vessels. In such cases, experience has shown that their owners may simply abandon the vessel on arrest[8]. Even so, it is doubtful that the vessel owners would simply ignore the possibility of arrest when deciding where to send their vessels to fish illegally, so the deterrent effect of the policy in this case would simply be diminished to some extent, rather than eliminated.

The legal certainty needed for application of the policies described above appears to be present only in coastal State EEZs and certain straddling stocks. Unfortunately, on the high seas, the certainty diminishes. Indeed, the IPOA notes that some activities that would nominally fall under the general umbrella of IUU fishing may not be in violation of any international law. To deter IUU fishing in such areas, other policies are needed, several of which are discussed in the following sections.

---

[7] *Note, however, that taking account of avoidance behaviour by fishers can also be important (see Charles et al., 1999).*

[8] *For example, the Belizian flagged vessel "Grand Prince" was arrested in December 2000 by France, which imposed a €1.7M fine, later reduced to a €206 149 fine on a ruling of the International Tribunal of the Law of the Sea. The company refused to pay this to obtain release of the vessel so it was sunk off La Reunion in April 2002. "Francia hunde el <Grand Prince>, apresado por faenar sin licencia." La Voz de Galicia, 13 Abril, 2002.*

FLAG OF CONVENIENCE VESSELS

As noted earlier, the IPOA proposes a range of measures designed to curtail the numbers and activities of IUU fishing vessels that have been re-flagged to open register States that do not exercise full control and responsibility over the vessels, as required by the Compliance Agreement and the FAO Code of Conduct for Responsible Fisheries. These include actions to be taken by Flag States, and potential actions to be taken by all States against their nationals who engage in IUU fishing. If these measures are taken up by all concerned States, then the problems posed by FOC vessels should be resolved. However, taking extraterritorial action against nationals is something many States are reluctant to do, and so far open register States do not appear to have been pushed into action on the basis of existing instruments alone. There are, however, encouraging signs that additional pressure from other directions may be having some effect.

The activities of FOC fishing vessels have been at the top of the agenda of many delegations to RFMO meetings in recent years, especially those concerned with the management of tuna stocks. Most of these have identified conservation problems for one or more tuna or tuna-like species within their jurisdiction and they have adopted management and conservation regimes, with corresponding Action Plans, aimed at restoring overexploited stocks to optimal levels and ensuring appropriate levels of fishing capacity. However, implementation of these measures and Action Plans has been greatly complicated by the operations of fishing vessels flying flags of States that are not Contracting Parties or Co-operating Non-Contracting Parties or Entities. The problem is not just that the fishing by these vessels is unregulated and often also unreported; in addition, Contracting or Co-operating Parties are understandably reluctant to curb fishing by their vessels if the Conservation Measures requiring that are completely ignored by FOC vessels.

The International Commission for the Conservation of Atlantic Tunas (ICCAT) has been at the forefront of moves to take action against countries deemed to have been acting in ways that undermine its conservation and management measures. In 1994, a Bluefin Tuna Action Plan was adopted by ICCAT that linked information gathered by the Bluefin Tuna Statistical Document Programme[9] with Contracting Party compliance and non-Contracting Party cooperation with ICCAT's conservation and management regime. Having identified in 1995 that Belize, Honduras and Panama had vessels that were fishing in a manner that diminished the effectiveness of ICCAT's conservation measures, ICCAT subsequently (1996) prohibited imports by its Members of bluefin tuna products from these three countries (effective from 1997 for Belize and Honduras and 1998 for Panama). This was successful in the case of Panama, which became a contracting party in 1998. Similar sanctions were extended to cover bigeye tuna taken by vessels flagged by Belize, Cambodia, Honduras, Equatorial Guinea and St Vincent and the Grenadines in 2000. Once again, this move seems to have been effective, and in 2001 (but still to come into force) ICCAT lifted the import ban on bigeye tuna from St Vincent and the Grenadines and the bluefin tuna ban from Honduras.

Other tuna Commissions have now introduced statistical document programmes modelled on the ICCAT programme. The Commission for the Conservation of Southern Bluefin Tuna (CCSBT) has introduced a southern bluefin tuna statistical document programme (effective from 2000) and the Indian Ocean Tuna Commission (IOTC) has

---

[9] *ICCAT resolutions 92-1 and 92-3, implemented in 1993.*

introduced a bigeye tuna statistical document programme (effective from July 2002). CCSBT has, like ICCAT, adopted a plan of action targeted at IUU fishing that will allow the prohibition of imports of southern bluefin tuna from non-members who allow their vessels to undermine conservation measures. It is likely that the IOTC statistical document scheme will, in time, lead to similar actions.

There are also now new moves within the EU to make progress on the FOC/IUU issue. Following written questions about FOC vessels in the European Parliament in early 2001, the EU has now promulgated a regulation that prohibits the use of subsidies ("structural assistance") to assist the transferring of EU vessels to registers of FOC countries[10].

One difficulty with the FOC/IUU issue has always been that of transparently identifying the vessels and States involved. Not all fishing vessels flying flags of FOC States are engaged in IUU fishing, and the actions of States and the behaviour of individual vessels change over time. Thus, developing reliable lists of offending vessels and States is fraught with difficulties. The EU has neatly circumvented this problem by making no reference to FOC or IUU at all. Rather, it relies on RFMOs to identify States acting to undermine their conservation measures by licensing fishing vessels, and then taking actions against those identified States.

This approach has naturally been followed by a drive to get RFMOs to adopt mechanisms to identify lists of vessels and flag states that are not complying with their regulations. In 2001, CCAMLR engaged in discussion of a possible list of vessels engaged in IUU fishing that would be required for the EU to act against them[11]. Prior to that, in 2000 CCAMLR adopted a resolution (13/XIX, later replaced by Conservation Measure 10-07 (2002)) against the re-flagging of vessels with a history of IUU fishing by Contracting Parties. Such a measure may not seem immediately necessary, until one recalls that many companies operating IUU vessels also operate legitimate vessels, and may wish to use these vessels as cover for their illegal activities. At its October 2002 meeting, CCAMLR brought in two important Conservation Measures regarding lists of vessels engaged in IUU fishing[12], and although both are "black" lists (CCAMLR chose not to create a "white" list other than its already existing list of vessels licensed by Members to fish in the Convention Area), they could be used by the EU legislation.

---

[10] *Regulation 2792/1999 laying down detailed rules and arrangements regarding Community structural assistance in the fisheries sector was amended by Regulation 179/2002 of 28 January 2002 as shown in italics below:*

3. The permanent cessation of vessels' fishing activities may be achieved by:

(a) the scrapping of the vessel;

(b) permanent transfer of the vessel to a third country, including in the framework of a joint enterprise within the meaning of Article 8, after agreement by the competent authorities of the country concerned, provided all the following criteria are met:

(iii) if the third country to which the vessel is to be transferred is not a Contracting or Cooperating Party to relevant regional fisheries organisations, that country has not been identified by such organisations as one which permits fishing in a manner which jeopardises the effectiveness of international conservation measures. The Commission shall publish a list of the countries concerned on a regular basis in the C Series of the Official Journal of the European Communities.

[11] *Paragraph 5.19 of the 2001 CCAMLR report states: "The Commission endorsed the advice .... on IUU fishing in the Convention Area .... and decided that: ... a list of Flags of Convenience should be compiled and maintained by the Secretariat together with a consistent process for identifying such flags".*

[12] *CCAMLR Conservation Measures 10-06 (2002) "Scheme to promote compliance by Contracting Party vessels with CCAMLR conservation measures" and 10-07 (2002) "Scheme to promote compliance by non-Contracting Party vessels with CCAMLR conservation measures"*

ICCAT has for some time been concerned about the activities of IUU vessels, particularly as most of them have crew from ICCAT Contracting Parties and there is considerable evidence of laundering of IUU catch either through links with legitimate vessels or through forging documentation[13]. In response to this concern, and following previous pressure from Japan (see Komatsu, 2001) and the EU, at its December 2002 meeting ICCAT enacted a series of resolutions[14] that create both "white" and "black" lists of vessels. Any vessel not on the white list that fishes, transships or otherwise engages in unregulated fishing is placed, following a series of review procedures, on the black list, and there are a number of punitive measures that are activated once a vessel is on this list.

Ideally, the end result of actions such as those above being taken by RFMOs and the measures envisaged in the IPOA against Flag States will be that all Flag States will become Contracting Parties or Cooperating Non-Contracting Parties to all relevant RFMOs. It should be noted, however, that if this occurs it will not be entirely painless, especially in those RFMOs that already have in place conservation measures that limit the allowable catches for certain species or stocks, or that limit the capacity of the fleets. The pain will arise from the very difficult issues of allocation among the Parties, an issue that is discussed by Butterworth & Penney (2003) in another keynote paper to this Symposium. This issue is potentially even more difficult for the IOTC, in which many of the Contracting Parties are developing coastal or small island States. These States, many of which currently have only coastal artisanal fisheries for tuna and tuna-like species in the Indian Ocean, are understandably concerned that actions to rein back the current levels of industrial fishing (purse-seine and longline) should not jeopardize their legitimate ambitions to be able to participate in this fishery in the future. Further sharing of the already shrinking quotas with new open register State Parties is likely to exacerbate these concerns.

These developments, in terms of identifying and acting against States that, through action or inaction, allow vessels flying their flag to engage in IUU fishing, are very encouraging. The fact that ICCAT has been able to use its Statistical Document Scheme to bring effective punitive action against States flagging vessels engaged in undermining bluefin, bigeye and swordfish conservation measures, points the way to other organizations, such as CCAMLR, to do the same to protect its stocks. As IUU fleets are also highly mobile and can easily move between oceans, it is also vital that there is close

---

[13] *The preambular paragraphs in ICCAT Resolution 01–19 make these concerns very clear: "RECALLING that the Commission makes yearly reviews of various trade and sighting data and based on that information prepares a list of IUU fishing vessels, RECOGNIZING that since IUU fishing vessels change their names and flags frequently to evade the sanction measures against them and that the lists of IUU fishing vessels based on the past trade data are still useful but should not be the sole tool to eliminate the IUU fishing vessels; EXPRESSING GRAVE CONCERN that a significant amount of catches by the IUU fishing vessels are believed to be transferred under the names of duly licensed fishing vessels; BEING AWARE that the majority of crew onboard the IUU tuna longline vessels are residents of the Contracting Parties, Cooperative Non-Contracting Parties, Entities or Fishing Entities; STRESSING THE NEEDS for Chinese Taipei, Japan and Parties concerned to investigate the relation between licensed vessel owners and IUU fishing activities and take necessary actions to prevent licensed vessel owners from being engaged in and associated with IUU fishing activities."*

[14] *Recommendations 02-22, "Recommendation by ICCAT concerning the establishment of an ICCAT record of vessels over 24 meters authorized to operate in the Convention Area" and 02-23, "Recommendation by ICCAT to establish a list of vessels presumed to have carried out illegal, unreported and unregulated fishing activities in the ICCAT Convention Area".*

cooperation amongst RFMOs, both in terms of developing lists of FOC/IUU vessels and in coordinating measures taken against them. It is also therefore also very encouraging that the Secretariats of the various RFMOs are maintaining close contact and consulting each other on these issues (FAO, 2002a).

The fishing industry itself is also showing increased concern about IUU fishing. As many of the best fishing masters are Spanish[15], both legitimate and IUU, the legitimate Spanish operators are becoming acutely embarrassed by their apparent association with IUU operations. They recognize that future access to international fishing agreements may well be jeopardized if Spanish nationals are involved in the IUU trade, even if Spanish flags are not. One manifestation of this concern is the move to convene a conference on deep-sea fishing in September 2003, which was initiated by the Spanish fishing industry, one of whose concerns is that IUU fishing is less easy to control on shelf/high seas stocks than in EEZ waters, so further undermining legitimate operators.

The actions described above by RFMOs and Contracting or Cooperating Parties to them are clearly forerunners to the wider trade-related measures outlined in the IPOA. Progress on that front is described in the next section.

TRADE-RELATED MEASURES

The IPOA identifies "internationally agreed market-related measures" (trade-related measures) as suitable mechanisms for combating IUU fishing. Such measures have, of course, been tried in a number of ways over the past 20 years.

The earliest and most well known attempt at using trade-related measures to force responsible fishing was the USA/Mexico Tuna-Dolphin case (see Schoenbaum, 1997). Very briefly, in the 1980s the USA became concerned about the large numbers of dolphins being killed during purse-seining operations for tuna. In 1990/91, the USA implemented tuna import embargoes on Mexico, Venezuela, Ecuador, Panama and Vanuatu, because it judged that the tuna fleets from these States did not meet its standards for dolphin protection during fishing operations. Mexico challenged this ban under GATT/WTO rules. In a similar case, the Shrimp–Turtle dispute, the USA banned imports from countries that did not require their vessels to use Turtle Exclusion Devices (TEDs) on their shrimp fishing gear (Schoenbaum, 1997).

The WTO panel ruled that both import bans were inconsistent with GATT rules, and should be removed. Schoenbaum (1997) states "WTO/GATT panels analyse Article XX(g) in terms of four requirements:

(1) that the policy of the measures for which the provision is invoked falls within the range of policies relating to the conservation of exhaustible natural resources;
(2) that the measures for which the exception is being invoked – that is, the particular trade measures inconsistent with the General Agreement – are related to the conservation of exhaustible natural resources;
(3) that the measures for which the exception are being invoked are made effective in conjunction with restrictions on domestic production or consumption; and

---

[15] *For instance, the captain and two officers on board the Russian flagged "Lena", intercepted and boarded by crew from the Australian Naval Vessel HMAS Canberra on 6 February 2002 after being sighted illegally fishing around Heard Island were Spanish nationals (Fisheries Information Service, 11 June 2002). See also "En el Índico se han apresado 37 pesqueros en seis años",.La Voz de Galicia, 9 Febrero 2003.*

(4) that the measures are applied in conformity with the requirements of the introductory clause of Article XX".

The reasons given by the panel in the Tuna–Dolphin case were that the USA could not embargo imports of+-- tuna products from Mexico simply because of the method of production, and that imported products must be treated no less favourably than domestic products. A further dissatisfaction of the panel was that the USA had not demonstrated that it had exhausted all other avenues for protection of dolphins, specifically international agreements (Schoenbaum, 1997). In the Shrimp–Turtle case, the USA was requiring that counties wishing to export shrimp to it had adopted a regulatory scheme essentially the same as their own, including using TEDs comparable in effectiveness to their own as used in the Gulf of Mexico, without considering whether individual vessels were using TEDs (IUCN, 1999).

These two cases indicate that, in disputes of this kind, the WTO is primarily concerned that trade measures should be non-discriminatory, should have been approached transparently in the context of an attempted multilateral resolution, and should be clearly directed at and connected to the policy of conservation of the resource in question. This is echoed now in paragraph 66 of the IPOA: "Trade-related measures should be adopted and implemented in accordance with international law, including principles, rights and obligations established in WTO Agreements, and implemented in a fair, transparent and non-discriminatory manner. ...[and] .. only used in exceptional circumstances, where other measures have proven unsuccessful to prevent, deter and eliminate IUU fishing".

The earliest successful trade-related measure aimed at curbing IUU fishing was that adopted by ICCAT for Atlantic bluefin tuna, as described earlier. The Bluefin Tuna Action Plan was used to prohibit imports from non-members whose vessels diminish the effectiveness of ICCAT conservation measures. In 1996, this was extended to allow the prohibition of imports from ICCAT Members who exceed their catch limits[16]. In this context, it should be noted that a statistical document scheme is not an essential precursor to the imposition of trade measures; sufficient information may already exist to provide evidence of undermining conservation measures. Thus, ICCAT maintains a Swordfish Action Plan, which together with its resolution 96–14 can be used to prohibit imports of swordfish from Member or non-Members. Also, ICCAT Resolution 98–18 is aimed at catches of tuna by large-scale longline vessels, and has been used to prohibit the importation of bigeye tuna from one ICCAT Member and four non-Members (FAO, 2002b). This is in the (then) absence of specific Statistical Document Schemes for these species.

There is only one trade-related scheme aimed at curbing IUU fishing that has been implemented for fish that are not tuna or billfish, and that is the CCAMLR scheme. This is described in detail by Sabourenkov & Miller (2003). It goes beyond the trade documentation schemes described for tuna, in that individual catches must be certified (Agnew, 2000). The difference between this scheme and the tuna schemes led the FAO Expert Consultation of Regional Fishery Management Bodies to define the CCAMLR catch certification scheme for toothfish as an amalgam of a *catch certification programme* (in which the fish are certified at point of harvesting or landing) and a *trade documentation programme* (in which documents are issued only for products that enter international trade) – FAO (2002a). The tuna schemes are trade documentation programmes. A further difference is that the trade document schemes simply identify the trade quantities, whereas

---

[16] *ICCAT Resolution 96–14.*

the CCAMLR catch certification scheme identifies whether the catch was caught illegally (IUU) or not.

CCAMLR has yet to take the next step and prohibit imports that are not accompanied by documents that state that the catch is legal. However, both this scheme and the ICCAT schemes are reported to have considerably assisted in tracing where IUU fishing is occurring, creating a disincentive to fish illegally (in the case of toothfish there is a 30–50% price difference between documented and undocumented catch) and encouraging compliance.

By themselves, documentation schemes are unlikely to be able to eliminate IUU fishing. Two of the critical limitations on the operation of such schemes, and the ultimate sanction of prohibition of imports arising from the schemes, are the type of fish and the concentration of markets among Parties to an RFMO.

First, for a trade documentation scheme to work there must be a sufficient incentive to export and import fish that are in recognizable form and that retain recognizable descriptions. Otherwise, fish would simply be imported or exported under a different name, one that is not recognizable in trade. While species-level genetic identification is usually possible (e.g. for abalone, Sweijd *et al.*, 1998), stock-level identification is more difficult and unlikely to be available routinely for customs officials (for toothfish, see Smith & McVeagh, 2000). Thus the only species that are currently subject to trade or catch documentation are the high value tuna and toothfish species.

Second, there must be a relatively concentrated market among Parties to RFMOs to make the threat of import prohibitions effective. If fish can simply pass in trade to a different country, one that is not a Party to the RFMO and therefore not bound by the import prohibition, then the whole scheme becomes ineffective. An example of this problem is the refusal of Canada, despite being an Acceding State to CCAMLR, to implement the toothfish Catch Documentation Scheme, so creating a market outside the CCAMLR Membership. If a fish has too many prospective markets, there is more likelihood of at least some of these being in States not Parties to the RFMO, or if not there will at least be more difficulty in implementing the measures with so many importing States to consider. ICCAT, for instance, acknowledges that the market for big-eye tuna and swordfish is much less concentrated than for bluefin tuna, so creating more challenges for implementation.

So far, no such RFMO scheme has come under challenge under the WTO rules. They have been negotiated openly, with considerable consultation with interested non-Parties, and they have been implemented gradually, usually with a simple tracing phase first. This points the way to the development of more such schemes on species that satisfy the trade identification and market conditions outlined above.

The findings by certain RFMOs that fishing by vessels of some Flag States has diminished the effectiveness of their conservation and management measures, and the subsequent imposition of trade-related sanctions, suggests at least superficially that the Convention on International Trade in Endangered Species (CITES) may also have a role to play in actions against IUU fishing. However, the issue of whether or not marine fish species should be considered for listing on the CITES Appendices, especially those species subject to extensive commercial harvesting, and if so using which listing criteria, has proved highly controversial and generated much polarized debate. As this would merit a paper at least as long as the current one, we will not attempt to rehearse these arguments here.

The toothfish situation is particularly interesting in this context, because the CCAMLR Convention Area does not cover the full range of occurrence of these species. Consequently, sometimes substantial catches of toothfish are reported as having been taken in high seas waters adjacent to but outside the Convention Area, so escaping the strictures of the CCAMLR Catch Documentation Scheme (CDS). While some of these fish may really have been caught on the high seas, there are strong suspicions that part of the reported catch was actually taken by IUU vessels inside the Convention Area. Issues surrounding the potential listing of toothfish on Appendix II of CITES have been discussed in Willock (2002). In 2002, following urgings from several conservation groups and possibly uniquely from an Australian fishing company involved in licensed toothfish fishing in CCAMLR waters[17], the Government of Australia tabled a proposal for the listing of toothfish under Appendix II of CITES[18]. The proposal specifically attempted to provide for the application of CCAMLR conservation and management measures for the species and for any other relevant measures or resolutions adopted by CCAMLR. However, there were a considerable number of difficulties with the proposal. These included problems of a jurisdictional nature (regarding the relative precedence of CCAMLR or CITES), and practical difficulties such as the method of setting catch limits for non-detriment findings, the practical application of these non-detriment findings, the fact that they would be unable to deal with any species other than the target species, and the difficult issues associated with introduction from the sea. In the end, the CITES parties adopted the solution proposed by Chile, which encourages all CITES parties to adopt the CCAMLR CDS, which of course is already in existence[19]. If CITES Parties comply, this should have the desired effect, which is to greatly increase the number of States adopting and implementing the CCAMLR CDS.

CONCLUDING REMARKS – OTHER POSSIBLE MEASURES

An increasingly widely used weapon in the armoury of monitoring, control and surveillance measures available to fisheries management authorities is the requirement that licensed fishing vessels should carry a functioning vessel monitoring system (VMS). It is therefore sensible to enquire whether a technological solution to deterring IUU fishing, such as VMS, could be identified. On its own, of course, VMS does nothing more than identify and track the activities of (presumably) legitimate fishing vessels; full-time specialist IUU vessels are hardly likely to install a properly functioning VMS. There are other possibilities, however.

---

[17] *"Unique partnership seek CITES listing of Patagonian toothfish". Press release by Austral Fisheries Ltd, Humane Society International, Traffic Oceania and the World Wide Fund for Nature, Sydney, Australia, 8 April, 2002.*

[18] *Government of Australia. Proposal for listing of Patagonian toothfish (Dissostichus eleginoides) on Appendix II of the [CITES] Convention. Proposal 12.39. June 2002.*

[19] *Cooperation between CITES and the Commission for the Conservation of Antarctic Marine Living Resources regarding trade in the toothfish. http://www.cites.org/eng/resols/12/12-4.shtml. This resolution includes the following recommendations (inter alia):*

"RECOMMENDS that, regarding these species, the Parties adopt the Dissostichus Catch Document used by CCAMLR for Dissostichus spp. and implement requirements for verification in all cases where specimens of these species are introduced into or exported from or transit through the territory under their jurisdiction;

INVITES all interested countries, the United Nations Food and Agriculture Organization (FAO) and other intergovernmental or international organizations active in this field to cooperate in efforts to prevent illicit trade in these species and transmit any relevant information to the Secretariat of CCAMLR; and

RECOMMENDS to the Parties that capture toothfish or that trade in toothfish products, and which have not yet done so, to adhere to the Convention for the Conservation of Antarctic Marine Living Resources and, in any case, to cooperate voluntarily with its conservation measures".

First, although it does not have the authority to require VMS to be installed on vessels that are not flying its flag, a Port State can require that fishing vessels of appropriate classes wishing to enter its ports have a functioning VMS and that they present their VMS records for inspection prior to port entry. Initially, these records may be used to determine whether the vessel had been fishing illegally in the State's EEZ or that of one of its Dependencies. However, there is certainly scope for extending this to allow refusal of entry to vessels for which there is evidence that they had engaged in any IUU fishing. Similar requirements would obviously be appropriate for Flag States when issuing authorizations to fish or Coastal States licensing vessels to fish in their EEZ. CCAMLR's Resolution 16/XIX requires that Flag States participating in its Catch Documentation Scheme for Patagonian toothfish ensure that their flag vessels maintain an operational VMS throughout the whole calendar year. The Government of South Georgia and the South Sandwich Islands has implemented this requirement to present VMS records for inspection for the whole year of operation for vessels wishing to fish within its waters. Any vessel submitting VMS records that demonstrate IUU fishing, or which are considered suspicious, will not be so licensed. In this context, it should be noted that, although VMSs can be tampered with, it is difficult to completely disguise illegal operations.

From an enforcement point of view, there is growing scope for using VMS in concert with satellite technologies. For example, Montgomery (2000) reviewed the potential use of wide-swath synthetic aperture radar (SAR) in international fisheries enforcement, and Agnew *et al.* (2002) noted the potential for using satellites to identify squid-jiggers fishing at night. With respect to SAR, when used on its own at present, there remains a difficulty in separating fishing vessels from other vessels, and in identifying whether a vessel is engaged in fishing or simply on innocent passage. For both technologies, even if they can identify a fishing vessel, there is no way of determining whether it is fishing legitimately or not. However, when used in conjunction with VMS records, they can be very effective in directing patrol vessels to areas of IUU fishing. Other possibilities that may be useful in some areas include the laying of passive sonar arrays that could be capable of picking up and identifying individual vessels.

Other technical solutions may be useful in specific situations. For instance, an excellent example of innovative thinking was given by the Spanish authorities in their approach to stopping IUU trawling in the Gulf of Cadiz. They scattered 610 artificial reefs in the area, and this is reported to have stopped all illegal trawling (Munoz-Perez *et al.*, 2000).

IUU vessels generally prefer not to fish in areas where there are licensed fishing fleets operating, especially if these vessels have observers on board who can report sightings of other vessels fishing nearby. Increasing the level of observer coverage and the temporal and spatial extent of licensed vessel operation will tend therefore to undermine the profitability of illegal or unregulated fishing, because the time available for such fishing with no licensed fleets around will be curtailed. It may even succeed in largely driving IUU fishing out of an area by making it unprofitable, a situation that is believed to have happened at South Georgia. Using data on the economics of illegal fishing obtained from industry sources, it has been shown, for instance, that for an illegal toothfish vessel to remain economic it must have access to fishing grounds for more than 30% of the year. Around South Georgia there is a 100% observer-covered licensed fishery for seven months of the year, and patrol vessels are present around South Georgia for 40% of the year. This leaves only about 10–20% of available fishing days free of legal or patrol vessel presence, which is not enough to make the area profitable.

While the argument that the presence of a legal fishing fleet in an area will have a deterrent effect on IUU fishing is generally accepted in principle, it can also create some controversies if there are other potential reasons for restricting the legal fishing season in time or space within a year. For example, in the early years of the toothfish fisheries in the Southern Ocean, the setting and hauling methods adopted for the deep longline fishing technique used to take this species also resulted in unacceptably high incidental catches of seabirds. A variety of mitigation techniques subsequently adopted by CCAMLR has greatly reduced the levels of incidental catches, but it remains true that incidental seabird catches are higher during the austral summer. Consequently, on these grounds there are good reasons to restrict the fishing seasons to avoid peak months of seabird bycatch. This presents a difficult dilemma: to what extent is it appropriate to extend the fishing season to cover periods of higher potential seabird bycatch in order to extend the period of deterrence of IUU fishing? South Africa has had to confront this directly in relation to its licensed fishery around the Prince Edward Islands, an area that has also been subject to very high levels of illegal fishing. They have opted to allow year-round fishing, despite opposition in CCAMLR, judging that any resulting added seabird bycatch taken by its vessels, which all use the full range of CCAMLR bycatch mitigation measures, will be much more than balanced by the reduced seabird bycatch by IUU vessels deterred from fishing in the area by the presence of the licensed fleet. As IUU vessels have little incentive to use seabird bycatch mitigation measures, this seems a reasonable judgement.

Enforcement is, of course, only effective within areas over which a State or organization has legal jurisdiction. Many acknowledge that the 200 nautical mile EEZ boundaries are too small to provide effective coverage of all important resources. For those resources outside EEZs, only RFMOs and the 1982 and 1995 Agreements apply. As we have seen, trade-related measures hold out some hope, but they are blunt tools. Perhaps motivated by the observation that 200 mile EEZs contain only a tiny fraction (though highly productive) of the world's oceans, occasionally the possibility of extending individual State jurisdictions has been canvassed. In most cases, these have been immediately overwhelmed by criticisms from those who recall the huge difficulties in negotiating the 1982 agreement that set up the existing 200 mile EEZs. However, one such proposal that did go a little further is the so-called "Mar Presencial" (Sea within which we are present) promoted by Admiral Jorge Martinez Busch, Commander in Chief of the Chilean Navy, in May 1991. The physical area covered by this was proposed as extending from Easter and Sala y Gomez Islands to the Chilean EEZ, and from there also to the Antarctic Continent (Dalton, 1993). The aim of this extension was to extend Chilean influence, but also to protect natural resources in the area that were important to Chile. It has thus been seen as running counter to the 1995 Agreement, and in any case it may not have added much because it did not suggest that Chilean patrols in the area would be able to arrest fishing vessels in high seas waters.

There being no appetite for addressing the high seas IUU fishing problem with extended state jurisdictions, the only ways in which direct enforcement measures can deter IUU fishing arise when such vessels are arrested by coastal States for fishing illegally in their current EEZ waters, or by the vessels' Flag States, if the vessels are found to have contravened conservation measures of RFMOs to which the Flag States are Contracting Parties. This further emphasizes the importance of indirect measures for deterring IUU fishing on the high seas contained in the IPOA. Of these proposed measures, the ones that are currently showing greatest promise are those involving introduction of catch certification or trade documentation schemes, and trade sanctions associated with them.

REFERENCES

Agnew, D. J. 2000 The illegal and unregulated fishery for toothfish in the Southern Ocean, and the CCAMLR Catch Documentation Scheme. *Marine Policy*, **24**: 361–374.

Agnew, D. J., Hill, S. L., Beddington, J. R., Purchase, L. V. & Wakeford, R. C. 2002. Sustainability and management of SW Atlantic squid fisheries. Paper presented at the World Conference on the Scientific and Technical Bases for the Sustainability of Fisheries, 26–30 November, 2002. University of Miami, USA.

Butterworth, D. S. & Penney, A. J. 2003. Allocation in high seas fisheries: avoiding meltdown (this volume).

Charles, A. T., Mazany, R. L. & Cross, M. L. 1999 The economics of illegal fishing: a behavioural model. *Marine Resource Economics*, **14**: 95–110.

Dalton, J. G. 1993. The Chilean mar presencial: a harmless concept or a dangerous precedent? *International Journal of Marine and Coastal Law*, **8**(3): 397–418.

Edeson, W. M. 1966 The code of conduct for responsible fisheries. An introduction, *International Journal of Marine and Coastal Law*, **233**.

FAO. 1999 Report of the Twenty-Third Session of the Committee on Fisheries, Rome, 15–19 February, 1999. *FAO Fisheries Report*, **595**. 70 pp.

FAO. 2000. World Review of Fisheries and Aquaculture. The State of World Fisheries and Aquaculture, **1**. FAO, Rome. 142 pp.

FAO. 2001 Report of the Twenty-Fourth Session of the Committee on Fisheries, Rome, 26 February – 2 March, 2001. *FAO Fisheries Report*, **655**. 87 pp.

FAO. 2002a Expert Consultation of Regional Fishery Management Bodies on Harmonization of Catch Certification, La Jolla, USA, 9–12 January 2002. FAO, Rome.

FAO. 2002b Implementation of the international plan of action to prevent, deter and eliminate illegal, unreported and unregulated fishing. *FAO Technical Guidelines for Responsible Fisheries*, **9**. 122 pp.

Freestone, D. & Makuch, Z. 1996. The new international environmental law of fisheries: the 1995 United Nations Straddling Stocks Agreement. *Yearbook of International Environmental Law*, **7**: 3–51.

Hart, P. J. B. 1997 Controlling illegal fishing in closed areas: the case of mackerel off Norway. In: *Developing and Sustaining World Fisheries Resources: the State of Science and Management. Proceedings of the Second World Fisheries Congress Brisbane, 1996*. Volume **2**. (Ed. by D. A. Hancock, D. C. Smith, A. Grant and J. P. Beumer), pp. 411–414. CSIRO Publishing, Collingwood, Australia.

IUCN, 1999 The catch documentation scheme under WTO rules. *CCAMLR Document*, **CCAMLR-XVIII/BG/48** (mimeo).

Komatsu, M. 2001 The importance of taking cooperative action against specific vessels that are diminishing the effectiveness of tuna conservation and management measures." Document IUU/2000/5 presented at the Expert Consultation on Illegal, Unreported and Unregulated Fishing, organised by the Government of Australia in cooperation with FAO, Sydney, Australia 15–19 May, 2000. *FAO Fisheries Report*, **666**: 59–87.

Montgomery, D. R. 2000 International fisheries enforcement using wide swath SAR. In: *SAR Symposium on Coastal and Marine Applications of Wide Swath SAR, Laurel, MD, March 1999. Johns Hopkins APL Technical Digest*, **21**(1): 141–147.

MRAG, 1993 Control of Foreign Fisheries. Report of a research project conducted under the ODA Fish Management Science Programme. 27 pp.

Munoz-Perez, J. J., Mas, J. M. G., Naranjo, J. M., Torres, E. & Fages, L. 2000. Position and monitoring of anti-trawling reefs in the Cape of Trafalgar (Gulf of Cadiz, SW Spain). *Bulletin of Marine Science*, **67**: 761–772.

Sabourenkov, E. N. & Miller, D. G. M. 2003 The management of transboundary stocks of toothfish, *Dissostichus* spp., under the Convention on the Conservation of Antarctic Marine Living Resources (this volume).

Schoenbaum, T. J. 1997. International trade and protection of the environment: the continuing search for reconciliation. *American Journal of International Law*, **91**: 268–313.

Smith, P. & McVeagh, S. M. 2000 Allozyme and microsatellite DNA markers of toothfish population structure in the Southern Ocean. *Journal of Fish Biology*, **57**(Suppl. A): 72–83.

Sweijd, N. A., Bowie, R. C. K., Lopata, A. L., Marinaki, A. M., Harley, E. H. & Cook, P. A. 1998 A PCR technique for forensic, species-level identification of abalone tissue. *Journal of Shellfish Research*, **17**: 889–895.

Willock, A. 2002 Unchartered waters: implementation issues and potential benefits of listing toothfish in Appendix II of CITES. Traffic International, Cambridge, UK and Traffic Oceania, Sydney, Australia. 35 pp.

Development of an estimation system for U.S. longline discard estimates of bluefin tuna

Carl M. O'Brien, Graham M. Pilling
CEFAS, Fisheries Laboratory, Pakefield Road, Lowestoft, NR33 0HT, UK
Craig Brown
NOAA Fisheries, Southeast Fisheries Center, Sustainable Fisheries Division, 75 Virginia Beach Drive, Miami, FL, 33149-1099, USA

ABSTRACT: Historical estimates of U.S. longline dead discards of bluefin tuna have been based on tallies of reported dead discards made by fishers on logbooks. Initiation of an observer programme allowed provisional estimates of bluefin tuna discards to be based on direct observation of the U.S. pelagic fleet for the period 1992–1999. The method applied provided a basis for characterizing uncertainty in estimates. However, sparse sampling in a number of spatial (e.g. geographical) and temporal (e.g. quarterly) strata meant pooling was required across strata to achieve a minimum number of observations per stratum. Further attention must be paid to the estimation of discards, so that its effects can be fully included in assessments of shared stocks. The paper presents an alternative approach to estimating dead bluefin tuna discards, which will improve both the estimates themselves and the calculation of uncertainty about them. Models are developed using observer data. A number of important model covariates are identified: quarter, year, grouped fishing area and set target species. The statistical concept of conditionality is applied, and a flexible mixture distribution – the negative binomial – proposed for modelling discard data. The results of applying these models to observer programme data to estimate total bluefin tuna discards are presented.

INTRODUCTION

North Atlantic United States (U.S.) fishing fleets primarily use longline gear to target tuna (*Thunnus* sp.) and swordfish (*Xiphias gladius*). Catch of these species may be discarded under certain market or regulatory conditions; for example, current regulations only allow the landing of one Atlantic bluefin tuna (*Thunnus thynnus*) per trip. Also, non-targeted by-catch, which may be hooked or entangled in the longline, are similarly discarded. A proportion of these discards will be dead. Efforts by U.S. National Marine Fisheries Service (NMFS) scientists have focused on improving estimates of dead discards, because this mortality must be included in estimates of the catch for stock assessment and management. This paper focuses on the estimation of dead discards of Atlantic bluefin tuna.

The year 2000 International Commission for the Conservation of Atlantic Tunas' (ICCAT) West Atlantic Bluefin Tuna Species Group recommended:

> *...that further attention needs to be paid to the collection of data on discards and their subsequent estimation so that the effect of discarding can be fully included in the stock assessment. The quality of the information is enhanced by Observer Programs. Observer sampling should be sufficient to quantify discarding in all months and areas and to avoid the need for pooling across time or area strata thought to be important to constructing estimates. Studies should be conducted to improve estimation of discards and to identify methods that would reduce discard mortality. Studies should also be conducted to estimate the subsequent mortality of bluefin discarded alive.*

Estimates of U.S. longline dead discards of bluefin tuna up to the mid–1990s were based on tallies of reported dead discards made by fishers on logbooks. Based on recommendations from the International Commission for the Conservation of Atlantic Tunas' Standing Committee on Research and Statistics (ICCAT SCRS), the U.S. implemented scientific observer sampling in selected U.S. fleets to pursue research on methods to address characterizing the total catch composition and disposition from the observed fleet. Initiation of the observer programme in 1992 permitted direct observations of discard rates by observers. Despite this, the West Atlantic bluefin tuna recovery plan allocation for dead discards was based on an historical standard that, in turn, was based on the logbook tally estimates. The corresponding consistent estimator for longline discards in the 1999 calendar year was 30 metric tonnes. Direct observations of dead discard rates are typically higher than self-reported rates from logbooks. This is not unexpected because, in general, it is believed to be more difficult for fishers to maintain accurate daily records of the numbers and condition of fish thrown back to the sea than for fish that are kept for sale.

Discarding in the U.S. Atlantic longline fleet has been estimated using several approaches. A simple proportional extrapolation of the observed catch rates to the logbook reported effort was used in 1993 (Cramer, 1995). This approach ignored self-reported information on catch rates from logbooks and did not provide a measure of uncertainty in catch estimates. Scott & Brown (1997) modified this simple approach to estimate marine mammal and turtle by-catch for the U.S. Atlantic pelagic longline fleet for 1994–1995 with a measure of uncertainty. Owing to sparse sampling in a number of spatial (e.g. geographical) and temporal (e.g. quarterly) strata, the estimation procedure also provided an option for pooling across strata, and the robustness of by-catch estimates from several different pooling schemes for by-catch rates were examined.

The application of two-stage models has been recommended within a fisheries context whenever a particular process under observation has a positive probability of not occurring (Pennington, 1983; Borchers *et al.*, 1997). Within this approach, presence/absence is first modelled as a binary process, and then the magnitude of the event under observation is subsequently modelled conditional upon presence. Yeung *et al.* (2000) developed the ideas of Scott & Brown (1997) further, by applying a two-stage approach and developing a specified criterion of a minimum number of observed sets to determine the level of pooling – set at 30 observations per stratum. Levels of quarter, year and grouped fishing area (NAREA – geographical zones used to classify observed and reported U.S. Atlantic pelagic longline effort – see Figure 1) were pooled in that order until the criterion was met. The method assumed

a log-normal distribution of the positive by-catch rate observations. Estimates were then constructed as a product of the proportion of occurrences of discarding and the average rate of discarding, given that there was discarding. The variance was a function of the variability of non-zero dead discard rates, as well as the number of discarding and non-discarding events. A slight modification of this approach was used to develop provisional estimates of U.S. longline discards of bluefin tuna based on direct observation of the U.S. pelagic fleet for the period 1992–1999 (Brown, 2001). The pooling order applied was based on an analysis that indicated smaller differences between years than between geographic or quarterly strata (unlike the analysis of Yeung *et al.*, 2000). The effect of the pooling assumption was compared and it was observed that, in recent years, estimates made without pooling were somewhat lower than with pooling, although this was not the case for all years. However, tests of normality were not presented, and the validity of the assumption that the natural logarithm of non-zero dead discard values was normally distributed was not commented upon.

The Centre for Environment, Fisheries and Aquaculture Science (CEFAS), in their review of the methodology used to estimate bluefin tuna dead discards, suggested a number of alternative models that could produce improved estimates of bluefin tuna dead discards. For example, the approach suggested by O'Brien & Fox (2003) for the estimation of annual egg production in the Irish Sea could be modified. In this approach, the probability of obtaining a zero observation (namely, no discarding) can be estimated first using a logit-link with a binomial error distribution. The mean number of bluefin tuna discarded dead, given that there was discarding, can then be estimated, using a log-link with a Gamma error distribution. The product of the two model components would give a bluefin tuna dead discard estimate whose variance may then be obtained through a computer-intensive technique such as bootstrapping (O'Brien & Fox, 2003).

Here, we examine discard rates of bluefin tuna from the U.S. Atlantic longline fleet observer programme and then examine the data available from the U.S. longline tuna fishery. Using available observer data, we develop methods to model the dead discards of bluefin tuna and present preliminary results. In the near future, these models will be applied and tested on logbook data from the U.S. longline fishery to estimate annual dead discards of bluefin tuna, which are included in the Total Allowable Catch (*TAC*) set by ICCAT. The approach developed should lead to improved estimates of bluefin tuna dead discards, stock assessment and management.

AVAILABLE INFORMATION

Two basic sources of data were identified – logbook data, completed by vessel skippers, and observer data, completed by observers placed on board the U.S. longline vessels.

LOGBOOK DATA

The NMFS pelagic logbook database contained information collected from 1986 through 2000. The current vessel logbook collects information in a number of general categories:

(1) Vessel identification details.
(2) Target species group details (assessment by skipper as to the species targeted by that set; may frequently contain multiple entries).

(3) Gear details (may contain multiple entries owing to the use of more than one gear during a trip).
(4) Set and haul details (date and time, location).
(5) Gear details (information on gear used, e.g. number of hooks, number of light sticks, length of gear).
(6) Catch (species of swordfish and tuna, shark, turtles and other species). Details are noted as number kept, number thrown back dead or alive, and estimated weight (lbs) kept.

OBSERVER DATA

Observer data for pelagic longline are available for the period May 1992 to December 2001. Observed sets represent some 5% of the total number of sets, and encompass 165875 individual biological records.

A number of data fields are common to both the logbook and observer logsheets (Table 1). The time of day at which a set is made may be important, because swordfish sets tend to be made at night, whereas sets targeting tuna tend to be during the day. Light sticks are often used at night, to attract swordfish to the line. However, if a relatively small number of light sticks are used compared with the number of hooks (e.g. <50%), tuna may also be being targeted by the set. Water temperature may also be important, because fishers may target areas of upwelling or thermal fronts when fishing.

Table 1 Fields common to both the longline logbook and observer logsheets. Concomitant information indicated with a * were influential in the GLM analysis performed by NMFS (Brown, 2001).

Factor	Database code
Year*	Year
Season/quarter/month/date*	[Calculated], quarter, [Calculated], esdate
Location/NAREA/Area/Lat and Long*	marea, NAREA, area, latd, lond
Time of day (am/pm/day/night/ twilight/dawn)	[Calculated from estimhr and estimmin]
Soak time	hours
Depth fished	hkdepmin.calc and hkdepmax.calc
Water temperature	estemp
Light sticks	litenum
Number of hooks	hookset
Bait type/status	Combination of baitkind and baittype. Live = baitkinds 1-3 and 5+ & baittype 3. Dead = 1-3 and 5+ & type 1-2 and 9, artificial = baitkind 4
Number of each species kept	nswok, nbftk, nalbk, nbetk, nyftk, nblkk, nbumk, nsaik, nwhmk, ndolk, nbshk
Number of each species discarded	nswod, [nbftd], nalbd, nbetd, nyftd, nblkd, nbumd, nsaid, nwhmd, ndold, nbshd

PRELIMINARY ANALYSIS

Graphical approaches were used to examine relationships between the number of dead bluefin tuna discard events and various factors. The geographical distribution of bluefin tuna dead discards in each year was also examined. A diagram of the NAREAs designated by NMFS for geographical analysis is presented in Figure 1. Specific species codes have been used throughout data analysis (Table 2) and represent the species or groups of species targeted by the set.

A plot of the numbers of bluefin tuna discarded dead against the numbers kept in the catch (Figure 2) presents the magnitude of the problem. In a single set, more than 30 bluefin tuna can be discarded dead. The highest level of discarding was where no bluefin tuna were retained. This may be due to either targeting issues or that a bluefin tuna was already retained on board the vessel, so requiring any additional fish to be discarded. The single event where three bluefin tuna were retained was before the regulation limiting the number of bluefin tuna to be kept to one was in place.

The number of dead bluefin tuna discarded was examined against the number of individuals of other species caught and retained in that set, for all areas combined (Figure 3). There are minor species-specific differences in the number and distribution of bluefin tuna dead discards relative to the number of each species retained in the catch. Overall, high numbers of dead discards are generally found where low numbers of other species are retained from the catch.

The level of dead discarding identified by the observer programme was examined by year and geographical location (Figure 4). The number of bluefin tuna discarded dead was displayed on a natural logarithmic scale (specifically $\log_e(n+1)$) to allow for extremely large discard events. There was considerable variation in the number of bluefin tuna discarded dead between years and NAREAs. Throughout the period, discarding was relatively common along the northern part of the eastern seaboard of the U.S. Discard levels in the Gulf of Mexico and off the east coast of Florida show more interannual variability. Discarding was particularly prevalent in 1995, confirmed by a plot of the numbers of bluefin tuna discarded dead each year (Figure 5). Analysing the 1995 discarding event by NAREA (Table 3), the high levels of discarding were all in the northeast sector (NAREA 'NEC').

The number of bluefin tuna dead discards were examined by 'target species' (species targeted by the set) and quarter of the year (Figure 6). A number of comments are worthy of note. First, where either tuna and swordfish mix (MIX) or all tuna except bluefin tuna (TUN) were the set target, most dead bluefin tuna discards were in the second quarter of the year. Second, where yellowfin tuna (YFT) were targeted, most dead discards were also in this second quarter, but discards in the first quarter were also high. In contrast, where swordfish (SWO) were targeted, dead discards occurred throughout the year.

Details of the frequency distributions of the number of bluefin tuna discarded dead by both target species and NAREA are listed in Table 4. During the period 1992–2001, no bluefin tuna were discarded dead in any location where dolphinfish or shark were targeted by the set, or in the Caribbean (CAR) or offshore (OFS) regardless of the target species. In many other combinations of NAREA and target species, bluefin tuna dead discarding was a binary event; i.e. a single bluefin tuna was either discarded dead or there was no dead discarding. In the remaining cases, the number of bluefin tuna discarded dead formed an obvious frequency distribution. Relatively high numbers of dead discard events were noted where swordfish or yellowfin and bigeye tuna were caught. Discarding was most

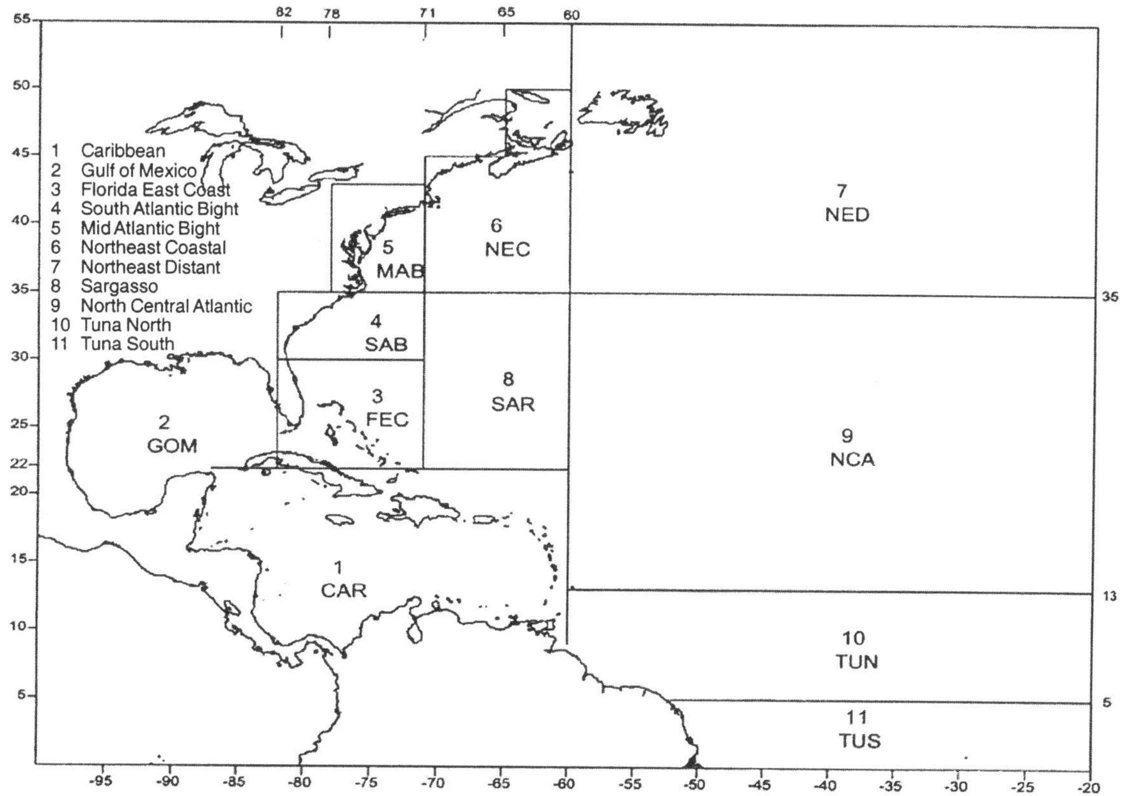

Fig. 1 Geographical zones (NAREAS) used to classify observed and reported U.S. Atlantic pelagic longline effort. For the purpose of estimation, several strata were combined. The Southeast Coast (SEC) stratum was defined as areas 3 and 4; the Northeast Coastal (NEC) stratum was defined as areas 5 and 6; and the Offshore South (OFS) was defined as areas 8, 9,10 and 11.

Table 2 Description of species codes used within the analysis.

Species code	Common name
BET	Bigeye tuna
DOL	Dolphinfish
SWO	Swordfish
YFT	Yellowfin tuna
SHX	Sharks (specifically Blue shark, BSH)
TUN	All albacore, bigeye and yellowfin tunas, but excluding bluefin tuna
MIX	Tuna (as for TUN) and including swordfish
BFT	Bluefin tuna

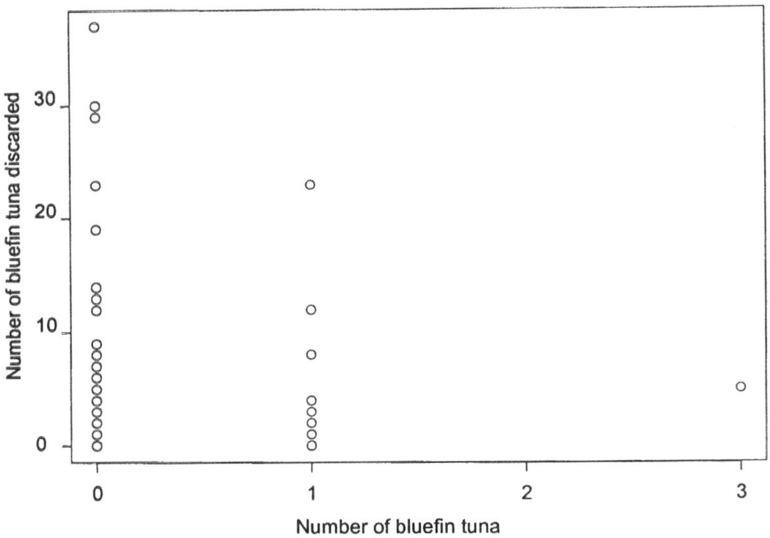

Fig. 2 Number of bluefin tuna discarded dead from a set against the number retained in the catch.

frequent in the Northeast Coast region (NEC) when targeting tuna (TUN), a result of the very high discard numbers seen there in 1995. Discarding of dead bluefin was common when swordfish were targeted, largely regardless of the NAREA fished.

These results suggest that year, NAREA, quarter and species targeted by a set may be important covariates for the model.

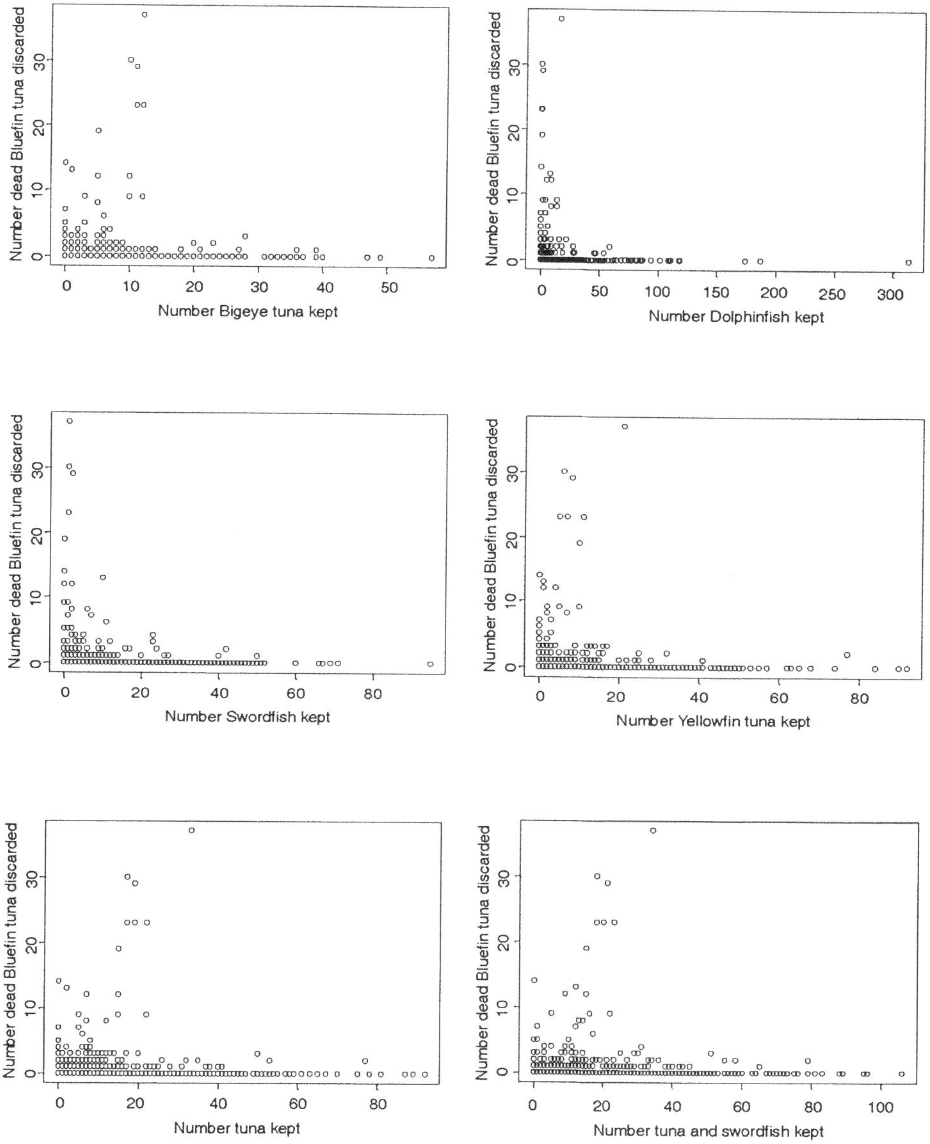

Fig. 3 Scatterplots of the number of bluefin tuna (BFT) discarded dead against the number of each species kept, all areas combined. In sets where blue shark were captured, no bluefin tuna were discarded dead.

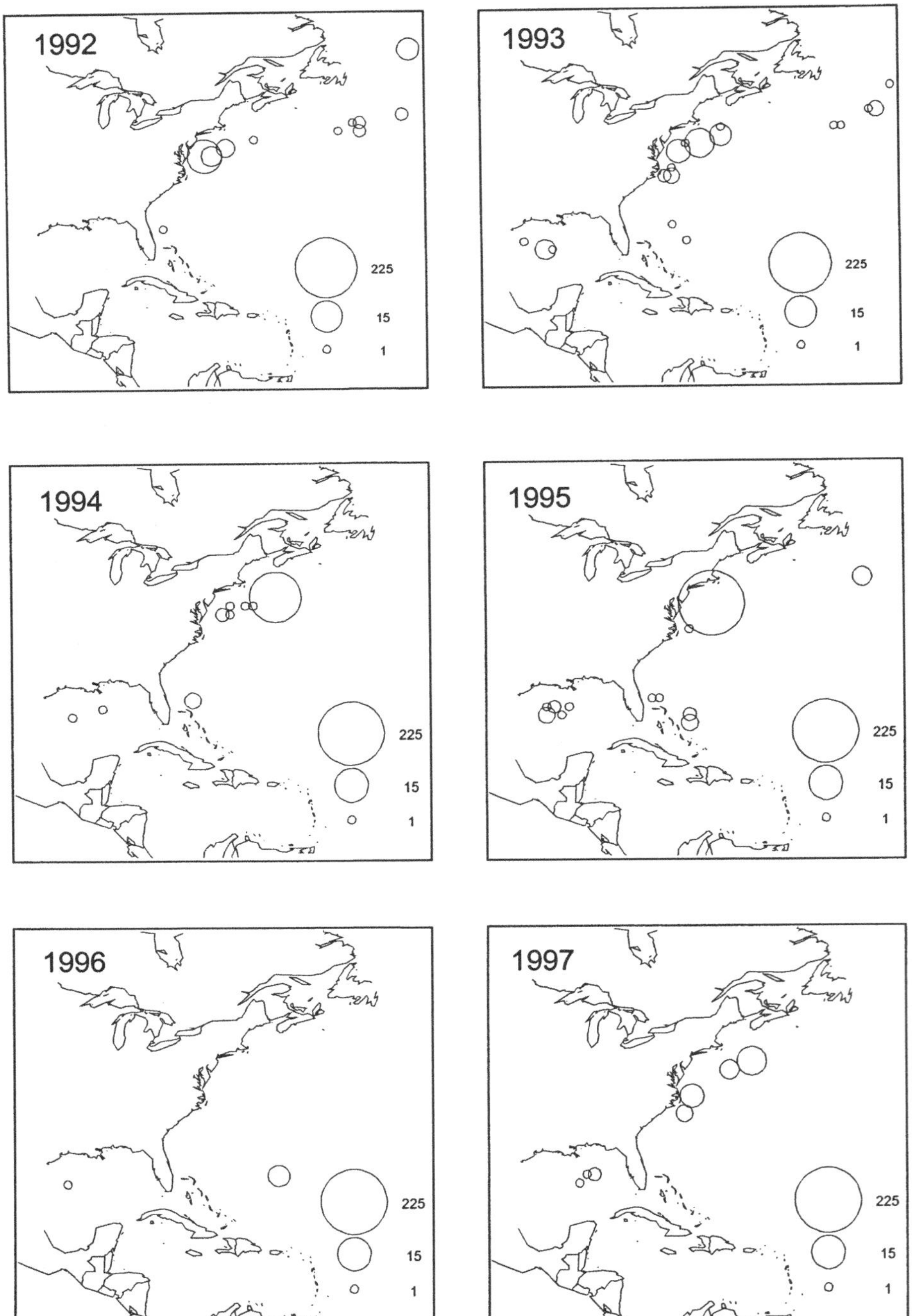

Fig. 4 Distribution and numbers of bluefin tuna discarded dead from the U.S. longline fishery, from observer logbooks, by year. Note that a logarithmic scale is used for number of discards ($\log_e(n+1)$). Data grouped by 1° latitude and longitude grid square.

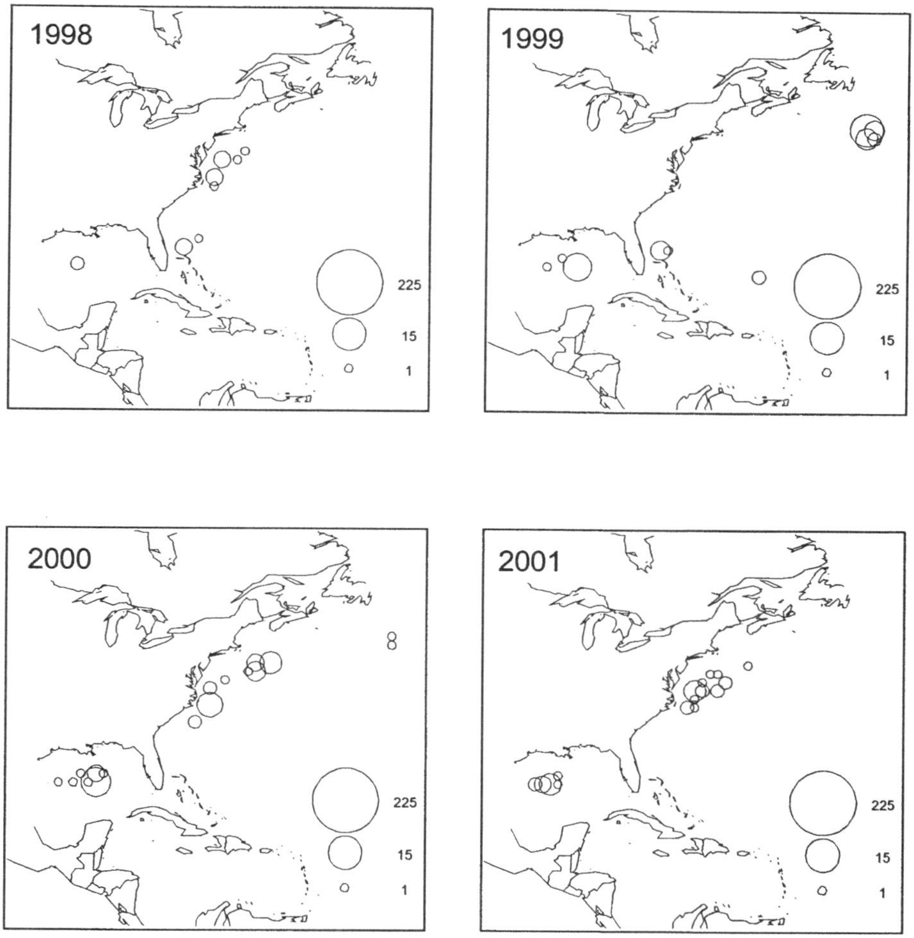

Fig. 4 continued.

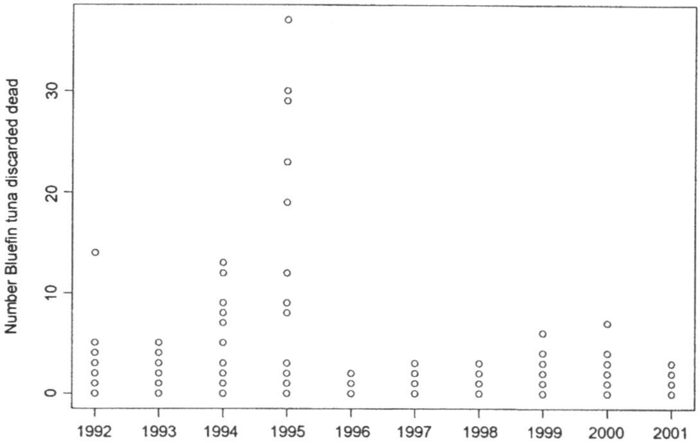

Fig. 5 Numbers of bluefin tuna discarded dead in a single set, by year.

Table 3 Frequency of bluefin tuna dead discard events in 1995 by magnitude (number discarded in a single set) and NAREA.

NAREA	Number of bluefin tuna discarded dead in single set											
	0	1	2	3	8	9	12	19	23	29	30	37
CAR	48	0	0	0	0	0	0	0	0	0	0	0
GOM	180	4	2	0	0	0	0	0	0	0	0	0
NEC	185	4	0	0	1	2	1	1	3	1	1	1
NED	62	2	1	0	0	0	0	0	0	0	0	0
OFS	82	0	0	0	0	0	0	0	0	0	0	0
SEC	89	2	1	1	0	0	0	0	0	0	0	0

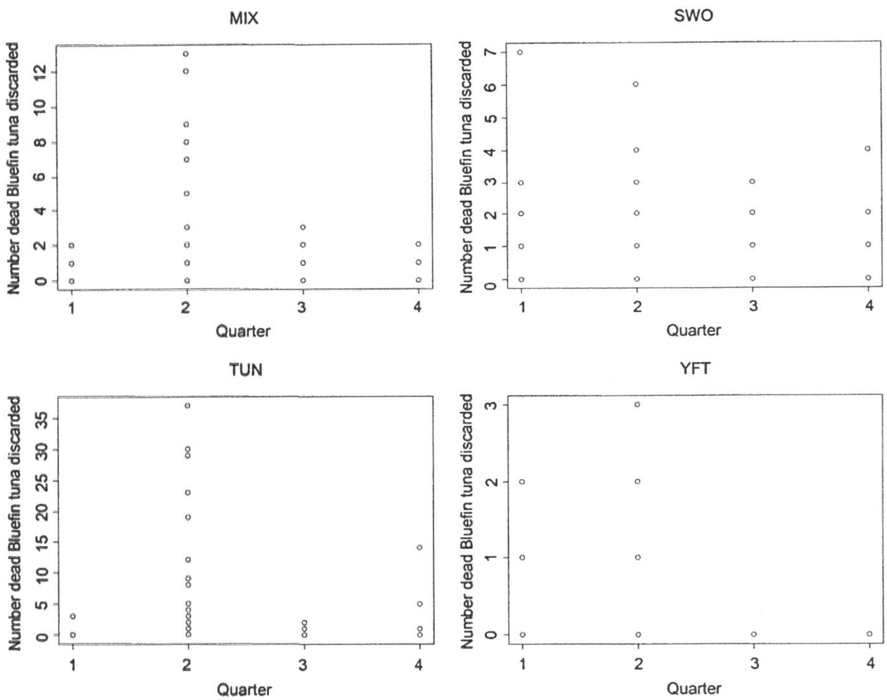

Fig. 6 Plots of the number of bluefin tuna discarded dead by quarter of the year, for four different target species (MIX, SWO, TUN, YFT). See Table 2 for details on target species codes.

Table 4 Number and distribution of bluefin tuna discarded dead during the period 1992–2001 by NAREA and target species. N = no discarding; B = binary [0,1], other events displayed as minimum and maximum numbers discarded [0,X].

Category	CAR	GOM	NEC	NED	OFS	SEC
BET		N	B			[0,4]
DOL						N
SWO	N	[0,2]	[0,7]	[0,6]	[0,2]	[0,3]
YFT		[0,3]	[0,3]		N	N
SHX			N			N
TUN	N	[0,3]	[0,37]	B		B
MIX		B	[0,13]	B		[0,3]

MODEL SPECIFICATION

This section describes the specification of the initial model of bluefin tuna dead discards.

EMPIRICAL DISTRIBUTION OF DISCARDS

The empirical distribution of discards per set can vary and depends on a number of different factors. In many cases, scientists assume the Poisson or related distributions as processes that may have generated the discard data (Taylor, 1953), but data transformations may seem appropriate; e.g. the non-linear natural logarithmic transformation in the case of an assumed log-normal distribution. Other fishery scientists, in contrast, remove zero catches and, after suitable transformation, treat both categories of data (zero and non-zero discards) separately (Pennington & Grosslein, 1978). Often, however, transformations are not without problems; conclusions representative of the original (untransformed) discard data might be difficult and at best *approximate*.

Aggregation, clumping or contagion characterizes most populations in the wild and counts of individuals $x_1, x_2, ..., x_n$ in small subareas of known size are among the oldest techniques in ecology employed to aid in the understanding of such processes (Du Rietz, 1929, 1930; Whittaker, 1962). The approach to statistical analysis of spatial patterns based on quadrat counts is simple and straightforward. For the fishery science application of this paper, the quadrat of interest is defined to be the sampled fishing event – each event corresponding to a single vessel trip.

OVER-DISPERSION AND CONDITIONALITY

Once over-dispersion or aggregation has been identified, there is a need to account for it, possibly by use of either a probabilistic distribution or model. Representing the over-dispersion in catch data by a specific model leads naturally to the theory of mixtures of Poisson distributions. The development of the theory of such mixtures has its roots in the early seminal works of Greenwood & Yule (1920) and Lundberg (1940), who introduced the *idea of conditionality* on a realization of a variable parameter. Using this approach, O'Brien *et al.* (1998) modelled daily catch rates of albacore (*Thunnus alalunga*) in the Bay of Biscay and adjacent waters; and O'Brien *et al.* (2000) accounted for spatial scale in research surveys by analysing trawl catches of 2-year-old cod from English, German and international groundfish surveys in the North Sea.

The idea of conditionality is of fundamental importance and leads to the development of models based on the identity:

$$P_n(t) = \int_0^\infty \{ e^{-\alpha t} (\alpha t)^n / n! \} \, u(\alpha) \, d\alpha$$

In such processes the parameter α appears as a random variable with a density $u(.)$, and the probability of exactly n events during time t is denoted $P_n(t)$. Typically, such processes are considered for a standard period of time and αt is replaced by a single parameter. Suppose, for the purposes of illustration, that the discrete variable Y representing collected catch data might be Poisson distributed with mean Z, and further, that this mean is itself a random variable which might be described by the Gamma distribution with mean μ and index $k\mu$; i.e. $E\{Z\} = \mu$ and $\text{var}\{Z\} = \mu / k$. Then, this mixture process leads directly to the negative binomial distribution (Plackett, 1981) with mean, $E\{Y\} = \mu$, and variance, $\text{var}\{Y\} = \mu (1 + \mu) / k$.

For a negative binomial distribution, the *index of cluster frequency* ICF gives the method of moments estimate of k (Anscombe, 1950), a parameter inversely related to the clumping of the population. Myers (1978) showed that such spatial indices based on k were strongly correlated with population density and that consequently, the use of ICF is generally recommended only once the suitability of a negative binomial distribution has first been assessed through the calculation of a goodness-of-fit test statistic such as the U-statistic.

U-STATISTIC GOODNESS-OF-FIT TEST

This test uses the observed and expected variances of the negative binomial distribution (Evans, 1953) and is defined by

U = sample variance – expected variance under a negative binomial distribution

$= s^2 - \{\bar{x} + \bar{x}^2 / k\}$

where the symbol s^2 denotes the sample variance and the symbol $\bar{x}$ the sample mean, both defined in the usual way. The standard error of U is defined as follows:

$$\text{s.e.}(U) = \sqrt{\frac{1}{n} \quad \{2\bar{x}(\bar{x} + g_1)g_2 \quad \frac{[g_2 g_2 \ln g_2 - g_1(1 + 2g_1)]}{[g_2 \ln g_2 - g_1]} + g_3\}}$$

where $g_1 = \bar{x} / k$

$g_2 = 1 + g_1$

$g_3 = g_2 \, g_1^4 \, [g_2^{(1+k)} - (\bar{x} + g_2)] / [g_2 \ln g_2 - g_1]^2$

and ln denotes the natural logarithm to the base e. The expected value of U is zero, so one approach to test if the observed value of U is significantly different from zero is to compare the calculated value with 2 standard errors of U. If a calculated value of U

exceeds 2 s.e.(U), then reject the null hypothesis that the negative binomial distribution is a suitable model for the observed data at the 5% level of significance. Once the suitability of the negative binomial distribution has been ascertained, the distribution and its estimated parameters may be used to simulate catch distributions (c.f. O'Brien *et al.*, 2000).

SPATIAL ABUNDANCE DISTRIBUTION AND RELATIONSHIP WITH COVARIATES

The negative binomial is a discrete probability distribution that includes both Poisson and geometric distributions as specific cases. The mode may be zero, with the probability of increasing values having monotonically decreasing probabilities. The negative binomial model has found increasing use in fisheries data, particularly for catch and catch rate data (Bannerot & Austin, 1983; Power & Moser, 1999) including for tuna (Lo *et al.*, 1992; Porch & Scott, 1993; Dong & Restrepo, 1996; Miyabe, 1997; Porch, 1997), where occurrence (or catch) can frequently be zero. It removes the requirement to decide *a priori* on the form of the catch distribution; only the nature of the mixing process needs to be considered. Generalized Linear Models (GLMs; Nelder & Wedderburn, 1972; McCullagh & Nelder, 1989) can then be used to apply a variety of error structures depending on the degree of aggregation, and predicted values estimated with confidence intervals based on, for example, boot-strapping (Efron & Tibshirani, 1993). All this suggests that the negative binomial might be appropriate within the context of bluefin tuna dead discard modelling. Possible relationships between the number of discarded dead bluefin tuna and one or more explanatory variable (*covariate*) can be investigated. Testing hypotheses about such relationships requires a statistical analysis.

There is a large and extensive literature on this problem in classical regression (see Thompson, 1978a, b): *given a number of possible covariates, find a subset of these which is in some sense optimum for constructing a set of fitted values*. Wolstenholme *et al.* (1988) have implemented a strategy for generalized linear models that is appropriate to the fishery application in this paper.

Implicit in the strategies proposed is that there is a balance to be struck between the improvement in the goodness-of-fit to data obtained by adding an extra term to a model, and the increase in complexity introduced by such an addition. A *combinatorial explosion* can result if all possible combinations of subsets of covariates are tested for inclusion in a model but approximate methods for generating a single *optimum* subset have been proposed. These include: (a) forward selection, whereby at each stage, the best unselected covariate satisfying a criterion is included next until none remain; (b) backward elimination, which begins with the full set and eliminates the worst covariates one by one until all remaining are necessary; and (c) stepwise regression, which combines the two procedures described under (a) and (b).

THE MODEL, ESTIMATION AND DIAGNOSTICS

Following this initial model specification, the rest of this paper will report on the methods developed to model the dead discards of bluefin tuna from the U.S. Atlantic longline fleet observer programme.

A generalized linear model was developed but conditioned on target species, i.e. yellowfin tuna (YFT), swordfish (SWO), all tuna except bluefin tuna (TUN), and tuna and swordfish (MIX). For each GLM, the calculated U-statistic goodness-of-fit test for the negative binomial distribution did not indicate that such a distribution was inappropriate.

Some notation and terminology is required first. The observations comprise an $n \times 1$ vector $\mathbf{y}$ of responses and an $n \times p$ matrix $\mathbf{X}$ of explanatory variables. The response variables Y_i are assumed independent with expectation $E(Y_i) = \mu_i$, where μ_i depends on the ith row $x_{i()}$ of $\mathbf{X}$ through the linear predictor $\eta_i = g(\mu_i) = x_{i()}\beta$. Here $g(.)$ is the link function and β is a $p \times 1$ vector of unknown parameters. The covariates comprise the p columns of $\mathbf{X}$; each covariate an $n \times 1$ vector $x_{()j}$. The distribution of Y_i is assumed to be a member of the exponential family with variance v_i, where v_i is the product of two functions; one depending only on a dispersion parameter Ø, the other a known function of μ_i. The probability density function of Y_i is denoted by $f(y_i; \mu_i; Ø)$.

The maximum likelihood (ML) estimates $\hat{\beta}$ of β are determined by iterated weighted least squares using weights $w_i = v_i(d\mu_i/d\eta_i)^2$. If $\mathbf{W}$ is the $n \times n$ diagonal matrix $\mathbf{W} = \text{diag}\{ w_i \}$, define h_i to be the ith diagonal element of

$$\mathbf{H} = \mathbf{W}^{1/2}\,\mathbf{X}\,(\mathbf{X}^T\,\mathbf{W}\,\mathbf{X})^{-1}\,\mathbf{X}^T\,\mathbf{W}^{1/2}$$

The Pearson residuals (standardized to have unit asymptotic variance) and deviance residuals (similarly standardized) are defined in the usual way.

In practice the values of w_i, h_i and v_i, all of which are functions of β, have to be determined at the ML estimates $\hat{\beta}$. Most of the standard statistical software packages currently in use will return estimates of the parameters, their standard errors and the goodness-of-fit deviance statistic.

Applying the stepwise procedure in Wolstenholme *et al.* (1988) leads to the identification of the covariates quarter, year and NAREA as important main effects. Repeating the stepwise procedure for first-order interaction effects does not indicate the presence of any significant interaction effects between covariates, and simple main effect models are adequate.

An important plot, after fitting, is that of standardized residuals against fitted values (with the latter transformed to the constant-information scale of the error distribution). The plot is capable of revealing isolated points with large residuals, or a general curvature, or a trend in the spread of residuals along the abscissa. Details of the technique are given in O'Brien & Kell (1996).

For the fishery application analysed in this paper, it is recommended to assess distributional form by the quantile-quantile (Q-Q) plot (Davison & Gigli, 1989) with simulation envelope (Atkinson, 1985). These appear to indicate a reasonable agreement between model and data for individual species YFT and SWO (Figure 7), but not for the two grouped target species MIX and TUN (Figure 8). An explanation of this result is discussed below, but with hindsight was merely to be expected.

DISCUSSION

Discard rates of dead bluefin tuna have been analysed based on data collected through the U.S. NMFS observer programme. The idea of conditionality has been applied, and a flexible mixture distribution – the negative binomial – proposed for the modelling of discard data. The application of generalized linear models (McCullagh & Nelder, 1989) to the modelling of discard data follows naturally (O'Brien *et al.*, 1998).

The need to decide whether discard data follow a log-normal, Poisson, gamma or some other distribution can be replaced by a general assumption about mixing processes and, therefore, alleviate the need to have different estimation methods and procedures for each separate distribution assumption.

Preliminary modelling has indicated that a number of factors affect the level of bluefin tuna dead discards, which may in turn be used to inform management approaches.

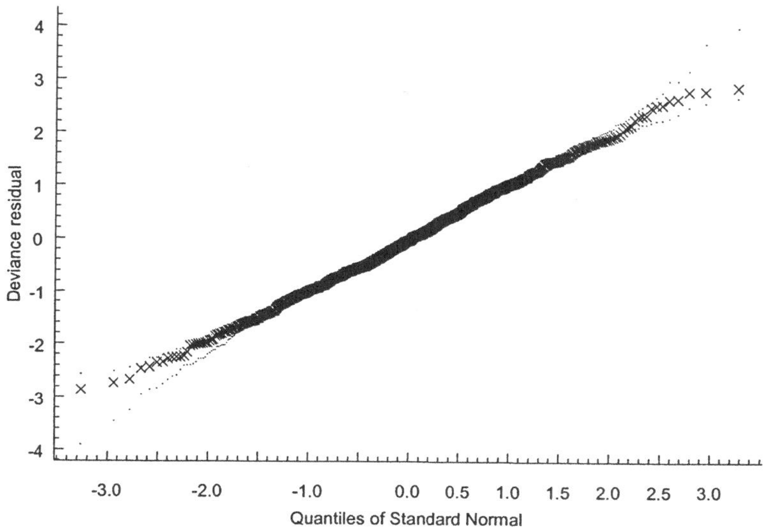

Fig. 7 Q-Q plot with envelope for the individual species SWO.

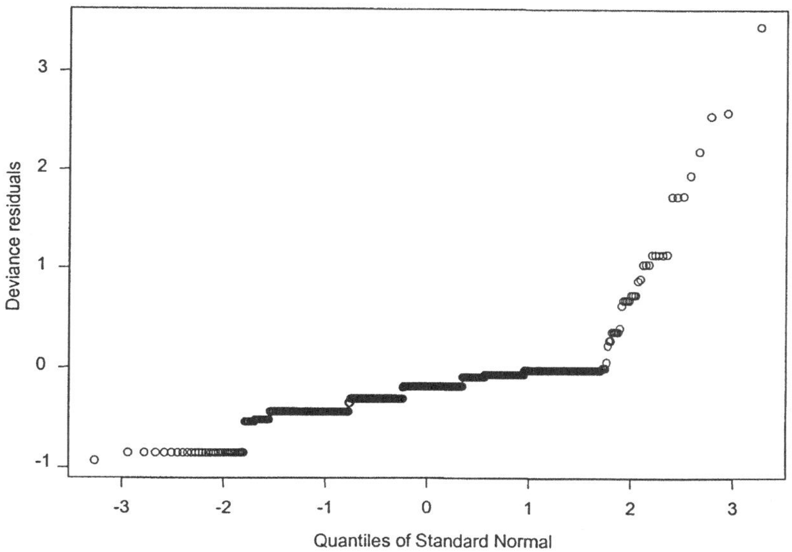

Fig. 8 Q-Q plot for grouped target species TUN.

Considerable variation in dead discard levels was found between years, quarters, NAREAs and target species, suggesting they are important covariates in modelling the level of bluefin tuna dead discards. In certain NAREAs, discarding is either a binary process, does not occur, or may be modelled using the negative binomial distribution. The negative binomial was inappropriate for grouped target species (MIX, TUN), because this was in effect fitting a mixture of mixtures. If the model were fitted separately to individual species in these complexes, the negative binomial may then apply but this conjecture requires further investigation and is outside the scope of the current paper.

The pooling system originally used in the estimation process (Brown, 2001) arose as a result of insufficient observer information on the level of dead discards by year, quarter and area. In order for the model developed using the observer data to improve the estimates of dead bluefin tuna discards, it must be applied to vessel logbook data, which have a significantly greater coverage of the fishery (cf. Walsh *et al.*, 2002). This will be the next stage in the modelling process and the subject of a further paper, but it should not lead to further modelling problems because the important covariates in modelling the level of bluefin tuna dead discards using the observer data are those recorded in the longline logbooks.

Development of this model may allow the impact of management approaches designed to limit dead discards to be modelled. For example, information on areas and periods of the year in which discarding is highest may assist management in the limitation of dead discarding through closed areas. Indeed, closures in the north-eastern area have been effective in reducing dead discard levels, confirming the influence of area on by-catch levels.

Identification of particular factors that result in a high level of by-catch may allow observer coverage to be weighted appropriately to monitor the influence of these factors, and optimize observer coverage. In turn, action can be taken to minimize the effect of these factors on by-catch.

In the recommendation to establish a re-building programme for western Atlantic bluefin tuna, point 12 of document ICCAT 98-7 (*Recommendation to Establish a Rebuilding Program for West Atlantic Bluefin Tuna*) notes that all Contracting Parties ... *shall minimize dead discards to the extent practicable* .... Refining the information on discards in space and time may allow further management measures, such as the current closed areas, to be used to improve both compliance with the recovery plan, and the management of these shared stocks.

ACKNOWLEDGEMENTS

We thank the Center for Independent Experts, US/RSMAS at the University of Miami, and the National Marine Fisheries Service (Southeast Fisheries Center) for funding this work, and Joe Powers and Jan Horbowy for helpful comments on the manuscript.

REFERENCES

Anscombe, F. J. (1950) Sampling theory of the negative binomial and logarithmic series distributions. *Biometrika*, **37**, 358–382.

Atkinson, A. C. (1985) *Plots, Transformations and Regression*. Oxford University Press, Oxford.

Bannerot, S. P. & Austin, C. B. (1983) Using frequency distributions of catch per unit effort to measure fish stock abundance. *Transactions of the American Fisheries Society*, **112**, 608–617.

Borchers, D. L., Buckland, S. T., Priede, I. G. & Ahmadi, S. (1997) Improving the precision of the daily egg production method using generalised additive models. *Canadian Journal of Fisheries and Aquatic Scences*, **54**, 2727–2742.

Brown, C. A. (2001) Revised estimates of bluefin tuna dead discards by the U.S. Atlantic pelagic long-line fleet, 1992–1999. *Collective Volume of Scientific Papers ICCAT*, **52**, 1007–1021.

Cramer, J. (1995) Large pelagic logbook newsletter – 1994. *NOAA Technical Memorandum*, **NMFS-SEFSC–378**, 33 pp.

Davison, A. C. & Gigli, A. (1989) Deviance residuals and normal scores plots. *Biometrika*, **76**, 211–221.

Dong, Q. & Restrepo, V. R. (1996) Notes on the Poisson error assumption made to estimate relative abundance of West Atlantic bluefin tuna. *Collective Volume of Scientific Papers ICCAT*, **45**, 158–161.

Du Rietz, G. E. (1929) The fundamental units of vegetation. *Proceedings of the International Congress of Plant Science, Ithaca*, **1**, 623–627.

Du Rietz, G. E. (1930) Classification and nomenclature of vegetation. *Svensk Botanisk Tidskrift*, **24**, 489–503.

Efron, B. & Tibshirani, R. J. (1993) *An Introduction to the Bootstrap*. Chapman & Hall, London.

Evans, D. A. (1953) Experimental evidence concerning contagious distributions in ecology. *Biometrika*, **40**, 186–211.

Greenwood, M. & Yule, G. U. (1920) An inquiry into the nature of frequency distributions representative of multiple happenings with particular reference to the occurrence of multiple attacks of disease or of repeated accidents. *Journal of the Royal Statistical Society, Series A*, **83**, 255–279.

Lo, N., Jacobsen, I. D. & Squire, J. L. (1992) Indices of relative abundance from fish spotter data based on delta-lognormal models. *Canadian Journal of Fisheries and Aquatic Sciences*, **49**, 2515–2526.

Lundberg, O. (1940) *On Random Processes and their Application to Sickness and Accident Statistics*. Almqvist & Wicksells, Uppsala.

McCullagh, P. & Nelder, J. A. (1989) *Generalised Linear Models*. Chapman & Hall, London.

Miyabe, N. (1997) Updated standardised CPUE of Atlantic bluefin caught by the Japanese longline fishery in the Atlantic. *Collective Volume of Scientific Papers ICCAT*, **46**, 102–117.

Myers, J. H. (1978) Selecting a measure of dispersion. *Environmental Entomology*, **7**, 619–621.

Nelder, J. A. & Wedderburn, R. W. M. (1972) Generalised linear models. *Journal of the Royal Statistical Society, Series A (Gen.)*, **135**, 370–384.

O'Brien, C. M., Adlerstein, S. & Ehrich, S. (2000) Accounting for spatial-scale in research surveys: analyses of 2-year old cod from English, German and International groundfish surveys in the North Sea. *ICES Document*, **CM 2000/K: 19**.

O'Brien, C. M. & Fox, C. (2003) Incorporating temporal information in ichthyoplankton surveys using a model-based approach: cod (*Gadus morhua* L.) in the Irish Sea. *ICES Journal of Marine Science* (in press).

O'Brien, C. M. & Kell, L. T. (1996) The use of generalized linear models for the modelling of catch-effort series. 1. Theory. *ICCAT Document*, **SCRS 96/173**.

O'Brien, C. M., Kell, L. T., Santiago, J. & Ortiz de Zarate, V. (1998) The use of generalized linear models for the modelling of catch-effort series. 2. Application to north Atlantic albacore surface fishery. *Collective Volume of Scientific Papers ICCAT*, **48**(1), 170–183.

Pennington, M. (1983) Efficient estimators of abundance, for fish and plankton surveys. *Biometrics*, **39**, 281–286.

Pennington, M. R. & Groslein, M. D. (1978) Accuracy of abundance indices based on stratified random trawl surveys. *ICES Document*, **CM 1978/D: 13**, 33 pp.

Plackett, R. L. (1981) *The Analysis of Categorical Data*. Griffin, London.

Porch, C. E. (1997) The implications of using the frequency of zero catches and other measures as indices of abundance. *Collective Volume of Scientific Papers ICCAT*, **46**, 237–241.

Porch C. E. & Scott G. P. (1993) A numerical evaluation of GLM methods for estimating indices of abundance from West Atlantic bluefin tuna catch per trip data when a high proportion of the trips are unsuccessful. *Collective Volume of Scientific Papers ICCAT*, **42**, 240–245.

Power, J. H. & Moser, E. B. (1999) Linear model analysis of net catch data using the negative binomial distribution. *Canadian Journal of Fisheries and Aquatic Sciences*, **56**, 191–200.

Scott, G. P. & Brown, C. A. (1997) Estimates of marine mammal and marine turtle catch by the U.S. Atlantic pelagic longline fleet in 1994–1995. *NMFS SEFSC Technical Report*, **MIA-96/97-28**, 14 pp.

Taylor, C. C. (1953) Nature of variability in trawl catches. *U.S. Fish and Wildlife Service, Fishery Bulletin*, **54**, 145–166.

Thompson, M. L. (1978a) Selection of variables in multiple regression. 1. A review and evaluation. *International Statistical Review*, **46**, 1–19.

Thompson, M. L. (1978b) Selection of variables in multiple regression. 2. Chosen procedures, computations and examples. *International Statistical Review*, **46**, 129–146.

Walsh, W. A., Kleiber, P. & McCracken, M. (2002) Comparison of logbook reports of incidental blue shark catch rates by Hawaii-based longline vessels to fishery observer data by application of a generalized additive model. *Fisheries Research*, **58**, 79–94.

Whittaker, R. H. (1962) Classification of natural communities. *Botanical Review*, **28**, 1–239.

Wolstenholme, D. E., O'Brien, C. M. & Nelder, J. A. (1988) GLIMPSE: a knowledge-based front end for statistical analysis. *Knowledge-Based Systems*, **1**, 173–178.

Yeung, C., Epperly, S. & Brown, C. A. (2000) Preliminary revised estimates of marine mammal and marine turtle bycatch by the U.S. Atlantic pelagic longline fleet, 1992–1998. *National Marine Fisheries Service Miami Laboratory PRD Contribution*. **99/00 –13**.

Problems of herring assessment and management in the Baltic Sea

Georgs Kornilovs
Latvian Fisheries Research Institute, Riga, Latvia

ABSTRACT: Hydro-meteorological conditions in the Baltic Sea vary from southwest to northeast. Herring have developed local populations as a means of adapting to conditions in the different ecological subsystems. The stock units have been grouped for assessment purposes in different combinations or assessed separately over the period 1974–2002. The unit changes generally resulted from new investigations into the distribution and separation of different stocks, but sometimes they were made for political reasons. The current assessment units were set in 1990 and were a compromise between actual knowledge of the stock structure of Baltic herring and the ability to separate the stocks in mixed catches. For the central Baltic, a large combined stock unit was created to eliminate the influence of the poorly documented migration of the species in the Baltic Sea. Management of the herring fishery in the Baltic Sea evolved in the mid-1970s. Since then the management scheme has remained virtually unchanged while the herring assessment units have been altered repeatedly. Management of the herring fishery did not encounter major difficulties until recently because of the good state of the stocks and the consequent high level of Total Allowable Catch, a level that surpassed the demands of the herring fishery. However, in the mid-1990s, many herring stocks declined and increasing uncertainty in the assessments became apparent. This change in perception of herring stock status was unexpected for managers and resulted in increasing disagreement with ICES advice and an unwillingness to follow it. A recommendation to manage on the basis of the existing assessment units was also ignored, and criticism of the herring assessment and its management units increased. In this paper, the validity of the current assessment and management units and their suitability for management is discussed.

INTRODUCTION

The herring *Clupea harengus* is commercially one of the most important fish in the Baltic Sea. Two seasonal races of Baltic herring have been distinguished since the 19th century (Heincke, 1898). Autumn spawners constituted an important part of the herring catches until the 1960s, but this component then declined rapidly. Spring-spawning herring populations have been discriminated partly on the basis of growth characteristics (Parmanne *et al.*, 1994). Hydro-meteorological conditions vary in different parts of the Baltic Sea. Salinity decreases gradually from southwest to northeast, and the size at age of the species also decreases. In the process of adapting to these different ecological conditions, herring developed local populations, and in routine sampling, herring stocks have been distinguished in the Baltic Sea on the basis of otolith morphology (Kompowski, 1969; Ojaveer *et al.*, 1981). In the mid-1970s, when the Working Group on Assessment of Pelagic Stocks in the Baltic was established by ICES, the results of these investigations were used for setting assessment units. However, there was never consistent agreement, so the assessment units and their borders were changed repeatedly (Sjöstrand, 1989). The proposed stock discrimination methods, based mainly on otolith structural differences, were not accepted by all countries involved in exploiting the stocks, and the result was complex stock separation procedures, e.g. for coastal herring in the southern Baltic (Anon., 1991; 2002a).

An opposing view is that apparent differences in morphological, meristic and biological characters are more likely to be a phenotypic reflection of the environmental gradient from southwest to northeast (Sparholt, 1994). This view is substantiated by the fact that there is no apparent genetic differentiation between herring stocks (Ryman *et al.*, 1984; Rajasilta *et al.*, 1999). Also, in summer and autumn, most Baltic herring undergo extensive feeding migrations (Parmanne *et al.*, 1994), so stock separation is unlikely. As a result, a single large herring assessment unit was established in the central Baltic (Subdivisions 25–29 + 32, including the Gulf of Riga), and currently most herring stocks are separated on the basis of their catch in a defined assessment area (Subdivision) of the Baltic Sea. The Gulf of Riga herring is the only stock that is distinguished and separated from other populations, both in the Gulf and outside the Gulf during the feeding migration of part of the stock. Gulf of Riga herring, although assessed separately until 2002, were also included in the herring assessment unit of the central Baltic. The results of the separate assessment of the Gulf of Riga herring and the respective ICES advice were used only in national fisheries regulation in the Gulf of Riga.

Since 1990 the following herring assessment units have been used in the Baltic Sea (Figure 1):

(1) herring in Subdivisions 22–24 and Division IIIa
(2) herring in Subdivisions 25–29 + 32 (including Gulf of Riga)
(3) Gulf of Riga herring.
(4) herring in Subdivision 30 (Bothnian Sea).
(5) herring in Subdivision 31 (Bothnian Bay).

In 2002 the assessment of herring in Subdivisions 25–29 + 32 excluded Gulf of Riga herring, and the ICES advice was given separately for the central Baltic and the Gulf of Riga (Anon., 2002c).

The International Baltic Sea Fishery Commission (IBSFC), which was established in 1974, regulates the Baltic Sea fisheries. The IBSFC has chosen to manage stocks chiefly by controls on total annual catch (Total Allowable Catches, TACs; Parmanne *et al.*, 1994). Since 1977, two herring management units have been used in the Baltic Sea: herring in the main basin (Subdivisions 22–29S + 32, including the Gulf of Riga), and Management Unit III (Subdivisions 29N, 30 and 31). Discrepancies between assessment and management units makes it difficult for the IBSFC to cope with differences in the stocks (Parmanne *et al.*, 1994). ICES has repeatedly asked that herring TACs for the Baltic be split and that individual TACs be applied to the stocks for which the assessments are made (Anon., 2002c). However, as this involves recalculation of national quotas, it is a difficult task.

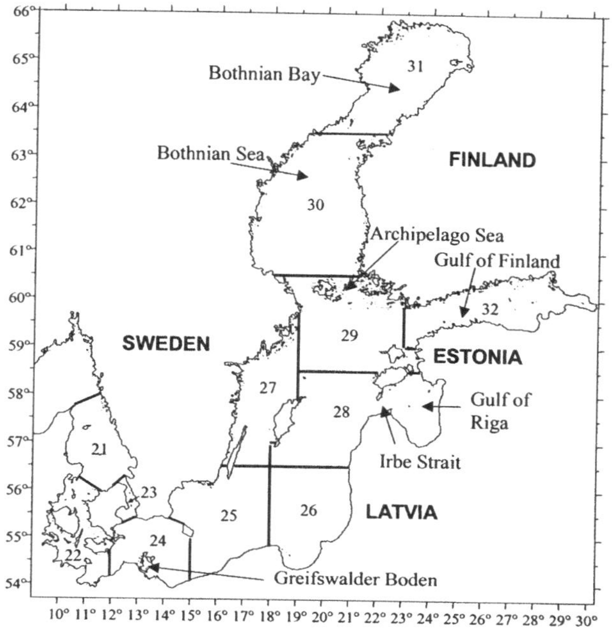

Fig. 1 ICES Subdivisions in the Baltic Sea.

This paper examines the present state and development of Baltic herring stocks on the basis of assessments by ICES Working Groups. It evaluates the reliability and consistency of the assessments and their development during the past decade, and examines the suitability of the management unit in providing for sustainable exploitation of herring stocks in the Baltic.

MATERIAL AND METHODS

During the period 1974–1994, assessment of Baltic herring stocks was performed at the Working Group on Pelagic Stocks in the Baltic, and since 1995 at the Baltic Fisheries Assessment Working Group. Since 1977, the major tool for stock assessment has been Virtual Population Analysis (VPA). Tuning programmes such as the '*ad hoc* tuning method' of Laurec and Shepherd (Pope & Shepherd, 1985) and Extended Survivors' Analysis (XSA; Shepherd, 1992) were applied (Parmanne *et al.*, 1994). Since 1996, assessment of herring in Subdivisions 22–24 and Division IIIa has been undertaken at the Herring Assessment Working Group for the area south of 62oN, applying Integrated Catch Analysis (ICA; Anon., 2002d). Abundance estimates from different surveys and fisheries are used for tuning VPA in addition to catch-at-age data. For herring in Subdivisions 25–29 + 32, the results of autumn hydro-acoustic surveys have been used for management advice since the beginning of the 1980s. For herring stocks in Subdivisions 30 and 31 and Gulf of Riga herring, catch per unit effort (cpue) data from different fishing fleets are used (Anon., 2002b). For herring in Subdivisions 22–24 and Division IIIa, several survey indices, including those from hydro-acoustic, bottom trawl and larval surveys, are used (Anon., 2002d). Since 1981, the ICES Working Group on Multispecies Assessments of Baltic Fish has estimated the natural mortality for herring and sprat *Sprattus sprattus* in the Baltic Sea, taking into account predation by cod *Gadus morhua*. The results of these analyses since 1991 have been incorporated into the assessment of herring in Subdivisions 25–29 + 32. For herring stocks in Subdivisions 30 and 31 and the Gulf of Riga, where cod are scarce or absent, a constant or slightly variable natural mortality rate of 0.2, 0.15 and 0.2–0.25 respectively, is used (Anon., 2002b). For herring in Subdivisions 22–24 and Division IIIa, natural mortality was assumed constant at 0.2 for all years and 2+ ringers, but a predation mortality of 0.1 and 0.2 was added to the 0- and 1-ringers (Anon., 2002d). The ICES advice was taken from the Reports of the Advisory Committee on Fishery Management (ACFM).

RESULTS

HERRING IN SUBDIVISIONS 25–29 + 32

This combined unit for the central Baltic, including the Gulf of Finland and the Gulf of Riga, was established in 1990. The justification for such a unit was that different herring stocks mix during their feeding migrations and that they are not distinguished in catches. During the early 1990s, the stock was considered to be within safe biological limits (Anon., 1994b). Fishing mortality F was, for most years, at or about $F_{0.1}$, i.e. 0.25. However, the spawning stock biomass from the XSA analysis then began to decline (Anon., 1997; Figure 2a). F gradually increased from the beginning of the 1990s and remained well above F_{pa} (0.17; Figure 2b). Recruitment after the 1980s was close to or below average (Figure 3a), and spawning stock by number was stable over the whole assessment period, decreasing only in the past 4–5 years (Figure 3b). A major cause for the divergent trends was thought to be the considerable changes in mean weight since the mid-1980s that continued through the 1990s (Figure 4; Anon., 1994a). In recent years, mean weight-at-age increased in some regions of the central Baltic. Also, assessments tended to overestimate spawning stock biomass and to underestimate fishing mortality. ICES scientists considered that, although the exact stock size was uncertain, there was confidence that spawning stock biomass had continued to decrease and was close to the historical low (Anon., 2001b). However, because the Gulf of Riga herring stock, which was included in the combined assessment unit, exhibited different stock dynamics, it was decided to perform an assessment for central Baltic herring excluding the Gulf of Riga. That assessment was carried out several times, but only in 2002 was it decided to give the advice separately for the central Baltic (Subdivisions 25–29 + 32) and the Gulf of Riga (Anon., 2002b). The exclusion of Gulf of Riga herring from the combined assessment did not change understanding of the dynamics of the herring stock in the central Baltic, because both assessments revealed a declining trend. Even in absolute figures, the spawning stock biomass was close for both assessments (Anon., 2002b).

GULF OF RIGA HERRING

Environmental conditions in the Gulf of Riga differ considerably from those in the main basin of the Baltic Sea and in the Gulf of Finland, which is situated farther north. The Gulf of Riga is part of Subdivision 28, and its water masses are separated from the waters of the Baltic proper by a hydrological front located in the Irbe Strait. Its salinity is in the range 2–6, so restricting migration of cod and sprat into its waters. Predation pressure on herring is therefore lower than in the Baltic Sea and herring are the dominant pelagic species in the Gulf, whereas sprat have dominated the pelagic fish community of the open Baltic since

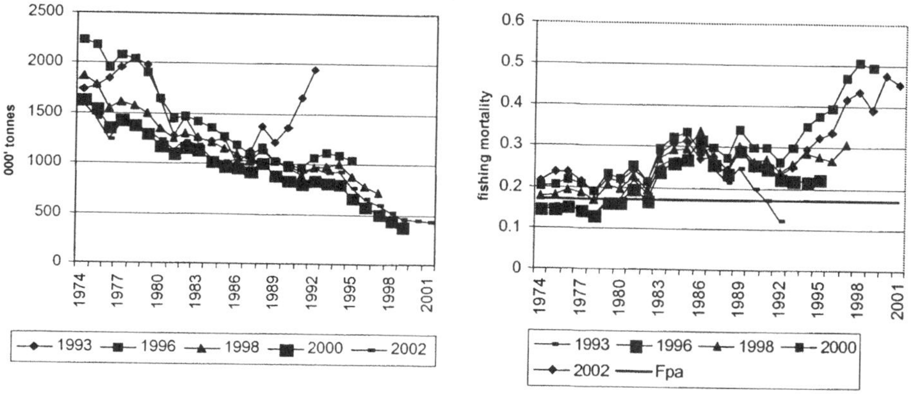

Fig. 2 (a) Spawning stock biomass and (b) fishing mortality (F_{3-6}) estimates of herring in Subdivisions 25–29 + 32 made during the years 1993–2002.

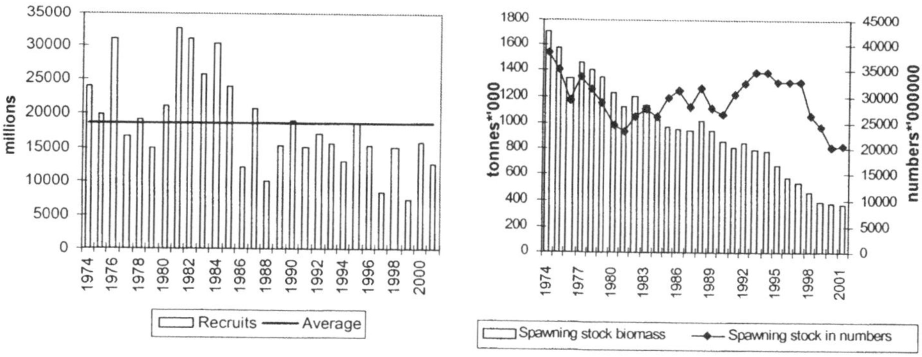

Fig. 3 (a) Recruitment (age 1) and (b) spawning stock biomass and numbers of herring in Subdivisions 25–29 + 32.

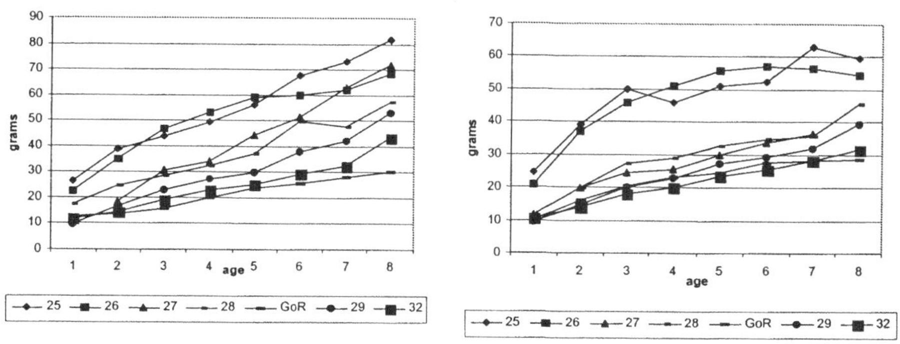

Fig. 4 Mean weight-at-age of herring in the main basin of the Baltic Sea by Subdivision in (a) 1993 and (b) 2001 (GoR – Gulf of Riga).

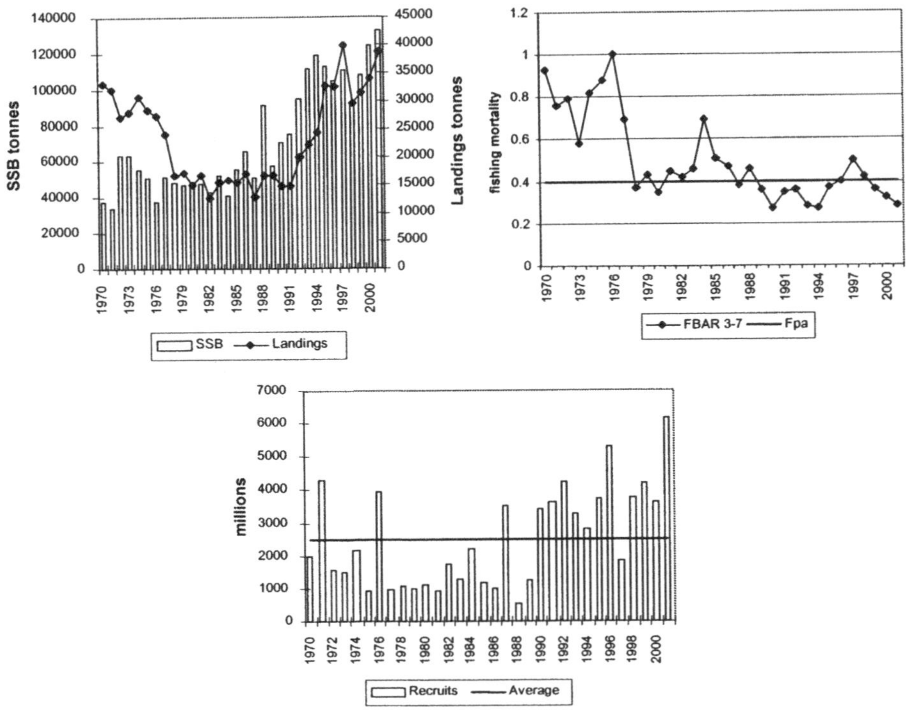

Fig. 5 (a) Spawning stock biomass and landings, (b) fishing mortality (F_{3-7}) and (c) recruitment (age 1) of Gulf of Riga herring.

the mid-1990s. In the 1970s and 1980s, spawning stock biomass of herring was low and fishing mortality was high (Figures 5a, b). Biomass has increased since 1989 as a consequence of the appearance of a- series of-abundant year-classes; catches have therefore increased (Figure 5c). Rannak (1971) noted that- strong year-classes of Gulf of Riga herring appeared after mild winters, and the same observation appears to be relevant for the latest period. Since 1989 -the only cold winter was in 1996;- in that year the herring year-class was below average, whereas all other year-classes were above average and some extremely abundant. There is a likely link with patterns of climate change, because such long periods of mild winters do not appear in the historical record. Long-term observations show that, in the Gulf of Riga in three years out of four earlier in the 20th century, there was permanent ice cover during winter (Pastors, 1967). In the past 15 years, such a permanent ice cover formed only once. The assessments since 1991 have been consistent and scientific perception of stock status has not changed. However, because only Latvia and Estonia prosecute a herring fishery in the Gulf of Riga, the assessment results were used only for purposes of national fishery regulation and quotas.

HERRING IN SUBDIVISIONS 22–24 AND DIVISION IIIA

Herring in Subdivisions 22–24 (western Baltic) are considered to be a separate stock of spring-spawning Baltic herring; Their spawning grounds are situated within the three Subdivisions, with the main distribution in Greifswalder Boden. The mature portion of the stock migrates to feed in the Kattegat and Skagerrak (Division IIIa) and eastern North Sea (Biester, 1979). During these migrations the fish become infected with *Anisakis simplex*, a nematode considered useful as a natural marker for this herring stock. In 2002, the assessment was accepted for the first time (Anon., 2002b). Spawning stock biomass decreased from the early 1990s and, although currently stable, has been low for the past six years (Figure 6a). Current fishing mortality is high, above F_{max} (0.37; Figure 6a).

Recruitment to the stock has been -stable for the past 10 years, but if compared with the assessment made during 1993, there has either been a scarcity of strong year-classes in the latest period or the stock-recruit relationship has changed (Figure 6b; Anon., 1993, 2002d). ICES considered that the uncertainty in the assessment could be related to problems in separating–this stock from North Sea autumn spawners (Anon., 2002c). This is particularly noticeable when comparing the assessments made in 1993 and 2002 (Figure 6a) for which different methodology was used to separate western Baltic spring spawners from North Sea autumn spawners. In the earlier assessment, the stocks were separated on the basis of otolith macrostructure or counts of vertebrae (Anon., 1998), but for the 2002 assessment, catches were separated on the basis of different otolith microstructure (Anon., 2002d).

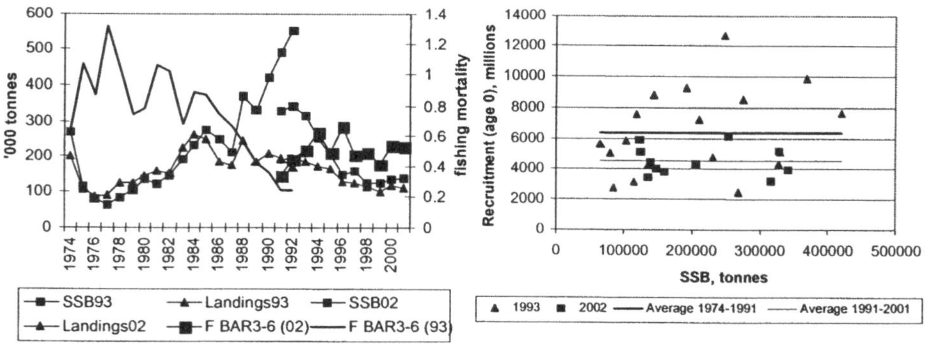

Fig. 6 (a) Spawning stock biomass, landings and fishing mortality and (b) a stock-recruitment plot of herring in Subdivisions 22–24 (assessments made in 1993 and 2002).

HERRING IN SUBDIVISION 30 (THE BOTHNIAN SEA)

The Bothnian Sea herring stock is thought to be distributed mainly within the assessment area (Aro, 1989; Parmanne, 1990). However, although immigration from other areas is negligible, emigration to the adjoining part of Subdivision 29 is common, possibly related to the utilization by Subdivision 30 stocks of spawning grounds in the Archipelago Sea (Subdivision 29N; Rajasilta *et al.*, 1999). Spawning stock biomass was -high at -the beginning of the 1990s, but it has now decreased, although it is still above B_{pa}, i.e. 200 000 t (Figure 7a). Fishing mortality increased after 1993 and was above F_{pa}, i.e. 0.21, from 1997 (Figure 7b). Towards the end of the 1990s, some assessments gave very high estimates of SSB and low estimates of F, and revealed partly differing trends in the stock dynamics. Repeated revisions of cpue data and the use of different tuning fleets caused these differences. In the 1990s, exploitation of the stock increased as the trawl

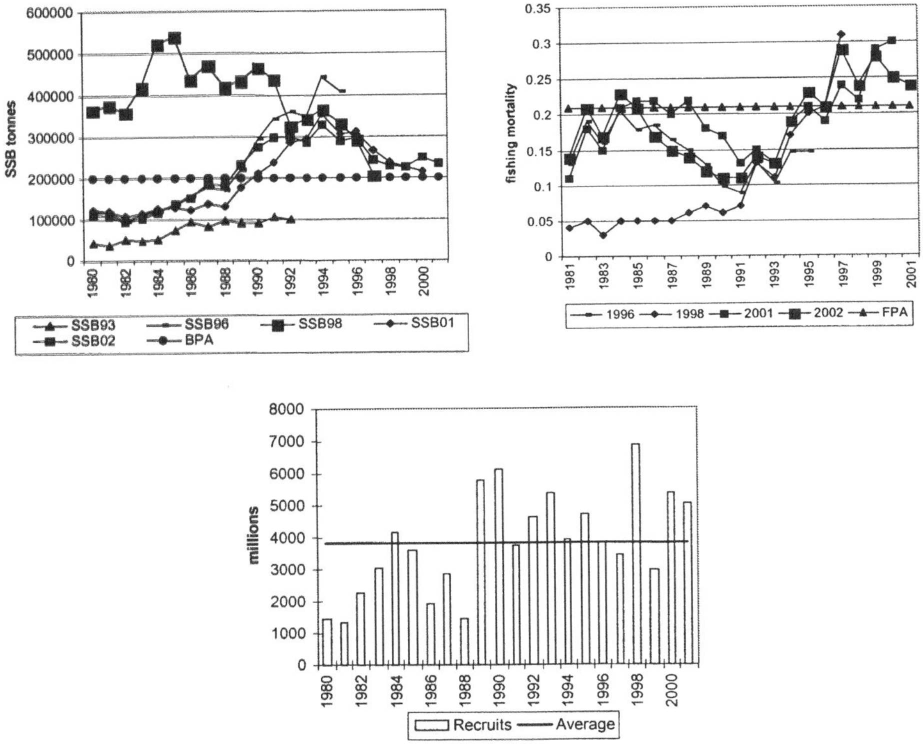

Fig. 7 (a) Spawning stock biomass and (b) fishing mortality estimates made between 1993 and 2002, and (c) recruitment (age 1) of herring in Subdivision 30.

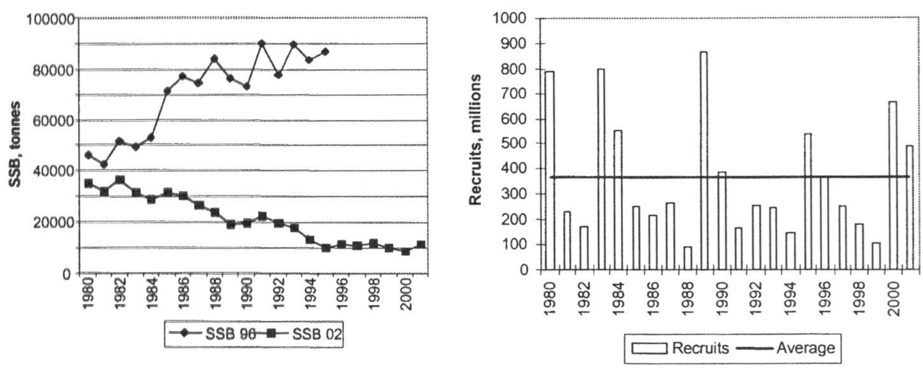

Fig. 8 (a) Spawning stock biomass estimates from the assessments made in 1996 and 2002 and (b) recruitment (age 1) of herring in Subdivision 31.

fishery became more effective and larger trawls were introduced. Recruitment to the stock has been high since the end of 1980s, as for the Gulf of Riga (Figure 7c).

HERRING IN SUBDIVISION 31 (BOTHNIAN BAY)

The status of herring stocks in Subdivision 31 is unknown, and the assessment was not accepted in 2002 (Anon., 2002c). However, the assessments made up to 1996 showed that the stock was only lightly exploited (Figure 8a; Anon., 1996), and the uncertainty in the stock estimates was explained by low fishing mortality F. Retrospective analysis revealed a -ssubstantial overestimate of spawning stock biomass and an underestimate of F. The revision of effort data in the tuning fleets during 1997 totally changed the perception of the state of the stock. Since then, assessments, although rather uncertain, have indicated that spawning stock biomass has declined substantially since the mid-1990s and that fishing mortality was high in the 1990s, both perhaps somewhat related to generally rather poor recruitment in the 1990s (Figure 8b). This herring stock inhabits the extreme northernmost area of the Baltic Sea and its reproductive success seems to depend heavily on some combination of environmental conditions that only rarely allow the appearance of strong year-classes (Anon., 2002b). Further, the difference between strong and weak year-classes is more striking than for other stocks in the Baltic Sea. The basis for treating this herring stock as a separate unit is that tagging results in the area show a small number of recaptures outside Subdivision 31, mainly in the neighbouring Subdivision 30 (Parmanne, 1990). Also, herring in Bothnian Bay differ morphologically from those in other parts of the northern Baltic Sea (Parmanne, 1990).

DISCUSSION

There are two opposing opinions on Baltic herring stock structure. The first is that several distinct herring populations have evolved as an adaptation to the diverse environmental conditions of the Baltic Sea (Ojaveer, 1988; Ojaveer & Elken, 1997). Therefore, merging possibly distinct Baltic herring populations, each with different basic biological parameters, resulted in increased uncertainty in the combined assessment (Anon., 2001a). It also increased the risk that exploitation would differentially impact the least productive and genetically distinct stocks (Anon., 2002a). Fisheries managers need to conserve intraspecific biodiversity in exploited fish species by avoiding eroding their population structure (Carvalho & Hauser, 1994) and should tend towards conservatism in an effort to manage as many potentially distinct stocks separately as practicable (Schweigert, 1991). For the Baltic Sea, Parmanne (1990) suggests using small management areas during the spawning season, and extending the areas during the autumn migration, when some herring are dispersed over vast areas of the Baltic Sea.

The second opinion contends that differences in morphological, meristic and biological characters are more likely to be a phenotypic reflection of the environmental gradient from south to north and therefore are not genetically inherited (Parmanne, 1990; Sparholt, 1994). Parmanne *et al.* (1994) point out that the stock units should ideally be chosen so that there is either no (or very little) overlap in their geographical distribution, or so that it is feasible to classify individual fish in mixed catches and to sort them according to stock. However, this requirement is difficult to satisfy because of the extensive herring feeding migrations and the consequent mixing of stock units, and because of the lack of common agreement on the separation of stocks for which some specific properties have been chosen. The latter were generally peculiarities of otolith morphology (Kompowski, 1969; Fetter *et al.*, 1990). Also, the different biological characters used to discriminate certain Baltic herring stocks -usually overlap. Until now, no genetic differentiation between Baltic herring stocks has been demonstrated (Ryman *et al.*, 1984; Rajasilta *et al.*, 1999). However, failure to demonstrate genetic differentiation does not necessarily mean that there is no stock structure (Hay & McCarter, 1997); it is more probably a failure of the genetic techniques used for stock identification. Since the early 1990s, microsatellite markers have been used with increasing success for stock identification, and some initial studies have recently commenced on Baltic Sea herring– (Anon., 2002a).

The difference between these two approaches is especially evident for the central Baltic herring in Subdivisions 25–29 + 32. Assessment of herring in this large area was started in 1990, and initial indications were of a healthy and safe stock state that, under conditions of decreasing demand and increasing abundance, made any need for restrictive management action less obvious (Parmanne *et al.*, 1994). However, after the mid-1990s, the scientific perception of stock status changed drastically. Assessments showed no increase in spawning stock biomass between the late 1980s and the early 1990s, likely because spawning stock biomass tended to be overestimated and fishing mortality underestimated (Anon., 2001c). The opinion of ICES was therefore that, although the exact stock size was uncertain, there was confidence that the spawning biomass had been decreasing and that it was then close to the historical low (Anon., 2001c). Eight countries exploit herring in Subdivisions 25–29 + 32 and several of them conduct their own national surveys of the stock, which do not always agree with the assessment estimates (e.g. Krasovskaya, 2002). ICES scientists concede that the stock structure of this assessment unit is complex, but stress that understanding the status of the different components of the combined unit is difficult. The only exception to this statement was for the Gulf of Riga stock. As that herring stock, which was included in the combined assessment unit, exhibited different dynamics, it was decided to perform an assessment for central Baltic herring excluding the Gulf of Riga component. That assessment was undertaken several times, but only in 2002 was it decided to give the advice separately for the central Baltic (Subdivisions 25–29 + 32) and Gulf of Riga herring (Anon., 2002c). Exclusion of the Gulf of Riga herring from the combined assessment did not change the perception of the stock dynamics of the central Baltic stock, because both assessments revealed declining trends. Even in absolute figures, the spawning stock biomass was close for both assessments.

As the reliability of the assessment of herring in Subdivisions 25–29 + 32 was -deemed dubious, it was decided to evaluate available information on herring stock components in the central Baltic and their migrations and to propose new assessment units on that basis. A Study Group on Herring Assessment Units in the Baltic Sea was established and it prepared a set of proposals (Anon., 2001a). Subsequently, assessments were made for two new assessment units: Subdivisions 25–27, and Subdivisions 28, 29 and 32 (Anon., 2001b). The results showed a good fit between the assessment of the combined stock and the proposed stock units (including Gulf of Riga herring). In both the southern and the northern part of the central Baltic, the herring stock size is at a low ebb. The assessments revealed poor stock status and a recent F of up to 0.7 in Subdivisions 28, 29 and 32. That result was supported by information on a massive decline in trap-net catches of herring in Subdivision 29 (Rajasilta *et al.*, 1999), but the assessments were not accepted by ICES because of insufficient scientific justification (Anon., 2001c).

It was considered that, for the western Baltic herring in Subdivisions 22–24, the main source of uncertainty in the assessment was the stock's separation from North Sea autumn-spawners in Division IIIa and the eastern North Sea. The degree of mixing of the latter stock with central Baltic herring from Subdivisions 25–29 + 32 was regarded as insignificant. However, since the mid-1980s, the abundance of herring infected with *Anisakis simplex* has increased considerably on the spawning grounds in Subdivisions 25 and 26, which are outside the assessment area of western Baltic herring (Podolska & Horbowy, 2001). One possibility could be that the spawning migration of western Baltic herring has extended eastwards. Another possibility is that local herring from Subdivisions 25 and 26 now perform longer feeding migrations to Division IIIa. For both options, some losses and gains in the catches underpinning the assessment of western Baltic and central Baltic herring are possible. In addition, the importance of local spring-spawning herring stocks in Subdivision IIIa should also be explored. Tagging experiments revealed another direction of migration for younger western Baltic herring, namely eastwards outside the assessment area (Jönsson & Biester, 1979); the magnitude of this migration was regrettably not estimated. It seems that such uncertainties in the distribution of herring in the southwestern Baltic and the stock to which they belong are partly responsible for current difficulties with the assessment.

Separate assessments of the Gulf of Riga herring have been made since the 1980s and continued after that component of the stock was included in the larger assessment unit of the central Baltic (Subdivisions 25–29 + 32). Those assessment results and the associated advice were used for national regulation within the national quotas. For instance, when the national quotas of Estonia and Latvia, the two countries that fish in the Gulf of Riga, were reasonably high during the 1970s and 1980s, exploitation levels were restricted on the basis of the Gulf of Riga herring assessments then. Both those countries separate herring in the catches by means of otolith structure. Mixing with central Baltic (open-sea) herring takes place in certain seasons both in the Gulf of Riga and in the open Baltic itself. Recently, some 7% of the Gulf of Riga herring catch has been taken outside the Gulf and some 12% of herring catches in the Gulf of Riga consist of open-sea herring from the main basin (Anon., 2002c). Analyses of herring length distribution at age revealed that annual changes support general understanding of the migration and mixing of stocks in the region (Kornilovs, 1994) and can also be used for distinguishing Gulf of Riga herring from other stock components.

The herring stock in the Bothnian Sea (Subdivision 30) is harvested outside safe biological limits, but spawning stock biomass is presently above B_{pa} (Anon., 2002c). During the 1990s, some assessments gave very high estimates of SSB and low estimates of F and therefore different understanding of stock dynamics. These differences were explained by repeated revisions of the cpue as a result of the introduction of more-effective and larger trawls. However, the extent of migration outside the assessment area could also have been having a significant influence on the assessment of that stock.

The state of the herring stock in Bothnian Bay (Subdivision 31) is unknown, but there are indications that at present it is very low. Migration outside the assessment area and mixing with other stock units is considered to be negligible (Parmanne, 1990), so any uncertainties in the assessment are unlikely linked to problems of stock separation.

Until the later 1990s, management of the herring fishery in the Baltic Sea did not encounter major difficulties because of the assumed healthy state of the stocks and the high TACs that always surpassed the needs of the herring fishery. In the main basin of the Baltic, the TAC was about twice the amount caught (Figure 9). However, in the mid-

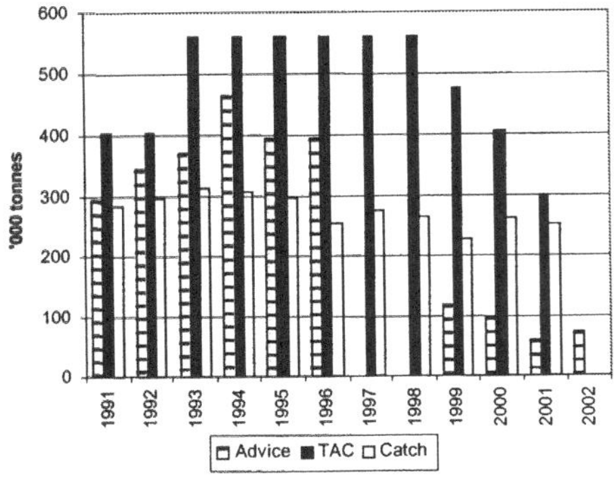

Fig. 9 ICES advice for herring in Subdivisions 25–29 + 32, and the agreed TAC and total catch of herring in Subdivisions 22–29 + 32 (no TAC agreed in 2002).

1990s, many herring stocks declined sharply and there was increasing uncertainty in the assessments. This changed perception of the health of the herring stocks was unexpected by managers, resulting in escalating disagreement with ICES scientific advice and a general unwillingness to follow it. A recommendation to manage on the basis of the existing assessment units was ignored and criticism of existing assessments and management units grew in intensity.

Overall management of the herring fishery was hampered by the discrepancy between management and assessment units. This problem was especially relevant for management units in the main basin, including Subdivisions 22–29S + 32, where management was governed by the state of the stock in the central Baltic. It was necessary to curb catches in Subdivisions 25–29 + 32, so restricting fishing in the western Baltic and Gulf of Riga. The states that fished those regions fought against the reduction in the TAC and therefore against the rebuilding of the stock in the central Baltic. Hopefully, such a struggle will end with the adoption of new management units that will be closer to the assessment units currently used by ICES. Such a change would certainly improve management of the herring fishery in the Baltic, because it would contribute to a more even distribution of fishing effort, in accordance with the ICES advice. However, there are significant obstacles to implementation of the new management units. These are associated mainly with the necessity to apply new allocation criteria that would lead to changes in the levels of fishing effort allowed for the various countries, mostly downwards. Of course, in the early 1990s when herring stock status was better, such a transition would have been easier.

The TAC for management unit III (Subdivisions 29N, 30 and 31) is drawn from the advice of three assessments: Subdivisions 25–29 + 32, Subdivision 30, and Subdivision 31, but it is obviously governed by the status of the stock in the Bothnian Sea (Subdivision 30). It has never been clear how managers calculate the TAC for Subdivision 29N. At present, with herring stocks in Subdivision 31 and the central Baltic, which includes Subdivision 29N, in such a poor state, management practices of this nature can only worsen the situation.

The present state of the herring stocks in the Baltic Sea indicates that management of large areas that include several stock units or fish stocks with different biological features can easily be unsuccessful. Current management practice simply does not ensure that fishing effort on each assessment unit is allocated according to the best possible advice.

Conversely, smaller management units, such as the Gulf of Riga and the Bothnian Sea, are easier to manage, especially when there is common understanding of the status of the stock and a common research policy.

REFERENCES

Anon. (1991) ICES Report of the Working Group on Assessment of Pelagic Stocks in the Baltic. *ICES Document*, **CM 1991/Assess: 18**.

Anon. (1993) ICES Report of the Working Group on Assessment of Pelagic Stocks in the Baltic. *ICES Document*, **CM 1993/Assess: 18**.

Anon. (1994a) Growth changes of herring in the Baltic. TemaNord 1994: 532. Nordic Council of Ministers, Copenhagen.

Anon. (1994b) ICES Report of the Working Group on Assessment of Pelagic Stocks in the Baltic. *ICES Document*, **CM 1994/Assess: 18**.

Anon. (1996) ICES Extract of the Report of the Advisory Committee on Fishery Management. Stocks in the Baltic, May 1996.

Anon. (1997) ICES Report of the Baltic Fisheries Assessment Working Group. *ICES Document*, **CM 1997/Assess: 12**.

Anon. (1998) ICES Report of the Study Group on the Stock Structure of the Baltic Spring-Spawning Herring. *ICES Document*, **CM 1998/H: 1 Ref. B**.

Anon. (2001a) ICES Report of the Study Group on Herring Assessment Units in the Baltic Sea. *ICES Document*, **CM 2001/ACFM: 10**.

Anon (2001b) ICES Report of the Baltic Fisheries Assessment Working Group. *ICES Document*, **CM 2001/ACFM: 18**.

Anon. (2001c) Report of the ICES Advisory Committee on Fishery Management, 2001. *ICES Cooperative Research Report*, **246**(3).

Anon. (2002a) ICES Report of the Study Group on Herring Assessment Units in the Baltic Sea. *ICES Document*, **CM 2002/H: 04. Ref. ACFM, D**.

Anon. (2002b) ICES Report of the Baltic Fisheries Assessment Working Group. *ICES Document*, **CM 2002/ACFM: 18**.

Anon. (2002c) ICES Extract of the Report of the Advisory Committee on Fishery Management on Stocks in the Baltic. Overview, June 2002.

Anon. (2002d) ICES Report of the Herring Assessment Working for the Area South of 62°N. *ICES Document*, **CM 2002/ACFM: 12**.

Aro, E. (1989) A review of fish migration patterns in the Baltic. *Rapports et Procès-Verbaux des Réunions Conseil International pour l'Exploration de la Mer,* **190**: 72–96.

Biester, E. (1979) Der Frühjahrshering Rügens. Doctoral thesis, Wilhelm-Pieck Universität, Rostock.

Carvalho, G. R. & Hauser, L. (1994) Molecular genetics and the stock concept in fisheries. *Reviews in Fish Biology and Fisheries*, **4**: 326–350.

Fetter, M., Groth, B., Kestner, D. & Wyszynski, M. (1990) Guide for the use of Baltic sprat and herring otoliths in fisheries studies. Fischerei-Forschung, Rostock, Index No. **31720**: 18–42.

Hay, D. E. & McCarter, P. B. (1997) Larval distribution, abundance, and stock structure of British Columbia herring. *Journal of Fish Biology*, **51**(Suppl. A): 155–175.

Heincke, F. (1898) Natugeschichte des Herings. Abhandlungen des Deutschen Seefisscherei-Vereins. Bd 2, H. 1, I-CXXXV: 1–128.

Jönsson, N. & Biester, E. (1979) Results of tagging experiments on the Rügen spring herring 1977/1978. *ICES Document*, **CM 1979/J: 29**.

Kompowski, A. (1969) Types of otoliths of southern Baltic herring. *ICES Document*, **CM 1969/H: 12**, 17 pp.

Kornilovs, G. (1994) Yearly length distribution of herring in the Gulf of Riga in relation to the populational structure of the stock. *ICES Document*, **CM 1994/J: 9, Ref. H**.

Krasovskaya, N. (2002) Spawning of Baltic herring in the Vistula Lagoon: effects of environmental conditions and stock parameters. *Bulletin of the Sea Fisheries Institute, Poland*, **1**(155): 3–25.

Ojaveer, E. (1988) Baltic herrings. Agropromisdat, Moscow: 3–204. (in Russian).

Ojaveer, E. & Elken, J. (1997) On regional subunits in the ecosystem of the Baltic Sea. In: *Proceedings of the 14th BMB Symposium, August 1995, Pärnu*, pp. 156–169. Estonian Academy Publishers, Tallinn.

Ojaveer, E., Jevtjukhova, B., Rechlin, O. & Strzyzewska, K. (1981) Results of investigations of population structure and otoliths of Baltic spring spawning herring. *ICES Document*, **CM 1981/J: 10**. 22 pp.

Parmanne, R. (1990) Growth, morphological variation and migrations of herring (*Clupea harengus* L.) in the nothern Baltic Sea. Finnish Fisheries Research, **10**: 1–48.

Parmanne R., Rechlin, O. & Sjöstrand, B. (1994) Status and future of herring and sprat stocks in the Baltic Sea. *Dana*, **10**: 29–59.

Pastors, A. A. (1967) Water and heat balance in the Gulf of Riga. In: *Sea Gulfs as Receivers of River Run-off*, pp. 8–33 (in Russian).

Podolska, M. & Horbowy, J. (2001) The analysis of infection of Baltic herring (*Clupea harengus* membras) with *Anisakis simplex* larvae using generalized linear models. *ICES Document*, **CM 2001/U: 12**.

Pope, J. G. & Shepherd, J. G. (1985) A comparison of the performance of various methods for tuning VPA's using effort data. *Journal du Conseil International pour l'Exploration de la Mer*, **42**: 129–151.

Rajasilta, M., Eklund, J., Laine, P., Jönsson, N. & Lorenz, T. (1999) Intensive monitoring of spawning populations of the Baltic herring (*Clupea harengus membras*). Final Report of Study Project 96-068, 1997-1999. University of Turku. 75 pp.

Rannak, L. (1971) On recruitment to the stock of spring herring in the north-eastern Baltic. *Rapports et Procès-Verbaux des Réunions Conseil International pour l'Exploration de la Mer*, **160**: 76–82.

Ryman, N., Lagercrantz, U., Andersson, L., Chacraborty, R. & Rosenberg, R. (1984) Lack of correspondence between genetic and morphologic variability pattern in Atlantic herring (*Clupea harengus*). *Heredity*, **53**(3): 687–704.

Schweigert, J. F. (1991) Multivariate description of Pacific herring (*Clupea harengus pallasi*) stocks from size and age information. *Canadian Journal of Fisheries and Aquatic Sciences*, **48**: 2365–2376.

Shepherd, J. G. (1992) Extended Survivors' Analysis: an improved method for the analysis of catch-at-age data and catch-per-unit-effort data. Working paper 11, ICES Multispecies Assessment Working Group, June 1992, Copenhagen, Denmark. 22 pp.

Sjöstrand, B. (1989) Assessment review: exploited pelagic stocks in the Baltic. *Rapports et Procès-Verbaux des Réunions Conseil International pour l'Exploration de la Mer*, **190**: 235–252.

Sparholt, H. (1994) Fish species interactions in the Baltic Sea. *Dana*, **10**: 131–162.

Relationships between fishing gear, size frequency and reproductive patterns for the kingfish (*Scomberomorus commerson* Lacépède) fishery in the Gulf of Oman

Michel R. G. Claereboudt, Hamed S. Al-Oufi, Jennifer McIlwain and J. Steven Goddard

Sultan Qaboos University, College of Agricultural and Marine Sciences, Department of Marine Science and Fisheries, Box 34, Al-Khod, 123. Sultanate of Oman.

ABSTRACT: A survey of fishing gears used in the traditional kingfish fishery along the Gulf of Oman coast of the Sultanate of Oman revealed significant differences in the type of gear and mesh size of nets used by fishers. In two regions of the coast, Al-Batinah and Muscat, different fishing gears (trolling, hooks and line, drift nets, set nets) were used in almost equal numbers, and the dominant mesh size of the nets ranged between 7 and 12 cm. In two other regions (Musandam and Ash-Sharqiyah), most fishing was with a single gear: either set nets or drift nets of large mesh size (>12 cm). An analysis of the maturity stages of the 778 fish landed indicated that, in these waters, females matured at a fork length (80.4 cm) smaller than males (84.7 cm). Further, significantly more females than males were caught in the two regions in which small-meshed nets and a higher diversity of gear types were used. Analysis of the reproductive cycle revealed a single spawning period, peaking in May/June, corresponding with a drop in gonadosomatic index and the appearance of post-spawning fish in the population. Length distribution and analysis of length at first maturity indicated that up to 90% of the fish captured were immature, adding to the risk of growth-overfishing in the two central regions of the coast studied. Improved fishery regulations are proposed and discussed.

INTRODUCTION

The kingfish *Scomberomorus commerson* (Lacépède) is one of the keystone species of the traditional fisheries along the Gulf of Oman. Since the introduction of fisheries statistics in Oman in 1988, landings of the species have fluctuated widely, but in recent years, a progressive tenfold decline in the number and biomass of landed kingfish has necessitated an urgent multidisciplinary study of the species (Al-Oufi *et al.*, 2002).

Understanding the spawning cycle of commercially exploited species is often an important consideration in the establishment of management strategies for their sustainable use. Kingfish constitute an important resource in many parts of the Indo-Pacific. Further, although several long-term studies have been carried out to evaluate its stock dynamics off South Africa (Govender, 1994, 1995, 1997) and Australia (McPherson, 1992, 1993), there have been few studies of the dynamics of kingfish off Oman (Dudley *et al.*, 1992; Siddeek & Al-Hosni, 1998). None of these have specifically set out to quantify the timing of spawning nor the various methods used in the species' traditional fishery. The objectives of this study are therefore to describe the different gear types used in the fishery and to determine some biological parameters for the species: sex ratio, length at first maturity, length distribution, spawning patterns. Within this framework we set out to (1) determine the sex ratio and sex-specific length distributions; (2) quantify the length at first maturity and describe the different maturity stages over a 12-month period; (3) identify the peaks in spawning activity for both sexes using maturity and gonadosomatic indices (*GSI*); (4) examine the relationship between gear size and length distribution of kingfish. The study was made at four locations along the Gulf of Oman.

METHODS

To collect data on fishing gear, traditional fishers were interviewed at several landing sites in four main regions along the Gulf of Oman (Musandam, Al-Batinah, Muscat, Ash-Sharqiyah; Fig. 1). Between January 2000 and January 2002, 360 interviews were conducted at landing sites immediately after the fish were landed. The interview included questions on different fishing methods and the gear employed to harvest the landed catch, the configuration of the gear and the way it was deployed. As part of the survey, the mesh size of the nets used to capture kingfish was also measured to the nearest mm (stretched mesh). All nets used in the Sultanate of Oman are of multifilament, polyamide twine, as per official regulations. Differences in the use of different gears between regions and differences in the mesh size of nets used in the four regions were tested statistically using contingency table analysis and χ^2 tests (Zar, 1984).

In all, 3650 kingfish were sampled between January 2000 and December 2001, representing the majority of kingfish landed at every visit to each landing site. Of these, 778 were purchased directly from the fishers at landing sites in each of the four regions so as to obtain the biological information only available on dissection of the fish. The other fish were simply measured and weighed to the nearest cm (fork length) and 0.01 kg respectively. By rotating the field trips to the various landing sites, it was ensured that monthly samples were available from each of the four regions during the two years of sampling.

For every fish purchased, the ovaries or the testes were removed and weighed to the nearest gramme. Each gonad was assessed macroscopically and assigned to one of six maturity stages based on size and appearance (Table 1). A gonadosomatic index (*GSI*) was calculated for each fish as:

$$GSI = \frac{\text{Mass of gonad (g)}}{\text{Fish mass (kg)}}$$

Data were pooled across locations, and the mean *GSI* was calculated for each calendar month. The length at first reproduction (the length at which 50% of the fish reach reproductive maturity L_{50}) was calculated by plotting cumulative maturity probability against fork length. Only those fish falling in a size range larger than the smallest mature fish and smaller than the largest immature fish were included in this analysis.

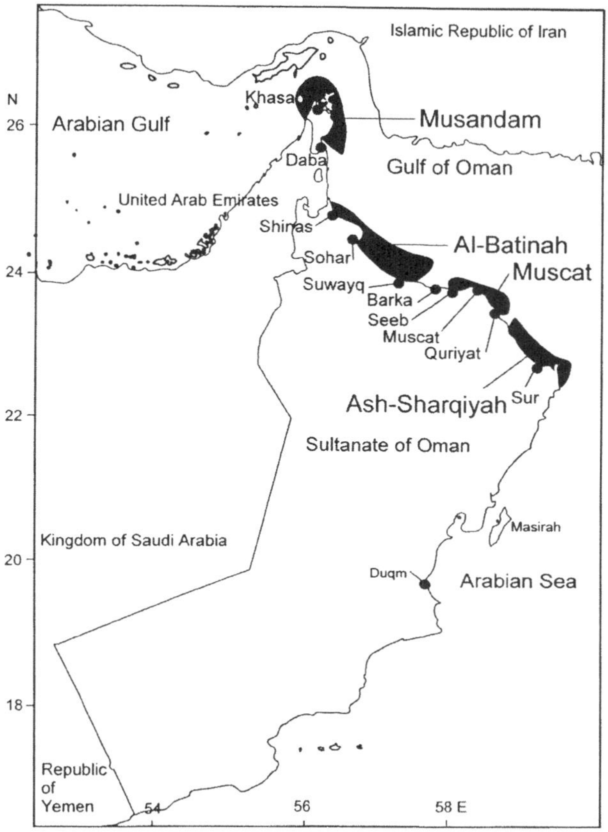

Fig. 1 Map of the study area showing the spatial extent of the coastal regions studied.

Table 1 Macroscopic criteria for assessing stages of reproductive development in *Scomberomorus commerson*.

Reproductive stage		Ovary	Testes
I	Immature	Ovary glassy, small, with compact wall; eggs invisible	Testes small, glass-like, transparent
II	Maturing	Eggs distinguishable.Ovary opaque, small, but rich in blood vessels	Testes opaque, reddish to white, small and compact
III	Mature	Ovary orange, compact and breakable, but no sexual products released when pressed	Testes opaque, white and releasing small amount of sperm when pressed
IV	Spawning	Ovary translucid, walls elastic. Eggs large, nearly transparent	Testes opaque, wall loose, and sperm released when pressed
V	Spent	Ovary looks like water- filled sac. Some oocytes may be found. Genital aperture inflamed	Testes short, dark reddish, no sperm released. Wall flaccid and rich in blood vessels
VI	Resting–maturing	Ovary opaque, pink. Similar to II	Testes opaque, creamy and compact

Median tests were used to compare the length at first maturity between sexes, because the size at maturity is defined as the median or L_{50} of the size distribution of mature fish (Zar, 1984). Deviations from the expected sex ratio of unity were analysed by one-dimensional Pearson's χ^2 tests with Yates correction for continuity (Zar, 1984).

RESULTS

FISHING GEAR

Interviews with fishers at the landing sites revealed that kingfish in the Gulf of Oman were exclusively exploited by traditional methods involving five different gear types. These gear types make different contributions to the total landings of kingfish and vary from one region to the next. The most commonly used gear was a drift gillnet (known locally as *Hayal*), accounting for nearly 39% of deployed gear, followed by a set or pen gillnet (known locally as *Mansab*), which constituted 35% of deployed gear. On two occasions, data collectors noted that beach-seine nets were used specifically to target juvenile kingfish (known locally as *Khabat*) in the Al-Batinah region (Fig. 2).

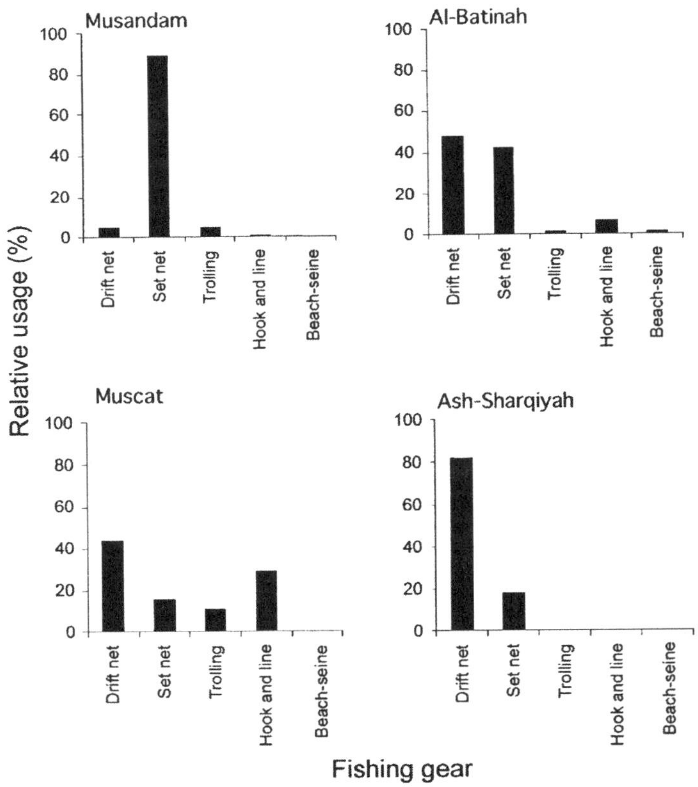

Fig. 2 Relative usage of different types of fishing gear deployed in the four regions studied.

The relative use of fishing gear and fishing methods varied significantly (χ^2 test, df = 9, χ^2 = 146.4, p < 0.0001) between regions (Fig. 2). Methods were more diverse in the Al-Batinah and Muscat regions, where drift nets accounted for approximately 40% of the total gear used. Hook-and-line gear and set nets were the second and third most common gear type used by fishers in the Muscat area. In contrast, set nets and drift nets accounted for more than 80% of the total gear used in the Musandam and Ash-Sharqiyah regions respectively.

The relative distribution of mesh sizes among gillnets and set nets varied significantly between the four regions (χ^2 test, df = 12, χ^2 = 67.2, p < 0.001; Fig. 3). Mesh sizes were smaller in the Al-Batinah and Muscat regions, 47 and 36% of nets having a mesh size in the range 7–12 cm respectively. In the Musandam and Ash-Sharqiyah regions, nets with mesh size <12 cm accounted for fewer than 20% of the nets deployed in the kingfish fishery (Fig. 3).

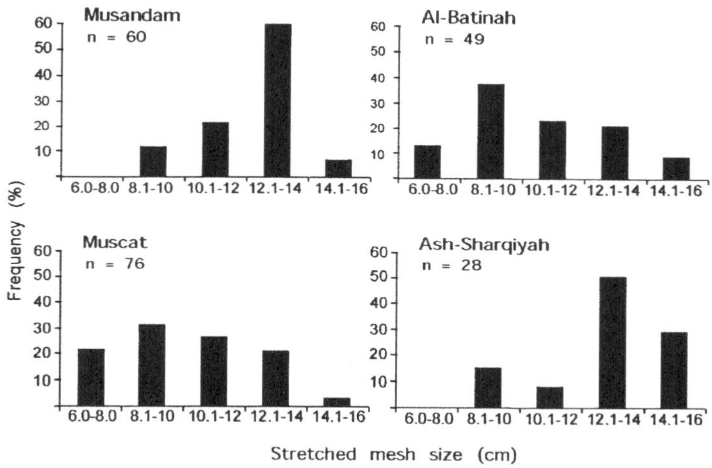

Fig. 3 Comparison of the mesh size used in drift nets and set nets in the four studied regions of the Gulf of Oman. The value n refers to the total number of sets of gear sampled in each region.

SIZE COMPOSITION AND SEX RATIO

Of the 3650 kingfish measured in the study, the smallest was 43 cm and the largest 201 cm (the latter from Musandam). There were clear differences in the size distribution of fish landed between regions (Fig. 4). Large fish (>100 cm) were in greater proportion in Musandam and Ash-Sharqiyah landings than in landings in the other two regions. At Muscat and especially at Al-Batinah, landings were dominated by fish <100 cm; fish >100 cm accounted for only 7% of the landings at Al-Batinah.

The sex ratio landed varied between regions (Table 2). Although females were caught in greater abundance in all four regions, only samples at Muscat and Al-Batinah revealed a sex ratio significantly different from unity (Table 2). At Musandam and Ash-Sharqiyah there was no significant difference in the sex ratio, with no deviation from the expected value of unity.

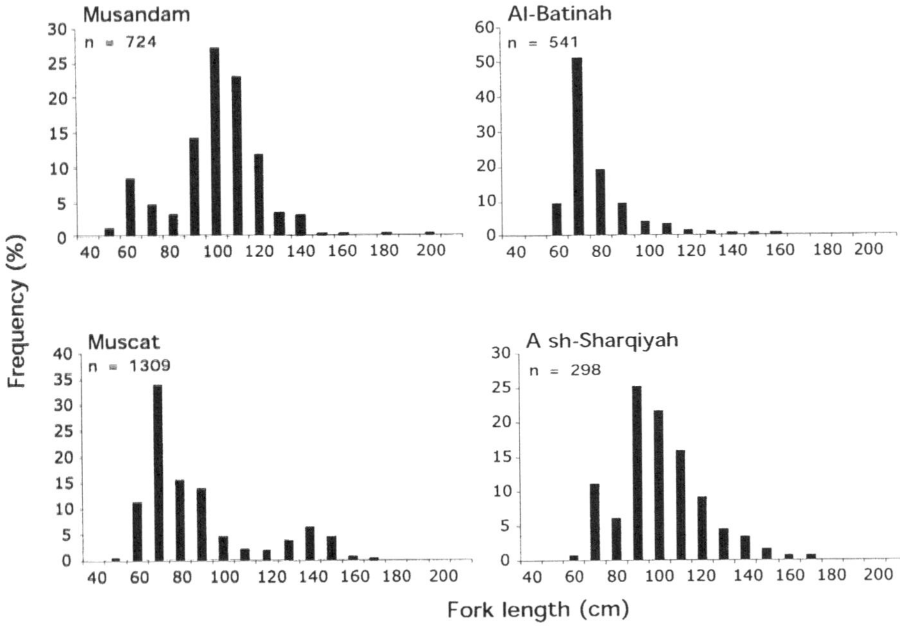

Fig. 4 Length frequency distribution of kingfish catches per region. The value n is the total number of fish measured in each region.

Table 2 Sex ratio and χ^2 test with Yates correction for deviation from the expected value of unity in four coastal regions of the Sultanate of Oman.

Parameter	Musandam	Al-Batinah	Muscat	Ash-Sharqiyah
Number of males	76	90	102	63
Number of females	92	121	151	83
Total number analysed	168	211	253	146
Sex ratio (male/female)	0.82	0.74	0.67	0.75
Expected number of males	84	105.5	126.5	73
Expected number of females	84	105.5	126.5	73
χ^2 (df = 1)	1.52	4.55	9.49	2.74
With Yates correction	1.39	4.26	9.11	2.47
p	0.24	0.039*	0.003*	0.116

* *test significant at 0.05 level.*

A plot of the relative cumulative proportion of mature fish against fork length revealed significant differences between sexes (Fig. 5). Female kingfish in the Gulf of Oman mature significantly smaller (80.4 cm) than males (84.7 cm; Median test, $\chi^2 = 18.8$, $p < 0.0001$).

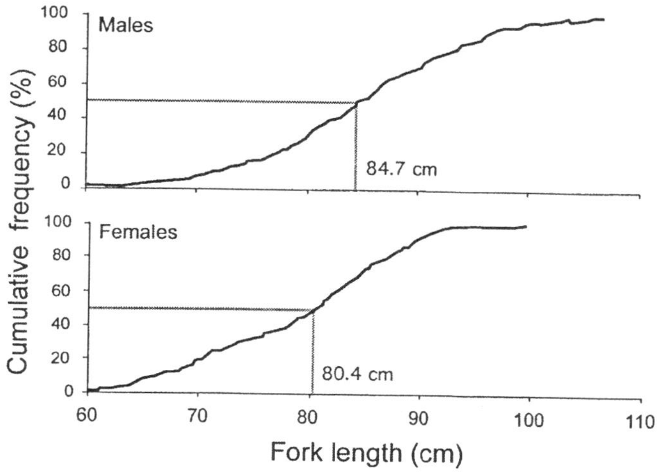

Fig. 5 Cumulative relative frequency of maturity in male and female kingfish from the Gulf of Oman. The lengths at 50% maturity (L_{50}) are indicated. Only fish larger than the smallest mature and smaller than the largest immature were considered.

MATURITY

Data on gonad stage development for the two years of sampling grouped by calendar month (Fig. 6) revealed that immature fish (Stage I) were present in nine of the 12 months, but were absent during summer (May – July). For the rest of the year (September – April), up to 30% of the fish landed in the northern Sultanate of Oman were immature.

The gametogenic cycle began in February with fish successively entering the developing stages (Stages III, IV) of the reproductive cycle (Fig. 6). Mature and spawning stages (Stages IV, V) began to appear in the May and June samples. However, no Stage IV fish were recorded from August to February. By July most kingfish (with the exception of a large number of Stage II fish) were in the post-spawning stages (Stage V, VI), indicating the end of the spawning season.

GONADOSOMATIC INDEX (GSI)

The monthly cycle in the gonadosomatic index (*GSI*) coincides with that of maturity (Fig. 7). Maximum gonad development was simultaneous in both sexes during May and June after a three-month period of steady increase. For females, spawning was synchronous in both years, and was characterized by a sharp decline in *GSI* in June. For males, *GSI* peaked one month earlier (in April) during the second year of sampling.

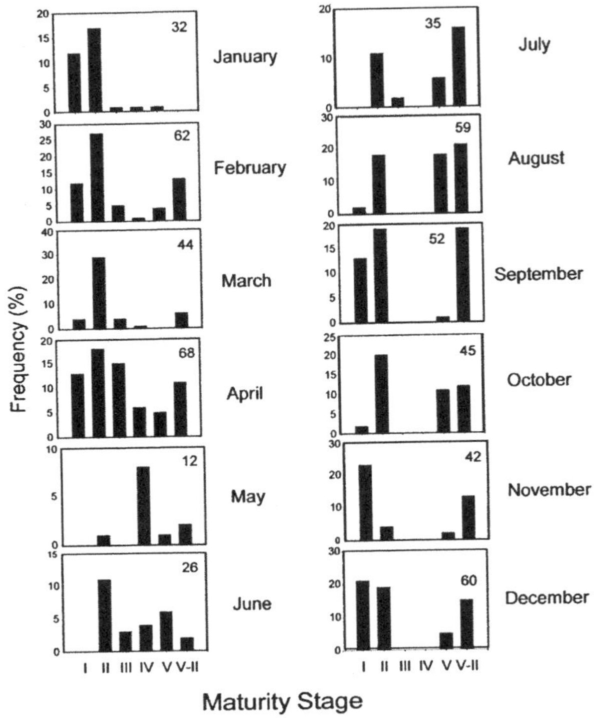

Fig. 6 Distribution of maturity stages in kingfish catches from the Gulf of Oman. Male and female data are grouped.

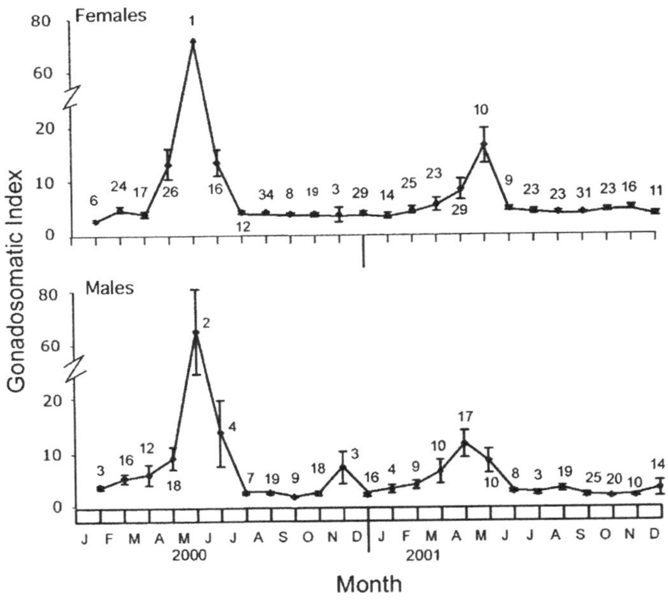

Fig. 7 *GSI* of female and male kingfish for two consecutive calendar years. Error bars represent standard errors, and numbers the sample size for each data point.

DISCUSSION

SEX RATIO

Females dominated the landings in two of the regions of the Gulf of Oman sampled. This finding could have resulted from differences in the actual sex ratio in the fish stock or from a bias introduced in the samples by the fishing methods used by traditional Omani fishers. Such fishery selectivity for females has been observed for another scombrid species (*Scomberomorus cavalla*) along the south coast of the USA, where recreational linefishing strongly favoured the catching of female fish (Trent *et al.*, 1987), although no biological explanation was given for this observation

In the present study, the hypothesis that fishing methods introduce a bias in sex ratio is supported by differences in fishing gear used in the four regions. In the Musandam and Ash-Sharqiyah regions, the use of both drift nets and set nets results in an equal number of male and female kingfish caught. On the other hand, in the Muscat and Al-Batinah regions, the addition of hooks and lines and trolling to the gear used possibly biased the number of females in the catches. This might be the case if females, particularly young ones, take bait more readily than males, or if females requiring more energy for gametogenesis feed more actively at dusk and dawn, when trolling most commonly takes place.

SIZE DISTRIBUTION

Differences in fishing gear may also explain some of the difference in size distribution between fish caught in the four regions. In the Sultanate of Oman, there are as yet no regulations on the mesh size of nets authorized for use in the kingfish fishery. In the Al-Batinah and Muscat regions, where small fish are captured in large numbers, small-meshed nets are common. Similarly, in areas where larger fish are landed, larger-meshed nets dominate (Figs 2, 3). The greater diversity of methods of fishing in areas where small fish are captured suggests perhaps a shift in the methods to capture fish at younger age. This is also supported by observations of beach-seine fishers targeting very young kingfish along the Al-Batinah coast.

SIZE AT FIRST MATURITY

Published values of length at maturity for *Scomberomorus commerson* in the western Pacific and Indian Oceans coincide with the present findings. Kingfish length at first maturity is attained at about 70–80 cm fork length off Madagascar, Papua New Guinea and Fiji (Collette & Russo, 1984), but not before 90–110 cm off South Africa (Govender, 1995). The eastern Australian stock first mature at 79 cm, whereas the northern population of the same country is slightly longer when it first matures (82 cm; McPherson, 1993). In the northern Indian Ocean, kingfish length at first maturity was estimated by Devaraj (1983) to be 75 cm.

The difference in length at first maturity between sexes found here could result from a slower growth rate of females. Combining the information on length at first maturity and frequency distribution of the landings reveals that the percentages of fish captured before maturity were 35, 89, 75 and 42% in the Musandam, Al-Batinah, Muscat and Ash-Sharqiyah regions respectively. A high incidence of juveniles in the catch, as demonstrated above, adds to the risk of growth-overfishing, which may ultimately impact on the status of stocks and compromise the potential sustainable yield that would accrue from these traditional fish.

SPAWNING

The variations in *GSI* and maturity stage demonstrate the existence of a single synchronous spawning period in May. This finding concurs with a preliminary analysis of the age of recruits following an analysis of the daily increments present on otolith sections (Dudley *et al.*, 1992). In the northern Indian and western Pacific Oceans, the reproductive biology of *S. commerson* is relatively well understood. Off north-eastern Australia, two distinct stocks of *S. commerson* have been identified. For the eastern stock, spawning peaks in late spring (October and November), but the more equatorial northern stock experiences a more protracted spawning season of approximately six months (August – December; McPherson, 1993). In the northern Indian Ocean, reproduction extends from January to September with a strong peak in April/May (Devaraj, 1983). Reports from Iran also suggest protracted spawning, from August to October (Kingfish Task Force, unpublished data, 1995), considerably later than off the Sultanate of Oman. Spring spawning has also been observed in other species of the genus, *S. queenslandicus* and *S. muroi*, in eastern Australia (Begg, 1998) as well as in *S. cavalla* in the southern USA (Collette & Russo, 1984).

The seasonality in the catches (most kingfish are caught during winter, and catches are very low in April and May) supports the idea of a migration (at least partial) out of Omani waters during the reproductive season. The traditional fishing community believes that *S. commerson* participates in a lengthy migration, moving north during summer to spawn in the Arabian Gulf (Al-Oufi *et al.*, 2002). However, the present results suggest that at least some of the population (in all regions) were engaged in localized spawning activity. Fully mature, spawning and spent fish were caught along the coast of the Gulf of Oman in April, May and June, supporting the theory that some spawning grounds at least are local. The issue of migration can only be resolved through a carefully planned tag-recapture programme for kingfish before, during and just after spawning.

MANAGEMENT ISSUES

Several indicators in the data suggest that the populations of kingfish are under heavy fishing pressure, particularly in the central part of the Gulf of Oman coast of the Sultanate, where high proportions of immature and small individuals constitute the catch as a consequence of the fishing gear being adapted to capture them. Preliminary data on the age of the fish sampled in the Sultanate (S. Zaki, pers. comm.) suggest that very few fish landed in the Gulf of Oman reach a third year of life (some 110 cm fork length) or a second reproductive period. Therefore, it would likely benefit the fishery if a larger proportion of fish were allowed to reach at least reproductive maturity and possibly reproduce once or twice before entering the fishery. Such an objective could be realized either by implementing input controls such as minimum size or possibly bag limits, although that could increase the risk of illegal landings and greater discarding.

Alternatively, measures such as mesh size limits for gillnets and set nets might be easier to regulate and enforce and so ensure an adequate level of protection of juvenile kingfish to allow rebuilding of the fish stocks. Off Australia, where the annual catch (some 3500 tons) is similar to that of Oman, there has been no significant decline in the catch for the 25-year period for which catch records are available (A. Tobin, pers. comm.). Setting such a minimum size might result in short-term decrease in yield, but it will likely generate long-term gains in terms of yield per recruit (Govender, 1995). The risks of stocks falling below a reproductive threshold, at least locally, are increasing, and decisions on management strategies should be taken rapidly if the Gulf of Oman kingfish fishery is to remain sustainable.

The results of this study have raised several questions that need to be answered before the implementation of a successful management strategy. For instance, to where do kingfish migrate during summer, and does their disappearance correspond to a spawning migration? Further, what is the selectivity of the different gear in terms of fish size and sex, and are differences in selectivity responsible for the greater number of juveniles caught in the central region of the north coast (Al-Batinah and Muscat). More research and, if possible, international collaboration are needed to provide answers to these fundamental questions.

ACKNOWLEDGEMENTS

We thank H. Al-Masroori, K. A. Al-Hashmi, H. N. Al-Habsi, G. V. Hermosa, S. M. Al-Barwani, A. Al-Nabhani, K. Al-Ryami, S. Al-Khusaibi and A. Ambuali for their assistance in data collection and processing, and A. S. Revill and R. P. van der Elst for constructive reviews of the draft manuscript. The financial resources for the project were provided by the Fisheries Research Funds of Oman, through the Ministry of Agriculture and Fisheries.

REFERENCES

Al-Oufi, H. S., McLean, E., Goddard, J. S., Claereboudt, M. R. G. & Al-Akhzami, Y. K. (2002) The kingfish, *Scomberomorus commerson*, (Lacépède, 1800) in Oman: reproduction, feeding and stock identification. In: *Contemporary Issues in Marine Science and Fisheries*. Ed. by E. McLean, H.S. Al-Oufi, Y.K. Al-Akhzami & Najamuddins, pp. 1-17. Hasanuddin University Press, Makassar.

Begg, G. A. (1998) Reproductive biology of school mackerel (*Scomberomorus queenslandicus*) and spotted mackerel (*S. munroi*) in Queensland east-coast waters. *Marine and Freshwater Research*, **49**: 261-270.

Collette, B. & Russo, J. L. (1984) Morphology, systematics and biology of the Spanish mackerels (*Scomberomorus*, Scombridae). *Fishery Bulletin U.S.*, **82**: 545–689.

Devaraj, M. (1983) Maturity, spawning and fecundity of the king seer, *Scomberomorus commerson* (Lacépède), in the seas around the Indian Peninsula. *Indian Journal of Fisheries*, **30**: 203-230.

Dudley, R. G., Aghanashinikar, A. P. & Brothers, E. B. (1992) Management of the Indo-Pacific Spanish mackerel (*Scomberomorus commerson*) in Oman. *Fisheries Research*, **15**: 17–43.

Govender, A. (1994) Growth of the king mackerel (*Scomberomorus commerson*) off the coast of Natal, South Africa – from length and age data. *Fisheries Research*, **20**: 63–79.

Govender, A. (1995) Mortality and biological reference points for the king mackerel (*Scomberomorus commerson*) fishery off Natal, South Africa (based on per-recruit assessment). *Fisheries Research*, **23**: 195–208.

Govender, A. (1997) Age and growth of two key exploited linefish species off Natal: the slinger (*Chrysoblephus puniceus*) and the king mackerel (*Scomberomorus commerson*), In: *Fish, Fishers and Fisheries*, Second Annual Marine Linefish Symposium, Durban, pp. 36–41.

McPherson, G. R. (1992) Age and growth of the narrow-barred Spanish mackerel (*Scomberomorus commerson* Lacépède, 1800) in north-eastern Queensland waters. *Australian Journal of Marine and Freshwater Research*, **43**: 1269–1282.

McPherson, G. R. (1993) Reproductive biology of the narrow barred Spanish Mackerel (*Scomberomorus commerson* Lacépède, 1800) in Queensland waters. *Asian Fisheries Science*, **6**: 169–182.

Siddeek, M. S. M. & Al-Hosni, A. H. S. (1998) Biological reference points for managing kingfish, *Scomberomorus commerson*, in Oman waters. *Naga*, **21**(4): 32–36.

Trent, L., Fable, W. A., Russell, S. J., Bane, G. W. & Palko, B. J. (1987) Variations in size and sex ratio of king mackerel, *Scomberomorus cavalla*, off Louisiana, 1977-85. *Marine Fisheries Review*, **49**(2): 91–97.

Zar, J. H. (1984) *Biostatistical Analysis*. Prentice-Hall Inc., Englewood Cliffs, New Jersey. 718 pp.

The Management of Transboundary Stocks of Toothfish, *Dissostichus* spp., under the Convention on the Conservation of Antarctic Marine Living Resources

Eugene N. Sabourenkov and Denzil G.M. Miller

The Secretariat, Commission for the Conservation of Antarctic Marine Living Resources (CCAMLR), P.O. Box 213, North Hobart, Tasmania 7002, Australia.

ABSTRACT: The Patagonian toothfish, *Dissostichus eleginoides*, is a deep-water species which occurs on the high seas as well as in the maritime zones of various Coastal States which fall within and beyond the area administered by the Commission for the Conservation of Antarctic Marine Living Resources (CCAMLR) – the 'Convention Area'. Commercial fishing for toothfish in the Convention Area developed rapidly from the early 1990s. With the fishery for a closely related species, the Antarctic toothfish (*D. mawsoni*), the fisheries for toothfish have posed a serious challenge to CCAMLR's efforts aimed at ensuring that their management is consistent with the precautionary and ecosystem conservation principles enshrined in the CAMLR Convention. The sustainable management of some of the stocks concerned has also been seriously compromised from about the mid-1990s by illegal, unregulated and unreported (IUU) fishing both inside and outside the Convention Area. Various management measures have been implemented to combat IUU fishing and these are described. Such measures include strict licensing requirements to control fishing, the requirement to deploy satellite monitoring systems (VMS) to monitor vessel location at sea in real time and both at-sea and in-port inspection schemes. The persistent high levels of IUU fishing have resulted in the recent implementation of an additional measure – the *Catch Documentation Scheme for Dissostichus* spp. (CDS) aimed at tracking the landings and trade flow of toothfish caught both inside and outside the Convention Area. The CDS has been operational since May 2000 and its impact is considered an important component in a suite of measures to combat IUU fishing along with its role in improving the participation of non-Contracting Parties in CCAMLR toothfish management measures. Some statistics associated with the operation of the CDS are discussed and an initial evaluation of its performance is provided. Despite its relatively short period of operation, the CDS can be shown to be an effective tool to combat IUU fishing in the CAMLR Convention Area when combined with other management measures. Details are also given of initiatives planned by CCAMLR to enhance performance of the CDS and the scheme's future is discussed in light of the entry into force of the United Nations Fish Stocks Agreement (UNFSA) and in relation to the FAO Compliance Agreement. Aspects of cooperation with the World Trade Organisation (WTO), the World Customs Organisation (WCO) and other international instruments are also discussed. The CCAMLR CDS possesses many features which make it a good model for recent FAO initiatives aimed at harmonising catch certification and trade documentation systems.

INTRODUCTION

The *FAO Fisheries Glossary* defines 'transboundary' stocks as 'those stocks of fish that migrate across international borders' (FAO, 2002a). The term 'transboundary stocks' is often used as a general term to describe both straddling and highly migratory fish stocks. Management of the latter stocks are now subject to the United Nations Fish Stocks Agreement (UNFSA)[1] which entered into force on 10 December 2001.

As Molenaar (2001) has emphasised, the 24-nation Commission for the Conservation of Antarctic Marine Living Resources (CCAMLR) manages fisheries of species which 'occur in various Exclusive Economic Zones (EEZs), even apart form those adjacent to Antarctica, and in an area beyond or adjacent thereto, whether or not this area constitutes the high seas'. This implies that the 'transboundary' stocks being regulated by CCAMLR are in fact 'straddling stocks' in the sense of Article 63.(2) of the 1982 United Nations Convention on the Law of the Sea (LOS Convention), and as a corollary, covered by UNFSA. It is in this context that CCAMLR's recent efforts to manage such fisheries will be discussed.

The management of transboundary stocks has been a focus of CCAMLR's attention for a number of years. However, neither CCAMLR nor its attached Scientific Committee have defined such stocks as being 'straddling' or 'highly migratory' in strict accordance with the principles enshrined in the UNFSA. Therefore, since the early drafting of the UNFSA, CCAMLR has preferred to describe 'transboundary' stocks as 'those which occur both inside and outside the CAMLR Convention Area' (CCAMLR, 1993).

The boundaries of the CAMLR Convention Area have been defined as falling between the Antarctic Polar Front (APF)[2] in the north and the area south of 60°S (the major part of the 'Southern Ocean'). Within the hydrographic barrier attributed to the APF, a key feature of Antarctic fish biogeography is the inherent endemism of the large number of species which inhabit shelf areas around the Antarctic Continent and close to the many oceanic islands of the Southern Ocean. However, such endemism is much less common in species which inhabit deep water, and these are often encountered in high-seas regions just to the south and north of the APF (Fischer & Hureau, 1985).

At present, stocks of Patagonian toothfish (*Dissostichus eleginoides*, family Nototheniidae) are the target of economically valuable fisheries, and using the CCAMLR definition the species could be described as 'transboundary'. Patagonian toothfish are distributed throughout much of the Convention Area and the species also occurs to the north of the APF, on the high seas and in the maritime zones of Coastal States along the continental margins of South America in particular. The distribution of the closely related Antarctic toothfish (*D. mawsoni*) is predominantly confined to the Convention Area, although some surveys conducted in the 1960s and 1970s reported encountering the

[1] *The United Nations Fish Stocks Agreement is the 1995 United Nations Agreement for the Implementation of the United Nations Convention on the Law of the Sea of 10 December 1982 relating to the Conservation and Management of Straddling Fish Stocks and Highly Migratory Fish Stocks.*

[2] *The Antarctic Polar Front (APF) is the zone where colder, fresher waters flowing north from the Antarctic meet the warmer, saltier waters flowing south from the Atlantic, Indian and Pacific Oceans. This term has effectively replaced the old term 'Antarctic Convergence' which was in use at the time of negotiating the CAMLR Convention.*

species in the Pacific Sector of the Southern Ocean to the north of 60°S (i.e. outside the Convention Area) (Yukhov, 1982). The two toothfish species look very alike and are often hard to distinguish, especially when mixed catches are landed (Fischer & Hureau, 1985). This has resulted in some recent work to develop genetic fingerprinting as a tool for separating the two species when attributing the source of fish products at points of landing (Smith *et al.*, 2001; Appleyard *et al.*, 2001).

In coastal waters around South America, fisheries for toothfish developed well before those in the Convention Area (Kock, 1992), with catches of Patagonian toothfish being reported from a number of fisheries in the Southern Ocean in the late 1970s (CCAMLR, 1990a). However, fisheries specifically targeting toothfish in the Convention Area only developed in the mid-1980s (CCAMLR, 1990b). Like other fisheries in the Convention Area, CCAMLR manages these fisheries on the basis of a precautionary and ecosystem approach (Constable *et al.*, 2000). In addition, CCAMLR has requested its Members to ensure that their vessels harvesting species which occur both within and in areas adjacent to the Convention Area, pay due respect to CCAMLR conservation measures for such species (CCAMLR, 1993).

The persistent encroachment of illegal, unregulated and unreported (IUU)[3] fishing for toothfish in the Convention Area from the mid-1990s has begun to seriously undermine CCAMLR's efforts to manage the exploitation of toothfish on a sustainable basis (Larson, 2000) as well those aimed at protecting species likely to be impacted by fishing activities (e.g. seabirds) (Kock, 2001). CCAMLR has implemented various measures to combat IUU fishing and these include mandatory license requirements to allow control of fishing, the monitoring of fishing vessel location in real time (so-called 'Vessel Monitoring Systems' – VMS), the marking of fishing gear, and at-sea and in-port inspection of vessels as well as catches (Agnew, 2000). At the same time, the presence of international scientific observers on board all vessels licensed to harvest toothfish in the Convention Area, acts as an extra deterrent. Despite such efforts, the continued persistence of IUU fishing has led to CCAMLR's implementation of a Catch Documentation Scheme (CDS) to monitor landings of, and trade in, both toothfish species.

CCAMLR's initial attempts to regulate IUU fishing in the Convention Area and the associated development of the CDS have recently been reviewed by Agnew (2000). This paper focuses on the implementation of the CDS over the past two years. It provides an assessment of the scheme's current performance and considers its possible future development. It also describes a current suite of CCAMLR measures that aim to combat IUU fishing and of which the CDS comprises a key element.

CCAMLR'S MANAGEMENT OF TOOTHFISH FISHERIES

ASSESSMENT OF TOOTHFISH YIELD

Both species of toothfish may live in excess of 30 years, exhibit slow growth rates and mature at a relatively late age (7–10 years) (VNIRO, 1985; Shust, 1998). Despite relatively high fecundity compared with other species of nototheniids (Shust, 1998), it is generally recognised that the life history characteristics of both species make them particularly vulnerable to overfishing (Agnew, 2000).

[3] *The terminology for IUU fishing is taken to cover fishing by vessels in defiance of areas of Coastal State jurisdiction within the Convention Area, fishing by vessels of CCAMLR Members in contravention of CCAMLR conservation measures, fishing by non-CCAMLR flagged vessels within the Convention Area outside areas of Coastal State jurisdiction, and fishing when catches are not fully reported or monitored (after Agnew, 2000).*

Since 1985, CCAMLR has employed a model-based approach to estimate toothfish yield and hence determine catch limits. Initially used to estimate precautionary catch limits for Antarctic krill (*Euphausia superba*) (Butterworth *et al.*, 1991; Butterworth & Thomson, 1995), this yield model has been developed into a more generalised formulation to be applied to toothfish and other species (the so-called 'Generalised Yield Model' – GYM) (SC-CAMLR, 1995; Constable & de la Mare, 1996). The method uses absolute estimates of recruit abundance and projects these forward in time. It allows discounting of known catches from population estimates, so providing estimates of long-term yield that can be assessed in terms of weight rather than as some proportion of the pre-exploitation biomass (B_0; Constable *et al.*, 2000). With attached decision rules, application of the GYM satisfies the precautionary management principles to which CCAMLR subscribes and serves to maintain spawning stock biomass at a sustainable level. A recent refinement allows estimates of IUU catches to be used to obtain better approximations of 'total removals' of toothfish, thus improving assessments of yield for existent fisheries.

STATUS OF TOOTHFISH FISHERIES IN THE CONVENTION AREA

The sustainable development of fisheries for toothfish in the Convention Area, along with reducing the extent of IUU fishing on the species, have been priority items on CCAMLR's agenda since 1996. A number of recent publications (e.g. Agnew, 2000; Kock, 2001; Molenaar, 2001) have addressed such concerns. These publications document toothfish catches derived from catch and effort statistics reported to CCAMLR by area and year. IUU fishing patterns have also been described mainly on the basis of information considered and published annually by CCAMLR in its reports. In addition, a number of studies sponsored by non-governmental organisations have analysed available toothfish trade statistics and have investigated the IUU fishing fleet, providing details on vessel flags, ownership and operations (e.g. Lack, 2001; Lack & Sant, 2001; Willock, 2002; see also information published on websites by Greenpeace International, 2002 and ISOFISH, 2002).

The estimated global catch of toothfish since 1984 has been summarised by Agnew (2000) for areas both inside and outside the Convention Area. In 1996, a large number of new and exploratory fisheries for toothfish were notified by CCAMLR Members and were opened in the Convention Area (CCAMLR, 1996). As noted in Agnew (2000), the juxtaposition of such notifications with the increase in IUU fishing observed in late 1996 and in 1997 may be entirely coincidental but there is a possibility that CCAMLR's notification process could itself have unwittingly triggered the outbreak of IUU fishing, particularly in the Indian Ocean (as shown in Fig. 1).

Under CCAMLR terminology, fisheries may be viewed as 'established', 'new' or 'exploratory'[4] (Constable et al., 2000). The extent of authorised toothfish fisheries in the

[4] *CCAMLR Conservation Measure 31/X defines a 'new' fishery as one for which specific information from an area has not been submitted to CCAMLR, or for which catch and effort information has never been submitted to CCAMLR, or catch and effort information from the two most recent seasons have not been submitted to CCAMLR. Conservation Measure 65/XI defines an 'exploratory' fishery as one previously defined as a 'new' fishery under Conservation Measure 31/X and one which continues to be defined as 'new' until such time that certain conditions (e.g. key data on which to base estimates of yield etc.) have been met. Both 'new' and 'exploratory' fisheries are subject to a rigid notification and review procedure before CCAMLR authorises fishing. All other fisheries are seen either as 'closed' (i.e. not specifically open to fishing as new or exploratory fisheries, or as closed by specific CCAMLR conservation measures) or are classed as no longer being exploratory (i.e. for which essential management and catch/effort information are available).*

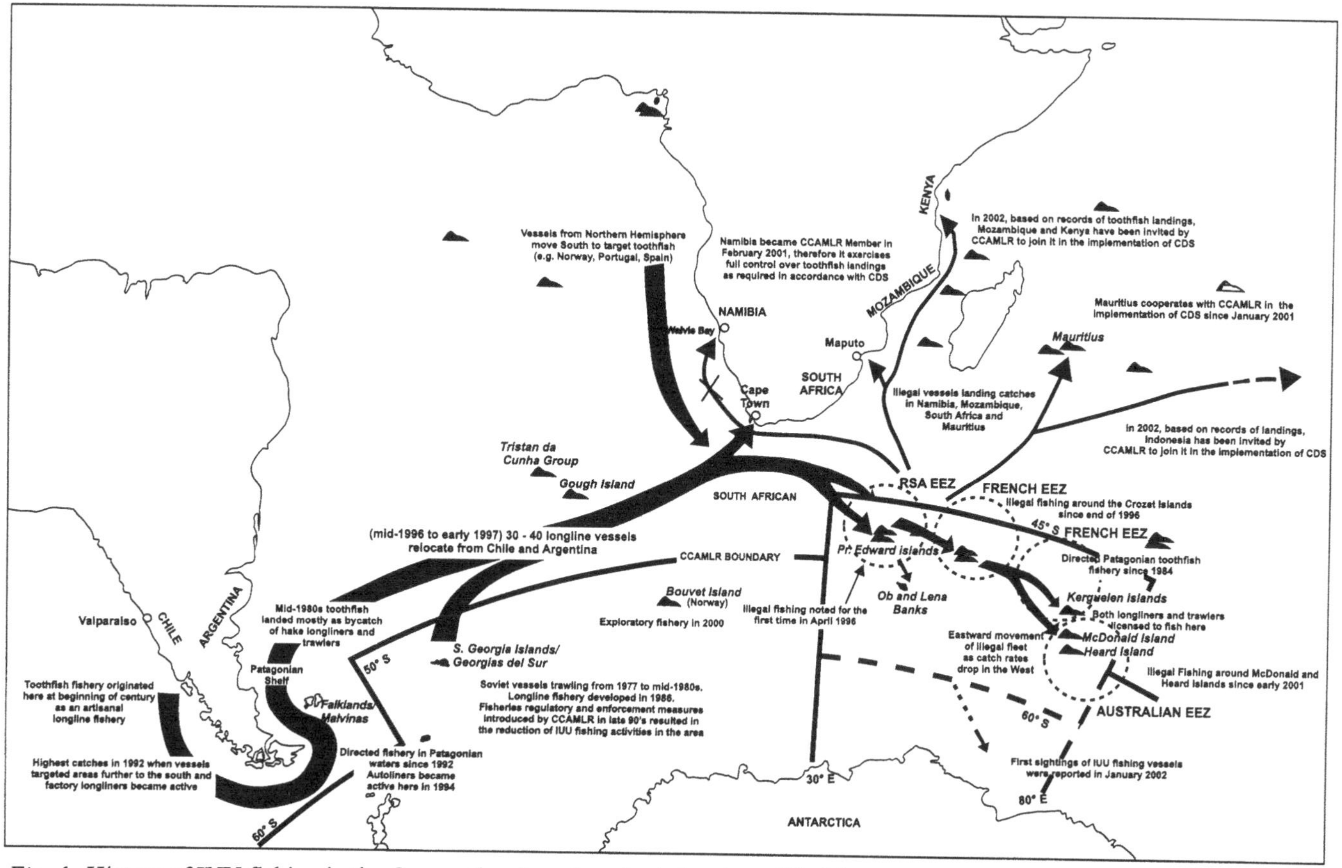

Fig. 1 History of IUU fishing in the Convention Area (as adopted and modified from a report on IUU fishing prepared by South Africa - D. Miller and B. Watkins, unpubl.).

Convention Area since the 1991/92 fishing season[5] is depicted in Fig. 2 where the number of fisheries authorised is compared with those actually prosecuted. It should be noted that, despite the large number of new and exploratory fisheries authorised by the Commission, only a small number were actually prosecuted. For example, out of 11 new and exploratory fisheries authorised in the 1997/98 season only one fishery was actually 'fished'. A similar situation is evident for the following seasons. Contrary to the view expressed in the previous paragraph, this trend would not suggest a general increase in fishing levels coincident with the increase of notifications to CCAMLR for the 1996/97 season. In this respect, the large number of new and exploratory fisheries authorised but not prosecuted in the pursuant seasons is more likely to reflect a growing realisation that there is no further potential for expanding the toothfish fishery or that impacts on available stocks, particularly by IUU fishing, are such that economically sustainable yields are no longer viable in certain areas.

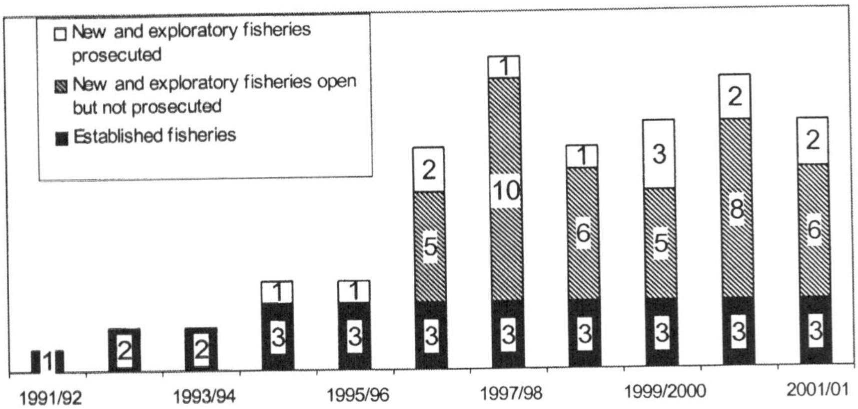

Fig. 2 Total number of established, new and exploratory fisheries for *Dissostichus* spp. (open and prosecuted, and open but not prosecuted) by fishing seasons.

Note: 1. Each fishery defined by a conservation measure is counted as one fishery irrespective of possible further subdivision of the TAC by fishing with different gear. It does not include fisheries conducted in EEZs' around Prince Edward and Marion Islands, Crozet and Kerguelen Islands which are exempt from management by CCAMLR conservation measures.

2. The 'fishery not prosecuted' is defined here as a fishery authorised by CCAMLR in any of the fishing seasons but for which no catch was reported or the reported catch was 1 tonne or less.

3. Information for the 2001/02 fishing season is not yet complete.

[5] *Since 2001, CCAMLR has defined the fishing season as the period from 1 December to 30 November of the following year. Before 2001, the fishing season for reporting purposes was defined as a split-year, i.e. a period from 1 July to 30 June of the following year.*

IUU FISHING IN THE CONVENTION AREA

Prior to 1996, CCAMLR determined IUU fishing activities from potential catches of unlicensed fishing vessels sighted in specific areas. However, with the rapid expansion of longline toothfish fisheries in the Convention Area, a standard methodology for the calculation of IUU catches has gradually evolved. This uses the following information:

(1) Type and size of CCAMLR licensed vessels, their catch and effort, and duration of fishing trips reported.
(2) Number, type and size of vessels sighted engaged in IUU fishing and reported by CCAMLR Members.
(3) Recovered illegal longline gear.
(4) Landings in ports of CCAMLR Members and known landings in ports of other States.
(5) Catch and effort information from vessels seized for IUU fishing by Coastal States in the Convention Area.
(6) Verified information from the international media.
(7) Known catch and trade statistics from a variety of sources (e.g. published trade statistics, customs declarations, etc.).

From the available information, estimates are made of the potential number of IUU fishing 'trips' per vessel by area and season, the duration of such trips, the days fished and the likely catch rates per day. Catch rates are based on those recorded in 'licensed' fisheries from the same area. Total IUU catch is then calculated as a product of fishing time and daily catch rate.

Much of the information used to calculate IUU catches is often incomplete as well as being subject to many underlying assumptions. Consequently CCAMLR's estimates of IUU-caught toothfish are not likely to be overly accurate and probably represent rather coarse approximations only (SC-CAMLR, 1999). However, as suggested below, estimates of the total removals of toothfish in both the Convention Area and elsewhere are likely to have improved as a consequence of the introduction of the CDS and the requirement to verify fishing position using satellite monitoring (VMS) of vessels licensed to fish by CCAMLR Members.

A number of recent publications have analysed available information on IUU fishing levels in the Convention Area (e.g. Agnew, 2000; Kock, 2001; Lack, 2001; Lack & Sant, 2001). Not surprisingly, and given the high levels of uncertainty attached to such estimates, these have differed considerably. Similarly, this is also true for CCAMLR estimates, which have often undergone revision as new information comes to hand. It is worth noting that estimates derived from world trade statistics are often higher than CCAMLR estimates (e.g. see Lack & Sant, 2001). Such a discrepancy could, at least partially, be attributed to the fact that the world toothfish trade statistics may reflect 'double' calculation with reported trade levels for some countries reflecting both imports and exports. For example, in 2000 and 2001, out of some 13 000 tonnes of toothfish imported by the People's Republic of China (including Hong Kong) about 3 000 tonnes were re-exported to Japan and the USA (from CDS data as described in the section below on the CCAMLR CDS). All available IUU catch estimates have been recently reviewed by the CCAMLR Working Group on Fish Stock Assessment (SC-CAMLR, 2001), together with new information on IUU fishing activities. These revised estimates are given in Table 1 for the period from 1996/97 to 2000/01. A decrease in the level of IUU fishing in the past three seasons is especially notable, although the underlying reasons for such a trend are not entirely clear (SC-CAMLR, 2001). It appears that the combined effect of CCAMLR enforcement measures and national measures implemented by Coastal States has resulted not only in a reduction in IUU fishing for toothfish but also in the identification of the remaining, and most persistent, Flags of Convenience, IUU vessels and their owners. It is that group of IUU perpetrators that is now the main focus of CCAMLR. The task of eliminating the remaining IUU operators is complicated as that group, when forced out of one area, relocates rapidly to another, often with changed flags and vessel names. This situation is evident from the recent relocation of IUU fishing operations to the Indian Ocean, both inside and outside the Convention Area (FAO Statistical Areas 51 and 57).

Table 1 Revised estimates of IUU toothfish catches (tonnes) in the Convention Area in the 1996/97 to 2000/01 fishing seasons (from SC-CAMLR, 2001).

	1996/97	1997/98	1998/99	1999/00	2000/01
Estimated IUU	40 000	22 415	4 872	6 546	7 599
Total catch	50 469	33 663	22 363	20 236	20 510
IUU as a % of total catch	79	66	21	32	37

In general, published studies have noted an eastward progression of IUU fishing from the Atlantic Ocean sector of the Convention Area (FAO Statistical Area 48) into the Indian Ocean Sector (Area 58) particularly around Prince Edward and Marion, Crozet, Kerguelen, Heard and McDonald Islands (e.g. Agnew, 2000). This pattern since 1996 is clearly illustrated in Figure 1, which is taken from a report on IUU fishing prepared by South Africa (D. Miller and B. Watkins, unpubl.). A similar progression is evident in Figure 3 which is derived from areal estimates of IUU catch undertaken by CCAMLR over the past five seasons. Most recently, information since 2000 has indicated further movement of IUU fishing into the higher latitudes of the Indian Ocean, in particular around Ob and Lena Banks (see Fig. 1) as well as further south to the Prydz Bay area.

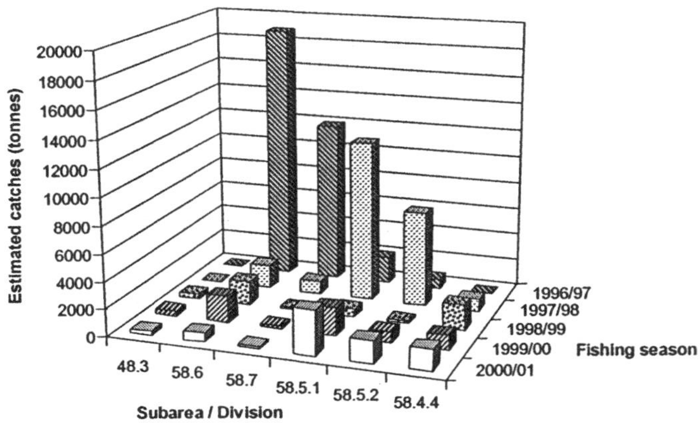

Fig. 3 Estimations of IUU catch in the Convention Area by area and fishing season.

CCAMLR has long been conscious of the fact that IUU fishing compromises the sustainability of toothfish stocks in the Convention Area and effectively undermines the organisation's conservation measures (CCAMLR, 1997). It has recognised that continued high levels of IUU fishing also compromise CCAMLR's long-standing objective of reducing incidental by-catch of seabirds during longlining. It should be specifically noted that, in CCAMLR's view, the continued take of seabirds by IUU longliners (as described by Kock, 2001) is likely to threaten a number of species for which there are already conservation concerns (CCAMLR, 2001a).

CONTROL OF IUU FISHING

The progressive development of specific conservation measures[6] has provided CCAMLR with a system to collect standard fisheries data as well as information on fish biology, demography, etc. Such information is essential for the monitoring of fisheries and in the assessment of the status of various stocks. Since 1989, CCAMLR has instituted a comprehensive System of Inspection where fisheries inspectors designated by the Member States are empowered to carry out at-sea inspections of Contracting Party vessels fishing in the Convention Area and to report sightings of fishing vessels flagged to non-Contracting Parties. Although the total number of at-sea inspections has never been large in any one year, these have been customarily concentrated in areas of most intensive fishing. Results from the CCAMLR System of Inspection have been summarised by Agnew (2000) for the period from 1989 to 1999. In 1992, the System of Inspection was augmented by the System of International Scientific Observation whereby observers are taken aboard vessels engaged in fishery research or commercial fishing in the Convention Area in order to collect essential scientific data and 'to promote the objectives of the Convention'. To preserve an element of scientific impartiality, observers designated under the Observation Scheme are required to be nationals of a CCAMLR Member other than the Flag State of the vessel to which they are assigned. The carrying of scientific observers under the CCAMLR International Scheme of Scientific Observation is a mandatory requirement for all CCAMLR-sanctioned toothfish fisheries, and a recent requirement has directed observers to provide factual data on sightings of activities by vessels other than those on which they are deployed (CCAMLR, 1998).

As clearly identified above, the initial expansion of IUU fishing for toothfish in the Convention Area coincided with an expansion in CCAMLR-sanctioned fishing activity, particularly in the 1997/98 fishing season when more than 40 IUU fishing vessels were sighted in the Indian Ocean alone (Agnew, 2000). Therefore, since 1997, CCAMLR has constantly developed and revised its conservation measures with the aim of eliminating IUU fishing in the Convention Area (see Table 2). In summary, these measures establish cooperative arrangements between CCAMLR Contracting Parties to improve compliance, at-sea inspections of Contracting Party vessels, compulsory identification markings of vessels and fishing gear, the introduction of a satellite-based VMS to verify catch location for all finfish fisheries, port inspections by Contracting Parties of all their vessels licensed to fish in the Convention Area and further development/promotion of ties with non-Contracting Parties. As already indicated, scientific observers collect and report factual information on fishing vessel sightings. CCAMLR has also established a vessel database to facilitate the exchange of information between its Members concerning all vessels known to have fished in contravention of the organisation's conservation measures.

[6] *The system of numbering Conservation Measures was changed by CCAMLR at its most recent meeting in November 2002. The numbering of Resolutions was not changed. A list of new and old numbers of Conservation Measures cited in the paper is given in the Appendix.*

Table 2 Conservation measures adopted by CCAMLR to manage toothfish fisheries and to eliminate IUU fishing in the Convention Area (measures have been developed since 1996/97 and are referenced to conservation measures currently in force – see CCAMLR, 2001c).

Measure	Current measures in force
Fishery Regulatory Measures	
• Prohibition of directed fishing for *Dissostichus* spp. in the Convention Area except in accordance with specific conservation measures.	218/XX
• Procedure for advance notification of new fisheries.	31/X
• Procedure for advance notification and conduct of exploratory fisheries, including data collection and research plans.	65/XII, 227/XX
• System for reporting catch and effort, and biological data, including reporting of fine-scale data.	51/XIX, 121/XIX, 122/XIX
• Placement of international scientific observers on vessels targeting *Dissostichus* spp.	A number of area-specific measures
• Reduction of seabird mortality in the course of longline fishing.	29/XIX
• Harvesting of stocks that occur both inside and outside the Convention Area with due respect for the CCAMLR conservation measures.	Resolution 10/XII
Flag State-based Measures	
• Licensing and inspection obligations of Contracting Parties with regard to their flag vessels operating in the Convention Area.	119/XX
• At-sea inspections of fishing vessels of Contracting Parties.	System of Inspection
• Marking of fishing vessels and fishing gear.	146/XVII
• Compulsory use of satellite-based VMS on all vessels licensed by CCAMLR Members to fish in the Convention Area.	148/XX
Port State-based Measures	
• Port inspections of vessels intending to land *Dissostichus* spp. in order to ensure compliance with CCAMLR conservation measures.	147/XIX
• Scheme to promote compliance by non-Contracting Parties vessels with CCAMLR conservation measures.	118/XX

In all aspects, the CCAMLR measures alluded to above are fully consistent with the provisions of Articles 116 to 119 of LOS Convention, Articles 21 to 23 of UNFSA and Articles III to VIII of the FAO Compliance Agreement. In keeping with Articles 8 (particularly paragraphs 3 and 4) and 17 of UNFSA, CCAMLR has also encouraged its Members to accept, as well as promote the entry into force of, UNFSA and the FAO Compliance Agreement respectively. Acceptance of the FAO Code of Conduct for Responsible Fisheries has also been encouraged. In addition, CCAMLR has noted on a number of occasions that the entries into force of both UNFSA and the Compliance Agreement are likely to contribute significantly to the reduction, and ultimately elimination, of IUU fishing in the Convention Area (CCAMLR, 1997). CCAMLR Members have actively participated in the work of FAO aimed at developing an

International Plan of Action to Prevent, Deter and Eliminate Illegal, Unreported and Unregulated Fishing (IPOA-IUU). The major purpose of the IPOA is to provide a comprehensive and integrated global approach to combat IUU fishing.

CCAMLR actively encourages cooperation at an institutional level with a number of regional fisheries organisations, especially those managing fisheries in waters adjacent to the Convention Area (e.g. the International Commission for the Conservation of Atlantic Tuna [ICCAT], the Indian Ocean Tuna Commission [IOTC] and the Commission for the Conservation of Southern Bluefin Tuna [CCSBT]) (see item 'Cooperation with Other International Organisations' in the CCAMLR annual reports). In particular, such cooperation includes exchanging information on IUU fishing on the high seas and on efforts to combat such fishing.

Levels of IUU fishing in the Convention Area have declined markedly over the past few years and are now confined mostly to vessels of non-Contracting Parties (Agnew, 2000). An apparent relocation of remaining IUU operators to waters of the Indian Ocean has been noted by CCAMLR. CCAMLR has called on all States participating in the CDS to ensure that catch documents relating to landings of toothfish caught in the Indian Ocean are verified, in collaboration with Flag States, using data reports from automated satellite-linked VMS (see Table 4, Resolution 17/XX). Despite common difficulties in establishing a 'genuine link' between the vessel and a Flag State (e.g. Vukas & Vidas, 2001), CCAMLR has a consistent procedure for identifying such flags and a policy for promoting their compliance with CCAMLR conservation measures (see Table 4, Conservation Measure 118/XX). This has been attributed to intensified pressure on IUU vessels arising from both stricter control over the activities of Contracting Party vessels, as well as the increased number of arrests and successful prosecutions of vessels engaged in IUU fishing. Key factors in the deterrent effect of the latter have been the levels of punitive fines imposed (in some cases in excess of US$1 million) and the seizure of vessels combined with increased likelihood and rate of capture.

CCAMLR CATCH DOCUMENTATION SCHEME FOR DISSOSTICHUS SPP. (CDS)

As already indicated, the persistence of IUU fishing for toothfish seriously undermines CCAMLR's conservation measures – a situation in clear contradiction to the principles set out in the UNFSA Articles addressing Flag State duties (Article 18) and the obligations of non-members or non-participants in regional fisheries arrangements (Article 17) as well as Articles 116 to 119 of the LOS Convention. As an economically valuable table fish, toothfish continues to attract high international demand and prices. Since stocks occur both inside and outside the Convention Area, fish taken during IUU fishing in the Convention Area have been difficult to trace and have enjoyed free access to international markets.

Therefore, in 1998, CCAMLR began to pursue additional measures as a means to monitor landings and the access to international markets of toothfish caught in the Convention Area by its Members and in waters under their jurisdiction. By that time, other international measures designed to trace international trade in specific fish species had been negotiated or were in the process of being negotiated. The most prominent of these was the Bluefin Tuna Statistical Document (BTSD) introduced by ICCAT in 1992. The BTSD monitors trade in fresh and frozen tuna and with a later measure requires ICCAT Members to deny landing in their ports of tuna caught outside ICCAT measures or in the absence of a BTSD (ICCAT, 1993). In contrast, the trade-related measures developed by CCAMLR for toothfish introduce a number of new and important elements.

The development of the CDS has been considered in detail by Agnew (2000). As he has stressed, the design, adoption and implementation of the scheme has constituted by far the most important step taken by CCAMLR to reduce and ultimately eliminate IUU fishing in the Convention Area. The following key principles clearly indicate the basis on which the CDS was developed (Agnew, 2000):

(1) IUU fishing by both CCAMLR Contracting and non-Contracting Parties should be addressed.
(2) The scheme should be non-discriminatory, fair and transparent.
(3) The scheme should also be practical and capable of entering into operation easily and rapidly.
(4) Fishing in areas both inside and outside the Convention Area should be included (e.g. recognition should be given to the 'transboundary' status of toothfish).
(5) The scheme should allow CCAMLR non-Contracting Parties to participate.
(6) The scheme should include sufficient validation and verification procedures to ensure ongoing confidence in the documentation.
(7) The scheme should clearly outline the responsibilities and obligation of all participants.

In addition to these principles, it should be stressed that the CDS was designed not as a stand-alone measure but rather as an integral component in a set of measures adopted by CCAMLR to combat IUU fishing in the Convention Area along the lines of those already considered in the section above 'Control of IUU Fishing'. With these considerations in mind, the CDS has two main objectives:

(1) To track landings of, and the world trade in, toothfish caught both inside and outside the Convention Area.
(2) To restrict access to international markets of toothfish taken by IUU fishing in the Convention Area.

The CDS was adopted by CCAMLR as a conservation measure and its key features are outlined in Table 3. From the table and an example of a CCAMLR toothfish catch document shown in Figure 4, it is clear that the tracking of landings through the CDS requires both identification and verification of catch origin. This enables CCAMLR, through records of either landing or transhipments, to identify the origin of toothfish entering the markets of all CDS Parties. Furthermore, it facilitates the determination of whether toothfish taken in the Convention Area were caught in a manner consistent with CCAMLR conservation measures. In contrast to other conservation measures that are essentially limited to the Convention Area, the CDS has worldwide applicability (see 'Discussion' below). It is also consistent with many of the provisions contained in Articles 7, 8 and 17 of the UNFSA.

The CDS establishes a 'probable cause' on which to deny toothfish landings, transhipments and exports or imports in the absence of the appropriate documentation and accompanying declarations. These provisions have been strengthened by the adoption of additional measures aimed at effectively and specifically prohibiting toothfish landings in the ports of CCAMLR Members unless it can be demonstrated that the fish have been caught either outside the Convention Area, or within it in conformity with CCAMLR conservation measures (e.g. Conservation Measure 147/XIX). Combined with

Table 3 Key features of the CCAMLR Catch Documentation Scheme (CDS) for *Dissostichus* spp.

Aims to ascertain the origin of catch for all transhipped/landed, imported/exported *Dissostichus* spp.
Has trade-related elements with worldwide application (i.e. not limited to the Convention Area). Aims at prohibiting entry of *Dissostichus* spp. into world markets without properly issued and verified catch documents.
Aims to determine whether *Dissostichus* spp. caught in the Convention Area were harvested in a manner consistent with CCAMLR conservation measures.
Requires vessels to have Flag State authorisation to fish for *Dissostichus* spp. for areas both inside and outside the Convention Area.
Requires every landing/transhipment to be accompanied by a valid catch document.
Requires every import or export of *Dissostichus* spp. to be accompanied by a valid export-validated or re-export validated document.
Requires submission of catch (export and re-export validated) documents to, and communication with, the CCAMLR Secretariat, including details of national authorities responsible for issuing/validating documents.
Open to participation of non-CCAMLR Contracting Parties under the same conditions as Contracting Parties.
Ensures examination of *Dissostichus* spp. shipments and catch documents by appropriate authorities in Port, Export and Import States.
Encourages cooperation between Flag State, Port State and Importing State on questions concerning the operation of the CDS.
Encourages any Party participating in the CDS to request additional verification of catch documents by Flag States, including the use of VMS, in respect of catches taken on the high seas outside the Convention Area.
Sets conditions for sale of *Dissostichus* spp. that were seized or confiscated as the result of IUU fishing investigations.

Conservation Measure 147/XIX *'Provisions to ensure compliance with CCAMLR conservation measures by vessels, including cooperation between Contracting Parties'*, CCAMLR Conservation Measure 118/XX *'Scheme to promote compliance by non-Contracting Parties with regard to their flag vessels operating in the Convention Area'* builds on the customary law rights enshrined in various Articles of the LOS Convention and expanded in Article 23 of the UNFSA which allow for Port States to inspect vessels in their ports and take action, including the prohibition of landings where it has been established that the catch has been taken in a manner that undermines the effectiveness of sub-regional, regional or global conservation and management measures on the high seas. Under certain conditions, non-compliance with CDS requirements could lead to confiscation of toothfish catches or imports. With the CDS, these measures clearly demonstrate CCAMLR's resolve to develop contemporary, and legitimate, measures aimed at avoiding the undermining of its conservation efforts by IUU fishing.

Together with the CDS, CCAMLR has adopted an *Explanatory Memorandum on the Introduction of the Catch Documentation Scheme for Toothfish (Dissostichus spp.)* and a *Policy to Enhance Cooperation between CCAMLR and Non-Contracting Parties* (CCAMLR, 1999). In 2000 and 2001, CCAMLR revised its existing measures aimed at combating IUU fishing and several new measures were adopted in support of the CDS. The complete suite of current CDS-related measures is shown in Table 4.

***DISSOSTICHUS* CATCH DOCUMENT** V 1.4

Document Number | **Flag State Confirmation Number**

PRODUCTION SECTION

1. Issuing Authority of Document

Name | **Address** | **Tel:** **Fax:**

2. Fishing Vessel Name | **Home Port & Registration Number** | **Call Sign** | **IMO/Lloyd's Number** (if issued)

3. Licence Number (if issued) | **Fishing dates for catch under this document** **4. From:** **5. To:**

6. Description of Fish (Landed/Transhipped)

7. Description of Fish Sold

Species	Type	Estimated Weight to be Landed (kg)	Area Caught*	Verified Weight Landed (kg)	Net Weight Sold (kg)	Recipient name, address, telephone, fax and signature.
						Recipient Name:
						Signature:
						Address:
						Tel:
						Fax:

Species: **TOP** *Dissostichus eleginoides*, **TOA** *Dissostichus mawsoni*

Type: **WHO** Whole; **HAG** Headed and gutted; **HAT** Headed and tailed; **FLT** Fillet; **HGT** Headed, gutted, tailed; **OTH** Other (specify)

8. Landing/Transhipment Information: I certify that the above information is complete, true and correct. If any *Dissostichus* spp. was taken in the Convention Area, I certify that it was taken in a manner which is consistent with CCAMLR conservation measures:

Master of Fishing Vessel or Authorised Representative (print in block letters) | **Signature and Date** | **Landing/Transhipment Port and Country/Area** | **Date of Landing/Transhipment**

9. Certificate of Transhipments: I certify that the above information is complete, true and correct to the best of my knowledge.

Master of Receiving Vessel | **Signature** | **Vessel Name** | **Call Sign** | **IMO/Lloyds Number** (if issued)

Transhipment within a Port Area: countersignature by Port Authority if appropriate.

Name | **Authority** | **Signature** | **Seal (Stamp)**

10. Certificate of Landing: I certify that the above information is complete, true and correct to the best of my knowledge.

Name | **Authority** | **Signature** | **Address** | **Tel.** | **Port of Landing** | **Date of Landing** | **Seal (Stamp)**

11. EXPORT SECTION

Description of Fish		
Species	**Product Type**	**Net Weight**

12. Exporter Declaration: I certify that the above information is complete, true and correct to the best of my knowledge.

Name | **Address** | **Signature** | **Export Licence** (if issued)

13. Export Government Authority Validation: I certify that the above information is complete, true and correct to the best of my knowledge.

Name/Title | **Signature** | **Date** | **Seal (Stamp)**

Country of export | **Export reference number**

14. IMPORT SECTION

Name of Importer | **Address**

Point of Unlading: | **City** | **State/Province** | **Country**

* Report FAO Statistical Area/Subarea/Division where catch was taken and indicate whether the catch was taken on the high seas or within an EEZ.

Fig. 4 Sample of the CCAMLR *Dissostichus* catch document.

With the entry into force of the CDS on 7 May 2000, CCAMLR was also endowed with a robust mechanism whereby toothfish catch data could be collected from areas within, and adjacent to, the Convention Area. Such data are vital for the estimation of 'total' toothfish removals for stock assessment purposes and provide a clearer perspective of the global levels of catch.

Table 4 CCAMLR conservation measures and resolutions that include provisions in direct support of CDS (provisions described should be read in conjunction with full texts of CCAMLR conservation measures and resolutions (CCAMLR, 2001c).

Reference	Provisions in Support of the CDS
Conservation Measure 147/XIX (as amended in 2001)	• Requests conduct of port inspections by Contracting Parties of vessels of both Contracting and non-Contracting Parties that intend to land or tranship toothfish. The purpose of inspections is to determine whether the catch is accompanied by the catch document required by the CDS, that the catch agreed with information contained in the document and, if the vessels fished in the Convention Area, that fishing was carried out in accordance with CCAMLR conservation measures. • Prohibits transhipment or landing to the vessel for which there is evidence that it fished in contravention of CCAMLR conservation measures.
Conservation Measure118/XX (as amended in 2001)	• Contains reference to Conservation Measure 147/XIX. • Prohibits transhipment by Contracting Party vessels from vessels of non-Contracting Parties that have been reported as sighted and engaged in fishing activities in the Convention Area. • Requests port inspections of such vessels. • Prohibits landings or transhipments in ports of all Contracting Parties from those vessels of non-Contracting Parties which were inspected under Conservation Measure 147/XIX and found fishing in contravention of Conservation Measures.
Conservation Measure 148/XX (as amended in 2001)	• Establishes a compulsory requirement for automated satellite-linked VMS to be used on all vessels of CCAMLR Members licensed to fish in the Convention Area (the krill fishery is currently excluded).
Resolution 13/XIX	• Urges Contracting Parties to avoid flagging or licensing a vessel of non-Contracting Party if such vessel has a history of IUU fishing.
Resolution 14/XIX	• Urges all Acceding States and non-Contracting Parties that fish for, or trade in, toothfish to participate in the CDS. • Reminds CCAMLR Members of their obligation under the CDS to prevent trade in toothfish in their territory, or by their flag vessels, with Acceding States or non-Contracting Parties when it is not carried out in compliance with the CDS.
Resolution 15/XIX	• Urges Contracting Parties not to use ports of those Acceding States and non-Contracting Parties that are not implementing the CDS for transhipments or landings of toothfish.
Resolution 16/XIX	• Urges all Parties to the CDS to ensure that their flag vessels authorised to fish for toothfish outside the Convention Area maintain an operational VMS throughout the year.
Resolution 17/XIX	• Urges all parties to the CDS to use VMS information for the verification of CDS catch data for areas outside the Convention Area, in particular, in FAO Statistical Area 51 (Western Indian Ocean).

DISCUSSION

CCAMLR CDS AND OTHER CATCH DOCUMENTATION SYSTEMS

Two main types of catch documentation systems exist in international practices – 'catch certification' and 'trade documentation'. During discussions at the recent FAO Consultation of Regional Fisheries Bodies on Harmonisation of Certification (FAO, 2002b) these terms were clarified. 'Trade documentation' refers to systems established by Regional Fisheries Management Bodies (RFMOs) that require documentation to accompany particular fish and fish products through international trade. Such documentation identifies the origin of fish for the purpose of ascertaining levels of unreported fishing. Trade documents are issued at the point of landing and only with respect to products that enter international trade. In contrast, 'catch certification' implies the issue of 'catch certificates' at the point of harvesting and covers all fish to be landed or transhipped.

As already indicated above, the first trade documentation system was adopted by ICCAT. Like the CCAMLR scheme for toothfish, the ICCAT BTSD (see the section above on the CCAMLR CDS) was developed to address problems caused by IUU fishing of bluefin tuna (ICCAT, 1993). Its immediate objective was to increase the accuracy of bluefin catch statistics by clarifying the origin of all catches, as well as to provide estimates of unreported catches by IUU fishing vessels. ICCAT requests all its Contracting Parties to require that any bluefin tuna imported into the territory of a Contracting Party be accompanied by a BTSD (Miyake, 2002). The document should contain information on the vessel that caught the tuna, where they were caught and quantities by product type. The document has to be validated by the national authorities of the Flag States of the catching vessel. The original validated document should accompany all the catch when traded. The BTSD only applies to catches that enter international trade and is not as encompassing as the CDS, which targets all harvested, landed, transhipped and traded catches of toothfish.

The ICCAT system has produced clear evidence that vessels flagged to particular States have been fishing for Atlantic bluefin tuna in a manner that undermines ICCAT's conservation and management measures for that species. That evidence, along with other related information, has led ICCAT to adopt binding recommendations that its Members prohibit the importation of bluefin tuna from such States (FAO, 2002b).

Since the adoption of the BTDS by ICCAT and the CDS by CCAMLR, other international fisheries organisations have taken steps to introduce similar systems. In general, these are trade documentation systems similar to the ICCAT system. At present, CCSBT and IOTC have adopted such systems.

It could be argued that each of the above systems has been tailored to meet its organisation's specific needs. Nevertheless, all such systems have addressed the problem of IUU fishing activities as one of their main objectives. However, in contrast to all the catch document systems identified, the CDS breaks new ground.

According to the FAO classification, the CDS is an amalgamation of a catch certification and trade documentation system (FAO, 2002b). While the CDS contains elements similar to the ICCAT BTSD, CCAMLR has recognised that to address uncertainties associated with estimating total toothfish removals, including IUU catches, Flag States should be required to certify the origin of any toothfish catch before it is landed or transhipped. For control purposes, certification is linked to a licence authorising fishing (CCAMLR Conservation Measure 119/XX), a requirement similar to that in Articles III and IV of the FAO Compliance Agreement. In these terms, CCAMLR's requirement to deploy VMS (Conservation Measure 148/XX) serves to improve the accuracy of the Flag State's determination of fishing or catch location. For catches from within the Convention Area, the CDS has been set up to also indicate whether, *inter alia*, toothfish were caught in accordance with CCAMLR conservation measures in force.

A second important CDS element aims to monitor, as well as limit, trade in IUU toothfish products. Thus, in order to address IUU fishing in the Convention Area, the

CDS essentially prohibits landings and/or shipments of toothfish by all Parties participating in the CDS from entering world markets unless accompanied by valid catch documents. In so doing, it relies on the participation of all CCAMLR Member States engaged in toothfish fishing both inside and outside the Convention Area. This requirement serves to separate toothfish taken within the Convention Area from those caught elsewhere (i.e. in areas under national jurisdiction or on the high seas outside the Convention Area) and thereby improves the estimation of total removals across the Convention Area's boundaries. The CDS also relies on participation of those non-Contacting Parties which conduct toothfish fishing outside the Convention Area or participate in the world toothfish trade. According to the CDS, each Contracting Party, and any non-Contracting Party participating in the scheme, may require additional verification of catch documents by Flag States, in particular, by using VMS, in respect of catches taken on high seas outside the Convention Area when landed at, imported into or exported from its territory.

Being open to both CCAMLR Parties and non-Parties, the CDS serves to draw in the latter in such a way that their catches can be monitored. Cooperation with non-Contracting Parties is encouraged in a way that does not discriminate between their obligations under the scheme from those of the Contracting Parties. This requirement essentially mirrors Articles 117 and 118 of the LOS Convention as well as Article 8 of the UNFSA.

CDS PERFORMANCE

CDS statistics for the period from May 2000 to June 2002 are summarised in Table 5. The information presented gives a general overview of the scope of CDS activities providing a perspective on the participation by Flag, Port and Import States as well as an indication of the number of CDS documents issued and processed. All CCAMLR Members have implemented the CDS, with only a small number having had to continue working on their national legal and administrative requirements to give it full effect. The European Community (EC) finalised its legislative procedures to implement the CDS fully on 1 June 2001. According to the EC fisheries legislation, the CDS is implemented on behalf of all Community Members. The EC membership includes, in particular, several States that are not Contracting Parties to CCAMLR, but are known to be involved in toothfish fishing or trade (e.g. Portugal).

Before the CDS was implemented by the EC, Spain, a Member of the EC and one of the main States engaged in fishing and trade of toothfish, implemented the CDS on a voluntary basis in anticipation of its adoption by the EC. In addition, the UK and France have fully implemented the CDS for vessels which are registered in their overseas territories and which are not subject to EC fisheries legislation. In 2001, Ukraine and Russia also put into place the remaining administrative procedures necessary to regulate imports of, and apply trade classification codes for, toothfish products as did Australia, New Zealand, South Africa and most other CCAMLR Members.

In addition, three of the seven Parties to the Convention that are not Members of CCAMLR (Peru, Netherlands and Greece) and which are involved in the fishing or trade of toothfish, have implemented the CDS. Netherlands and Greece have done this as part of the EC. Of the remainder involved in toothfish trade, only Canada has not yet implemented the CDS. In March 2001, Canada advised that 'the Canadian Government is currently reviewing the feasibility of implementing the CCAMLR Catch Documentation

Table 5 CDS operational statistics from May 2000 to June 2002.

Contracting Parties (1)	Vessel Flag State (2)	Port State (3)	Documents Issued by Flag States in respect of Catches		Certifications Issued in respect of Exports and Re-exports (5)
			Inside the Convention Area (4a)	Outside the Convention Area (4b)	
Argentina	⚐	⚐	2	176	210
Australia	⚐	⚐	15	3	121
Brazil					
Chile	⚐	⚐	18	3750	5230
EC					
Belgium					
France					
Germany					
Greece	⚐				
Italy					
Portugal	⚐				
Spain	⚐	⚐	6	52	67
Sweden					
UK					
France (overseas territories)	⚐	⚐	101		283
India					
Japan			1		16
Republic of Korea			12	140	307
New Zealand	⚐	⚐	13		123
Namibia		⚐			22
Norway					
Peru	⚐	⚐	154		158
Poland					
Russian Federation	⚐		17	2	
South Africa	⚐	⚐	26	1	100
Ukraine	⚐		4	2	
UK (overseas territories)	⚐	⚐	15	96	3
US					8
Uruguay	⚐	⚐			272
Non-Contracting Parties participating in the CDS					
People's Republic of China					246
Mauritius		⚐			12
Seychelles	⚐		44		253
Singapore		⚐			11
Non-Participating Parties identified for attention					
Belize	⚐				
Bolivia	⚐				
Canada					
Indonesia		⚐			15
Kenya		⚐			
Mozambique		⚐			
Sao Tome and Principe	⚐				
St Vincent & the Grenadines	⚐				
Togo	⚐				
Totals			**428**	**4222**	**7457**

scheme' (CCAMLR, 2001b). CCAMLR has continued to pursue the matter with the Government of Canada, urging it to become a Member and immediately participate in the CDS. Bearing in mind that a failure to participate in the CDS by a Contracting Party could be construed as acting contrary to the objectives of the Convention or to undermine its conservation measure, it is anticipated that Canada will eventually become a CDS Party.

The database administered by the CCAMLR Secretariat allows the checking and cross-referencing of all CDS documents submitted. It also gives CCAMLR Members access, via password-protected pages on the CCAMLR Website, to scanned images of CDS documents. This facilitates searches for documents according to a pre-defined number of data fields as well as their verification. The system developed by the CCAMLR Secretariat to process, store and access CDS information not only takes into account the scheme's immediate objectives, it also attempts to address future needs for its potential integration with information on the implementation and status of related compliance and enforcement measures. As such, the system has considerable potential as an effective, real-time, enforcement tool for use by all CDS Parties.

As already mentioned, the CDS is one of a suite of CCAMLR measures aimed at eliminating IUU fishing in the Convention Area. Table 6 shows how, during the period it has been operational, the CDS has combined with other measures to detect and address violations of CCAMLR conservation measures. It is clear that the efforts of all CDS Parties have increased, especially in conducting in-port inspections of vessels from both CCAMLR Contracting and non-Contracting Parties. This trend has continued in 2002.

Table 6 Performance of CDS-related conservation measures for the period the CDS has been in force.

	2000	2001	First Half 2002
Inspections of fishing vessels reported (both of CCAMLR Contracting and non-Contracting Parties)			
• At port	4	26	8
• At sea	10	6	1
Fraudulent or otherwise unauthorised catch documents		18 (May 2000–June 2002)	
Sightings of IUU fishing vessels in the CCAMLR Convention Area	5	1	4
Apprehensions of IUU fishing vessels by CCAMLR Members	2	5	2

Several of the non-Contracting Parties identified in Table 5 are engaged in fishing for, and/or the trading of, toothfish. CCAMLR has communicated with these Parties and informed them about the introduction of the CDS. They have also been invited to cooperate with CCAMLR in the scheme's implementation. The Republic of Namibia became a full Member of CCAMLR in 2001 and subsequently introduced the CDS. The Republic of Seychelles (June 2001) and the Republic of Singapore (September 2000) have both recently become CDS Parties along with the People's Republic of China (July 2001). The Republic of Mauritius (since January 2001) implements most of the CDS elements, but is still considering whether toothfish transhipments and/or landings in the free port area of Port Louis should be considered as imports into the Mauritius territory. CCAMLR maintains ongoing contacts at various levels with other non-Contracting Parties identified in Table 5.

The CDS-derived statistics indicate that the main importers of toothfish are Japan, USA, EC and the People's Republic of China. Based on trade statistics provided by Canada for 2000, that country should also be included in the list of main importers of toothfish (CCAMLR, 2001b). While all the States listed above, with the exception of Canada, are CDS Parties, recent reports prepared by TRAFFIC (Trade Records Analysis of Flora and Fauna in Commerce) International (Lack & Sant, 2001; Willock, 2002) show that the market share of these countries in the global toothfish market is of the order of 90%. In excess of 90% of the products are provided by Argentina, Australia, Chile, France, New Zealand, South Africa and the UK, all of which are CCAMLR Members. CCAMLR works actively in order to involve those non-Contracting Parties which are responsible for the remaining 10% of the toothfish world market in the implementation of the CDS.

CCAMLR has gone to considerable effort to create the CDS such that it is consistent with the provisions of the World Trade Organisation (WTO) and the General Agreement on Tariffs and Trade (GATT) (Agnew, 2000). As Larson (2000) has indicated, the CDS employs similar conservations measures to those used for other marine species and which have been found to be in conflict with WTO-GATT obligations. However, unlike these invalidated conservations measures, the CDS strikes a careful balance, allowing CDS Parties to meet CCAMLR's conservation needs without violating the legal rights of fellow WTO members. In particular, the CDS framework ensures that any discrimination on the basis of national origin is minimised. This means that all three of the key elements of concern to the WTO (i.e. non-discrimination, transparency in multilateral resolution and clear linkage to a policy of conserving the resource(s) in question) are expressly addressed by the CDS.

The CCAMLR CDS has attracted widespread interest within the WTO, particularly its Committee on Trade and the Environment (CTE). According to the WTO Secretariat, the CDS, along with ICCAT's BTSD, 'can be considered to provide examples of appropriate and WTO-consistent (i.e. non-discriminatory) use of trade measures in multilateral environmental agreements' (WTO, 2000a). However, the CTE as a whole has not yet reached consensus on the issue (WTO, 2000b). The CCAMLR Secretariat has been regularly invited by the WTO to participate as an observer at special CTE meetings addressing cooperation with Multilateral Environmental Agreements (MEA) and, although not yet able to attend, has already submitted background information on CCAMLR measures to eliminate IUU fishing and on the implementation of the CDS.

APPRAISAL OF CDS PERFORMANCE

By June 2002, the total number of catch documents issued for toothfish landings was about 5 000. These documents were issued by all CDS Parties (see Table 5) for areas both inside and outside the Convention Area. About twice this number of export-validated documents has been received and processed by the Secretariat before being entered into the CDS database. In appraising the performance of the CDS, four questions present themselves and these are addressed in turn:

Does the CDS work? The answer is a definite 'yes'. The 27 CCAMLR Contracting Parties and four non-Contracting Parties involved are responsible for more than 90% of the global trade in toothfish. They have deployed considerable (human and financial) resources to put the CDS in place and to maintain its efficient operation. CCAMLR works actively in order to involve those non-Contracting Parties that are responsible for the remaining 10% of the toothfish world market in the implementation of the CDS. Furthermore, the CCAMLR Secretariat has been provided with significant funds to develop and maintain a comprehensive CDS database. This has enabled all CDS Parties to access catch documents 'online' to verify their authenticity and has provided other essential information necessary to certify landings or to authorise exports or imports.

Does the CDS work effectively in tracking international trade in toothfish? Again the answer is resoundingly affirmative. The CDS Parties have introduced specific trade custom codes for toothfish products and such trade is being monitored across national borders. In addition, the trade statistics collected through the CDS are more timely, complete, precise and detailed than those previously available.

Does the CDS work effectively to prevent toothfish trade not accompanied by catch documents? For the three following reasons, the answer is again 'yes'. First, CCAMLR Members have implemented the CDS for toothfish fisheries by their flagged vessels within the Convention Area, in areas under their national jurisdiction and on the high seas. The CDS has served to identify better catch location and has required Flag States to assume responsibility for verifying authorisation of their vessels to fish for toothfish. In combination with other CCAMLR conservation measures, it has required Flag States to ensure that toothfish taken in the Convention Area or transhipped at sea have been caught in a manner consistent with CCAMLR conservation measures. Secondly, the CDS extends these requirements to Port States with respect to landings of toothfish subject to export and import operations. This provision is applied irrespective of whether the catch originated from inside or outside the Convention Area. Finally, the CDS is open to participation by all non-Contracting Parties engaged either in fishing for, or in trade in, toothfish, or both. Since about 90% of toothfish world fisheries are under control of CCAMLR Members and more than 90% of the world toothfish market is limited to markets of CCAMLR Members and CDS Parties, the CDS has provided some incentive for other Parties to join. Such incentive has become more compelling in light of the fact that toothfish products accompanied by the necessary CDS documents consistently fetch a higher price in most of the current markets. For example, in August 2000 it was estimated that in excess of 1 000 tonnes were being offered for sale without catch documents. Such fish fetched prices of US$3.00/kg lower than fish with catch documents, then trading at around US$8.40/kg (CCAMLR, 2000). Some future expansion of the CDS is anticipated, particularly if CCAMLR's policy on cooperation with non-Contracting Parties succeeds in moving the remaining identified and relevant non-Contacting Parties to take up CCAMLR's invitation to join it in the implementation of the CDS. The task of eliminating the remaining IUU operators is complicated because that group, when forced out of one area, relocates rapidly to another area, often with changed flags and vessel names. CCAMLR has a consistent procedure for identifying such flags and a policy for promoting their compliance with CCAMLR conservation measures.

Does the CDS effectively address IUU fishing in the Convention Area? To a large extent, it appears that the CDS has been effective in reducing levels of IUU fishing (see Table 1 and Fig. 3 above). The CDS has also enabled CCAMLR to identify the remaining, and most persistent, IUU fishing operators that require immediate and concentrated attention of all CDS participating States. However, this having been said, it has not achieved this in its own right alone, but in combination with a suite of associated measures. Despite some success, IUU fishing activities persist by moving from one area, where CCAMLR control has intensified, to other areas where such controls are less robust. Furthermore, the use of flags of 'third Parties' (i.e. 'Flags of Convenience' as defined by Vukas & Vidas, 2001) by vessels or their owners wishing to avoid compliance with CCAMLR conservation measures or CDS provisions, has made CCAMLR's task of eliminating the remaining IUU fishing that much more difficult. The measures already taken by CCAMLR Members to improve Port State control, and even to deny landings of toothfish by vessels outside the CDS, have come some way in addressing the issue when used in combination with limitations on toothfish trade (especially imports into the territories of CCAMLR Parties) which is not accompanied by the necessary CDS documentation. CCAMLR continues to work on improving identification of Flags of Convenience (CCAMLR, 2001b) and on its dealings with States that provide such flags both on a case-by-case basis on behalf of CCAMLR as well as through the 'moral persuasion' attached to bilateral and multilateral political demarches.

FUTURE CDS DEVELOPMENTS

Despite the obvious merits of the CDS, CCAMLR remains concerned with the levels of IUU fishing in the Convention Area despite the observed decrease over the past year or so. It is hard to assess whether such a decrease is the result of more effective enforcement or reduced availability of fish. However, on the assumption that IUU fishing will persist and will continue to undermine its conservation measures, CCAMLR has embarked on a number of priority initiatives in order to enhance the performance/effectiveness of the CDS. The more prominent of these relate to the following issues:

(1) Non-participation by Canada (a Contracting Party) in the implementation of the CDS.
(2) Use of Flags of Convenience (see discussion above) associated with reduced Flag State participation and responsibility.
(3) Use of ports of non-Contracting Parties to land toothfish without a catch document.
(4) Insufficient monitoring of transhipments at sea – a difficult task given the geographic extent of the Convention Area.
(5) Misreporting of catch origin – either deliberate or accidental.
(6) 'Laundering' of fish either using false CDS documents or by the issue of documents that are not carefully monitored by the Flag and/or Port State.
(7) Inconsistency in verification of catch position, either by 'patchy' Flag State monitoring/verification of VMS data, or by non-deployment of VMS, or by at-sea tampering with VMS units. Outside the Convention Area, the situation is exacerbated by non-mandatory deployment of scientific observers that may prejudice objective verification of catch position.

Over the past two years, CCAMLR has attempted to address several of these concerns, resulting in some significant revisions of the CDS. These revisions have been mainly aimed at clarifying the CDS provisions, particularly deadlines for the exchange of CDS information between participants at all stages in the trade cycle, commencing with landing of the catch through actions taken by the Flag, Port and Export States to final import and sale. Such work continues and is currently being conducted by a specially convened CDS Intersessional Task Group set up by CCAMLR. This includes among its members, representatives from CDS Parties with expertise in the fishery, trade, fisheries enforcement, law and information management.

A prominent development initiated in 2001/02 by CCAMLR is the establishment of an 'online' facility to augment the timely processing of CDS documents using a central and secure database archived at the CCAMLR Headquarters in Hobart. The proposed system will enable all CDS Parties to issue and process CDS documents in quasi 'real time', so creating an effective network for information exchange on the certification and verification of landings, exports and imports. It will also contain security features to eliminate fraud at any stage of the trade cycle, including 'watermarks' to ensure authenticity of original documents.

With the entry into force of the UNFSA, particularly the applicability of its Articles 8 and 17, the impact of the CDS has been considerably strengthened. As a consequence, CCAMLR's competency as a RFMO has been boosted in relation to the provision that non-CCAMLR Parties or participants are not discharged from their obligation to cooperate in the conservation and management of 'transboundary' stocks such as toothfish. While CCAMLR may be unlike any other RFMO, Article 1.(1)(d) of the UNFSA does not rule out that an RFMO may have purposes other than fisheries conservation or management (Molenaar, 2001). Even though not essential, the unique sovereignty situation and other characteristics of the CAMLR Convention could be used to justify the use of non-Flag State powers under UNFSA Article 21 and 22 in the interest of the international community.

Another prominent, and potential improvement, of the CDS is likely to arise from global standardisation of catch certification and trade documentation systems as initiated by FAO in collaboration with various regional fisheries organisations, including CCAMLR. While still relatively novel, the FAO consultations are likely to include consideration of cooperation with the WTO and WCO. As the CDS combines features of both catch certification and trade documentation, and since it is consistent with the provisions of current international fisheries arrangements such as UNFSA, it appears to offer a useful prototype on which to build future standardisation.

CONCLUSIONS

The CDS has been in operation for a little over two years. It is still too early to assess whether, in combination with other related CCAMLR measures, the CDS is an 'indispensable' tool for addressing IUU fishing on toothfish and whether it is specifically so with respect to the management of transboundary stocks in the Convention Area. For the former to happen, all toothfish trading will need to be limited internationally to fish taken legally with fish taken illegally being prevented from entering the market. Given recent initiatives to list toothfish under the Convention on International Trade in Endangered Species of Wild Fauna and Flora (CITES), the issue of marrying the competencies of RFMOs like CCAMLR, which uphold the long-held legal precedent of

'Flag State control', with more widespread trade-based agreements such as CITES, poses some interesting questions and is unlikely to be easily resolved. Most importantly, CCAMLR has expanded the role of 'Port' and 'Market' States in its effort to discourage trade in IUU-caught toothfish. Without diminishing the Flag State responsibilities, CCAMLR's recent efforts have brought into focus the need for other States to assume heightened responsibility for combating trade in toothfish taken in a manner which undermines CCAMLR conservation measures. This development is the key to improving CCAMLR's ability to combat IUU fishing directly and is fully at the centre of ensuring that the obligations of UNFSA Articles 20, 21 and 23 are effectively met.

In should be noted that a system like the CDS must remain dynamic in order to respond to changing circumstances and needs. Therefore the CDS requires maintenance and a high level of support along with comparable tightening of associated measures to ensure that the objectives of the overall package of measures will be successful. As such, the CDS is an essential component in a 'toolbox' of measures, and cannot be implemented and evaluated in isolation.

Despite the obvious misgivings outlined here, an initial evaluation of the CDS is encouraging. Not only is the CDS unique in its scope and application, it became fully operational in a relatively short time (less than two years). It has also drawn in a number of CCAMLR non-Contracting Parties and its overall coverage extends to more than 90% of the global world trade in toothfish.

At an operational level, the introduction of the CDS has led to its Parties denying a number of toothfish landings and/or shipments in the absence of the required documents. It has also allowed an improved appreciation of toothfish global catch levels and focused on incidents of malpractice or even fraudulent catch documents. Given the evidence that introduction of the CDS has made trading in illegally caught fish less profitable, it has also served to restrict unfettered market access to IUU-caught products. While some of the highlighted improvements of the CDS will undoubtedly make it more effective in combating IUU fishing in the Convention Area, the question remains – 'what would have happened in its absence?' On the basis of the information presented in this paper, the answer appears too horrible to contemplate.

ACKNOWLEDGEMENTS

The authors are grateful to Davor Vidas and Geoff Rohan for constructive and helpful comments on the manuscript.

REFERENCES

Agnew, D. J. (2000) The illegal and unregulated fishery for toothfish in the Southern Ocean, and the CCAMLR Catch Documentation Scheme. *Marine Policy,* **24**: 361–74.

Appleyard, S. A., Ward, D. R. & Williams, R. (2001) Population structure of the Patagonian toothfish (*Dissostichus eleginoides*) in Australian waters. Document *WG-FSA-01/38*. CCAMLR, Hobart, Australia: 14 pp.

Butterworth, D. S. & Thomson R. B. (1995) Possible effect of different levels on krill fishing on predators – some initial modelling attempts. *CCAMLR Science*, **2**: 79–98.

Butterworth, D. S., Punt, A. E. & Basson, M. (1991) A simple approach for calculating the potential yield from biomass survey results. In: *Selected Scientific Papers*, 1991 (SC-CAMLR-SSP/8). CCAMLR., Hobart, Australia: 201–15.

CCAMLR (1990a) *Statistical Bulletin*, Vol. 1 (1970–1979). CCAMLR, Hobart, Australia: 61 pp.

CCAMLR (1990b) *Statistical Bulletin*, Vol. 2 (1980–1989). CCAMLR, Hobart, Australia: 109 pp.

CCAMLR (1993) Resolution 10/XII. In: *Schedule of Conservation Measures in Force, 1993/94.* CCAMLR, Hobart, Australia: p. 35.

CCAMLR (1996) *Report of the Fifteenth Meeting of the Commission (CCAMLR-XV).* CCAMLR, Hobart, Australia: 23–32.

CCAMLR (1997) *Report of the Sixteenth Meeting of the Commission (CCAMLR-XVI).* CCAMLR, Hobart, Australia: 8–12, 24–8.

CCAMLR (1998) *Report of the Seventeenth Meeting of the Commission (CCAMLR-XVII).* CCAMLR, Hobart, Australia:12–22.

CCAMLR (1999) *Report of the Eighteenth Meeting of the Commission (CCAMLR-XVIII).* CCAMLR, Hobart, Australia: 11–21.

CCAMLR (2000) Report of the Standing Committee on Observation and Inspection. In: *The Report of the Nineteenth Meeting of the Commission (CCAMLR-XIX)*, Annex 5. CCAMLR, Hobart, Australia: 105–146.

CCAMLR (2001a) *Report of the Twentieth Meeting of the Commission (CCAMLR-XX).* CCAMLR, Hobart, Australia: 1–69.

CCAMLR (2001b) Report of the Standing Committee on Observation and Inspection. In: *Report of the Twentieth Meeting of the Commission (CCAMLR-XX)*, Annex 5. Hobart, Australia: 121–170.

CCAMLR (2001c) *Schedule of Conservation Measures in Force, 2001/02.* CCAMLR, Hobart, Australia: 138 pp.

Constable, A. J. & de la Mare, W. K. (1996) A generalised model for evaluating yield and the long-term status of fish stocks under conditions of uncertainty. *CCAMLR Science*, **3**: 31–54.

Constable, A. J., de la Mare, W. K., Agnew, D. J., Everson, I. & Miller, D. (2000) Managing fisheries to conserve the Antarctic marine ecosystem: practical implementation of the Convention on the Conservation of Antarctic Marine Living Resources (CCAMLR). *ICES Journal of Marine Science,* **57**: 778–91.

FAO (2002a) *Fisheries Glossary* (www.fao.org/fi/glossary/default.asp).

FAO (2002b) The Report of the Expert Consultation of Regional Fisheries Bodies on the Harmonization of Catch Certification. Committee on Fisheries, Sub-Committee on Fish Trade. Document COFI: FT/VIII/20002/Inf.13: 19 pp.

Fischer, W. & Hureau, J.-C. (Eds) (1985) *FAO Species Identification Sheets for Fishery Purposes. Southern Ocean (CCAMLR Convention Area Fishing Areas 48, 58 and 88)*, Vol. II: 232 pp. Prepared and published with the support of the Commission for the Conservation of Antarctic Marine Living Resources. FAO, Rome.

Greenpeace International (2002) *Greenpeace Gallery of Toothfish Vessels* (www.greenpeace.org/~oceans/southernoceans/expedition2000/gallery/pirates.html).

ICCAT (1993) Recommendations adopted by the Commission at its Eighth Meeting (Madrid, November 1992), Report for Biennial Period, 1992–93, Part 1.

ISOFISH (2002) *Rogues Gallery* (www.isofish.org.au).

Kock, K.-H. (1992) *Antarctic Fish and Fisheries.* Cambridge University Press, Cambridge: 359 pp.

Kock, K.-H. (2001) The direct influence of fishing and fishery-related activities on non-target species in the Southern Ocean with particular emphasis on longline fishing and its impact on albatrosses and petrels – a review. *Reviews in Fish Biology and Fisheries,* **11**: 31–56.

Lack, M. (2001) Antarctic toothfish: an analysis of management, catch and trend. TRAFFIC Oceania, Sydney, Australia: 27 pp.

Lack, M. & Sant, G. (2001) Patagonian toothfish: are conservation and trade measures working? *TRAFFIC Bulletin*, **19**(1): 18 pp.

Larson, K. (2000) Fishing for a compatible solution: toothfish conservation and the World Trade Organization. *The Environmental Lawyer*, **7**(3): 123–58.

Miyake, P. (2002) Catch Certification and the Feasibility of Harmonizing Certifications among Regional Fisheries Management Bodies. FAO Expert Consultation of the Regional Fisheries Management Bodies on the Harmonization of Catch Certification, La Jolla, 9–11 January 2002. Document FI:HCC/2002/Info 2: 90 pp.

Molenaar, E. J. (2001) CCAMLR and Southern Ocean Fisheries. *The International Journal of Marine and Coastal Law*, **16**(3): 465–99.

SC-CAMLR (1995) Report of the Workshop on Methods for the assessment of *Dissostichus eleginoides*. In: *Report of the Fourteenth Meeting of the Scientific Committee (SC-CAMLR-XIV)*, Annex 5, Appendix E. CCAMLR, Hobart, Australia: 387–413.

SC-CAMLR (1999) *Report of the Eighteenth Meeting of the Scientific Committee (SC-CAMLR-XVIII)*. CCAMLR, Hobart, Australia: 1–107.

SC-CAMLR (2001) Report of the Working Group on Fish Stock Assessment. In: *Report of the Twentieth Meeting of the Scientific Committee (SC-CAMLR-XX)*, Annex 5. CCAMLR, Hobart, Australia: 195–558.

Shust, K. (1998) *Fishes and Fish Resources of Antarctica*. VNIRO, Moscow: 163 pp. (in Russian).

Smith, P. J., Gaffney, P. M. & Purves, M. (2001) Genetic markers for identification of Patagonian toothfish and Antarctic toothfish. *Journal of Fish Biology.,* **58**: 1190–4.

VNIRO (1985) Age, growth rate and length-age population structure of abundant species of neritic and mesopelagic fish on the Southern Ocean. VNIRO Publishing, Moscow: 31 pp. (in Russian).

Vukas, B. & Vidas, D. (2001) Flags of convenience and high seas fishing: the emergence of a legal framework. In: Stokke, O.S. (Ed.). *Governing High Seas Fisheries. The Interplay of Global and Regional Regimes.* Oxford University Press: 53–90.

Willock, A. (2002) Unchartered waters: implementation issues and potential benefits of listing toothfish in Appendix II of CITES. TRAFFIC International. Cambridge, UK and TRAFFIC Oceania, Sydney, Australia: 35 pp.

WTO (2000a) Environmental benefits of removing trade restrictions and distortions: the fisheries sector (note by the Secretariat). World Trade Organization, Committee on Trade and Environment, Document WT/CTE/W/167 (16 October 2000).

WTO (2000b) Report of the meeting held on 24–25 October 2000. World Trade Organization, Committee on Trade and Environment, Document WT/CTE/M/25 (12 December 2000).

Yukhov, V. L. (1982) *Antarctic Toothfish*. 'Nauka' Publishing, Moscow: 114 pp (in Russian).

APPENDIX

Old and new numbers for the Conservation Measures cited in the paper

Readers may also refer to a current "Schedule of Conservation Measures in force, 2002/2003", which is available on the CCAMLR Website (http://www.ccamlr.org/pu/e/pubs/cm/drt.htm).

Old number, as cited in the paper	New number, since November 2002
29/XIX	25-02 (2002)
31/X	21-01 (2002)
51/XIX	23-01 (2000)
65/XII	21-02 (2002)
118/XX	10-07 (2002)
	Resolution 13/XIX - rescinded
121/XIX	23-05 (2000)
122/XIX	23-04 (2000)
146/XVII	10-01 (1998)
147/XIX	10-03 (2002)
148/XX	10-04 (2002)
218/XX	32-09 (2002)
227/XX	41-01 (2002)

On the management of shared fish stocks: critical issues and international initiatives to address them

Gordon Munro
University of British Columbia, Vancouver, Canada
Rolf Willmann and Kevern L. Cochrane
Department of Fisheries, Food and Agriculture Organization of the UN, Rome

ABSTRACT: In this chapter we analyse, from an economics standpoint, the management of shared fish stocks, with the assistance of the theory of cooperative and non-cooperative games. Particular emphasis is given to the question of allocation criteria, and to the fundamental importance of ensuring that cooperative fisheries management arrangements have the robustness and resilience to withstand, through time, unexpected shocks, be those shocks environmental, economic or political in nature.

INTRODUCTION

The Government of Norway, in cooperation with the FAO, held an Expert Consultation on the management of shared fishery resources in Bergen during October 2002[1]. The Consultation was convened in recognition of the fact that the management of these resources stands as one of the great challenges on the way towards achieving long-term sustainable fisheries. The objective of the Expert Consultation was to assist countries in improving their efficiency and performance in meeting the challenge.

Among the background documents prepared for the Expert Consultation was a discussion paper entitled "On the Management of Shared Fish Stocks" (Munro, 2003). The paper attempted to outline the scope and magnitude of relevant resource management issues. In so doing, the paper undertook to review the academic, or theoretical, aspects of resource management issues in a non-technical manner, and went on to illustrate the key points arising therefrom, by presenting brief case studies from the real world. This paper draws heavily upon that discussion paper.

The FAO definition of shared fish stocks includes all fish stocks that cross the boundaries of a coastal state EEZ -into neighbouring EEZs and/or the adjacent high seas, along with those stocks found exclusively on the high seas. The Expert Consultation in Norway was not called upon to examine all shared stocks. Rather it was to confine itself to:

(a) *Transboundary stocks*: fish stocks crossing the EEZ boundary of one coastal state into the EEZ(s) of one, or more, other coastal states.

[1] *The Norway–FAO Expert Consultation on the Management of Shared Fish Stocks. Bergen, Norway, 7–10 October 2002 (FAO, 2002)*

(b) *Straddling stocks*: fish stocks crossing the EEZ boundary into the adjacent high seas, other than highly migratory and anadromous/catadromous stocks.

This paper also confines itself to transboundary and straddling stocks. Having said this, the scope of the management problem posed by this restricted set of shared fish stocks is immense. Caddy (1997) estimated that there may be as many as 1500 transboundary fish stocks alone. We can only guess at the number that would have to be added to this total if straddling stocks were taken into account. Caddy also notes, with reference to transboundary stocks, that the number of these stocks subject to effective cooperative management is very modest in relation to the global total.

LEVELS OF COOPERATION IN FISHERY RESOURCE MANAGEMENT

The management of shared fish stocks as a major, worldwide, resource management issue owes its origins to the UN Third Conference on the Law of the Sea, 1973–1982 (LOSC), the UN Convention on the Law of the Sea (United Nations, 1982) to which the Conference gave rise, and the resultant EEZ regime. Appropriately, the FAO began analysing this resource management issue several years prior to the close of the LOSC.

Among those writing on the issue in the FAO during those years was John Gulland (1980). In his still widely cited 1980 paper, *Some Problems in the Management of Shared Stocks*, Gulland raises the question of the appropriate level of cooperation between/among states sharing a fishery resource, given that a *prima facie* case for cooperation exists. He deems a *prima facie* case to exist if the harvesting activities of one state sharing the fishery resource impinges on the harvesting opportunities of one or more other states sharing the resource.

There are, Gulland (1980) points out, at least two levels of cooperation. The first, what we might term the primary level, consists of cooperation in research alone, without reference to coordinated management programmes. As all parties should stand to benefit from improved information and data, the cooperation should be *relatively* easy to achieve. If it is not possible to achieve cooperation at this primary level, it certainly will not be possible to achieve cooperation in active management of the resource. In actual cooperative management regimes that have proven successful, cooperation in research alone is often seen, in retrospect, to have been the precursor to cooperation in active management.

What we might call secondary cooperation – "active management" – involves, almost by definition, the establishment of coordinated joint management programmes. As Gulland (1980) informs us, this will require:

(a) Determination of an optimal management strategy through time, including, *inter alia*, the determination of optimal global harvests over time.
(b) Allocation of harvest shares among the participating states (or entities).
(c) Implementation and enforcement of coordinated management agreements.

Obviously, this is a much more formidable undertaking than the primary level of cooperation. To begin, even cooperation in research may lose its benign character. Research findings can influence harvest allocations, and thus can easily become "tools of combat" in negotiations between and among relevant states. Second, as recognized by the FAO as early as 1979 (FAO, 1979), there is no necessary reason why states sharing a fishery resource should be expected to have identical resource management goals. If management goals are not identical, one is faced with the burden of developing a mutually acceptable compromise resource management programme, or so it would seem (FAO, 1979).

Therefore, establishing cooperative management at the secondary level can prove to be frustrating and costly. Each relevant state could conclude that the aforementioned gross benefits of cooperation are not substantial by taking the view that, if it and its fellow states sharing the resource manage their respective segments of the resource in a rational manner, the overall resource management results, while not ideal, will be adequate. One of the central questions to be addressed in the discussion of the analytical aspects of this resource management issue is whether or not this comfortable view of the world is, in fact, reasonable.

BASIC ECONOMICS OF THE MANAGEMENT OF SHARED FISH STOCKS – A REVIEW

The economics of the management of shared fish stocks has been developed in two stages. The first, which dates back to the late 1970s (Munro, 1979), consisted of developing the economics of the management of transboundary fish stocks. This is a reflection of the fact that, at the close of the LOSC in 1982, management of transboundary fish stocks was recognized as an important problem, whereas management of straddling fish stocks was not. The true magnitude of the problem of managing straddling stocks was to become manifest over the ensuing decade.

It is also a reflection of the fact that management of transboundary fish stocks is considerably less complex than management of straddling fish stocks. In the case of transboundary fish stocks, the states involved are, with few exceptions, fixed through time, and the shared, or joint, property rights to the relevant resources are reasonably straightforward (McRae & Munro, 1989). The same does not pertain to straddling stocks. Furthermore, the number of states involved is usually relatively small. In an economic analysis of the management of transboundary resources, one can often make do with models consisting of just two countries.

The second stage, consisting of the development of the economics of the management of straddling (and highly migratory) fish stocks, dates back only to the early 1990s (Kaitala & Munro, 1993). In this stage, the economics of management of transboundary fish stocks is used as a foundation. The question is then asked what modifications to, and what extensions of, the analysis are required, in light of the special problems arising from, and issues raised by, the management of straddling fish stocks.

ECONOMICS OF THE MANAGEMENT OF TRANSBOUNDARY FISH STOCKS

The economics model used in analysis of the management of transboundary fishery resources is a blend of two components. The first consists of the now standard bioeconomics model used for the analysis of fisheries confined to waters of a single coastal state (see, for example: Clark, 1990; OECD, 1997). The second component consists of the theory of games. The reason for incorporating game theory into the analysis is the realization that, without game theory, analysis of the economics of shared fish stock management becomes incomprehensible.

On the assumption that most readers are not familiar with the theory of games, we now give a brief review of the essentials of the theory. The theory is designed to analyse strategic interaction between and among individuals, be they persons, firms, nations or others. The theory of games is relevant when the actions of one "individual" have a clearly perceived impact upon other "individuals", so inviting a reaction from those other "individuals".

Cooperative resource management between, or among, coastal states sharing a fishery resource becomes worthy of consideration, it can be argued, if the harvesting activities of one coastal state have a significant impact on the harvesting opportunities open to the other state(s) sharing the resource. If this condition is met, then strategic interaction between "individuals", in the form of states sharing the resource, becomes virtually inescapable. It is for this reason that it was very difficult to make significant progress in developing the economics of the management of transboundary fishery resources until the analytical tools provided by the theory of games were brought to bear.

In the terminology of game theory, the "individuals" are referred to as "players". The "players" are assumed to be rational and to have various courses of action open to them, which are referred to as "strategies". The expected return to a player, in following a particular strategy, is then referred to as a "payoff". The size of the expected return or "payoff" will, needless to say, be dependent upon the expected reactions of other "players". The interaction between, or among, the players, as they execute their strategies, is the game. The stable outcome of a game, if it exists, is termed the "solution" to the game. Finally, the game may be a "once only" affair, or it may be repeated.

There are two broad categories of games, competitive, or non-cooperative, games, and cooperative games. In a cooperative game, the players are assumed motivated entirely by self-interest, but have some incentive to endeavour to cooperate. Of critical importance is the fact that players are able to communicate with one another effectively. In competitive, non-cooperative, games, the lines of communication between and among players are faulty or simply non-existent.

In analysing the economics of the management of shared fishery resources, economists have asked themselves two fundamental questions, the first being: what are the consequences of coastal states sharing a fishery resource refusing to cooperate in the management of the resource? The implication is that, in the absence of cooperation, each coastal state will simply go its own way and manage its segment of the resource as best it can. If the answer to the question is that the negative consequences of non-cooperation are trivial, then one need proceed no further. If, on the other hand, the answer to the question is that the negative consequences of non-cooperation are severe, then cooperation does matter, and the second fundamental question must be asked – what requirements must be met for a cooperative resource management regime to be stable and sustainable over the long run?

The first question, that of the consequences of non-cooperative management of a shared fishery resource, is addressed, not surprisingly, by bringing to bear the theory of non-cooperative games. Consider a two "player" (coastal state) game.

A stable solution to a non-cooperative game was defined by Nash (1951) as a situation in which each player has no incentive to change, given the strategies being followed by the other player(s). All investigations of the non-cooperative fisheries game, of which these authors are aware, come to the same conclusion. A stable solution to the game would involve, except in unusual circumstances, mismanagement of the resource from society's point of view. One example is provided by Clark (1980), who argues that, if the players are symmetric, i.e. identical in all respects, the outcome will be similar to that encountered in an unrestricted, open access domestic fishery, referred to in the economics literature as Bionomic Equilibrium (Gordon, 1954). Bionomic Equilibrium is characterized by overexploitation of the resource, from society's point of view, and by fleet capacity far in excess of that required if the resource were to be exploited optimally.

The overall outcome to the game is an example of what is probably the most famous of all non-cooperative games, known as the "Prisoner's Dilemma". The essence of the "Prisoner's Dilemma" is that the players are driven inexorably to adopt strategies which they know to be harmful[2]. With respect to fisheries, the predictive power of the theory has been high.

The implication of the analysis is straightforward. Even if coastal states sharing a resource have the capability of managing fishery resources within their domestic waters effectively, one has no justification to assume, in the absence of cooperation, the resource management outcome will be "adequate". The risk exists that the outcome will be disastrous. Other than in exceptional cases, cooperation does matter. Moreover, cooperation is a prerequisite for effective management, not merely a useful supplement to resource management by individual states.

Consider the following example. The FAO has in place an International Plan of Action for the Management of Fishing Capacity (FAO, 1999). The IPOA – Capacity does, *inter alia*, emphasize the importance of addressing the problem of excess fleet capacity in the management of shared stocks (FAO, 1999, p. 2). One can be confident that, if shared stocks plagued by excess fleet capacity are managed non-cooperatively, the excess capacity problem will continue indefinitely, IPOA or no IPOA.

Cooperation does matter, except in unusual circumstances. In examining cooperative management of transboundary fishery resources, one brings to bear the theory of cooperative games, which is to be seen as a theory of bargaining. It is assumed that each player is motivated by self-interest alone, so if cooperation does occur, it does so because each player is convinced that cooperation will pay.

In developing a cooperative resource management programme, it is not sufficient, as we have noted, to be concerned only with the allocation of benefits from the fishery, in harvests or other forms. One must also be concerned with the possibility that the players will have different management goals, which means that compromise management programmes will have to be negotiated.

One must also ask whether or not the cooperative arrangement is binding. Binding arrangements are obviously more stable than non-binding ones. Next, one must ask whether so-called side payments are allowed for.

A side payment, in its simplest form, is a type of transfer that may be either monetary or non-monetary in nature. We shall define, for our purposes, a fisheries cooperative game, without side payments, as one in which one coastal state's return from the shared fishery is determined solely by the harvests of its fleet(s) within its own waters. The importance of side payments, although practitioners will seldom use this term, has become increasingly recognized over the past few years (see Caddy, 1997). One role that side

[2] *The name of the game, "Prisoner's Dilemma", comes from a story told by the author of the game to illustrate the point (Tucker, 1950). Two men are arrested on suspicion of having committed a major theft. The suspicions are, in fact, entirely valid. The two suspects, A and B, are kept separated from one another. A is interviewed by the chief prosecutor, who admits that the evidence he has is limited. A is told that, if both he and B plead not guilty, they can each expect to receive a six-month sentence on a lesser charge. If both A and B plead guilty, they will each receive a five-year sentence. If A pleads guilty, but B pleads not guilty, A will be released for having assisted the prosecution. If A pleads not guilty, but B pleads guilty, then it will go very hard with A, and A will get ten years. The chief prosecutor then holds exactly the same interview with B.*

A and B are the players. Each player has two alternative strategies: to plead guilty, or to plead not guilty. If A and B could communicate, and enter into a binding agreement, they would both plead not guilty, and would look forward to being out of prison in six months time. They cannot communicate, however. The best strategy for A, regardless of which of the two strategies B might choose, is to plead guilty. What is true for A is true for B. Hence, both plead guilty and end up with the decidedly inferior outcome of serving five year sentences.

payments can play is that of helping to resolve the problem that arises when the relevant coastal states have differing management goals.

There are two fundamental conditions that must be met if the solution to a cooperative game is to be stable. Both appeal to common sense, although one has no difficulty in finding cases in the real world where these common sense conditions are ignored. The first implies that there exists no other outcome or agreement that could make one player better off, without damaging the other(s). To introduce some economist's jargon, this condition is referred to as Pareto Optimality[3]. The second, sometimes referred to as the Individual Rationality Constraint, is that, under cooperation, each and every player must be at least as well off as it would be under conditions of non-cooperation. Potential solutions to the game that meet both conditions are said to constitute the "core" of the game. There is no guarantee that the "core" will be other than empty. If the "core" is empty, then attempts to achieve cooperation will prove futile. Consider now Figure 1, which is now found well outside the realm of academia (e.g. OECD, 1997).

In this example, it is assumed that the two players (coastal states) have different management goals. The axes show the "payoffs" to the two respective players. A given

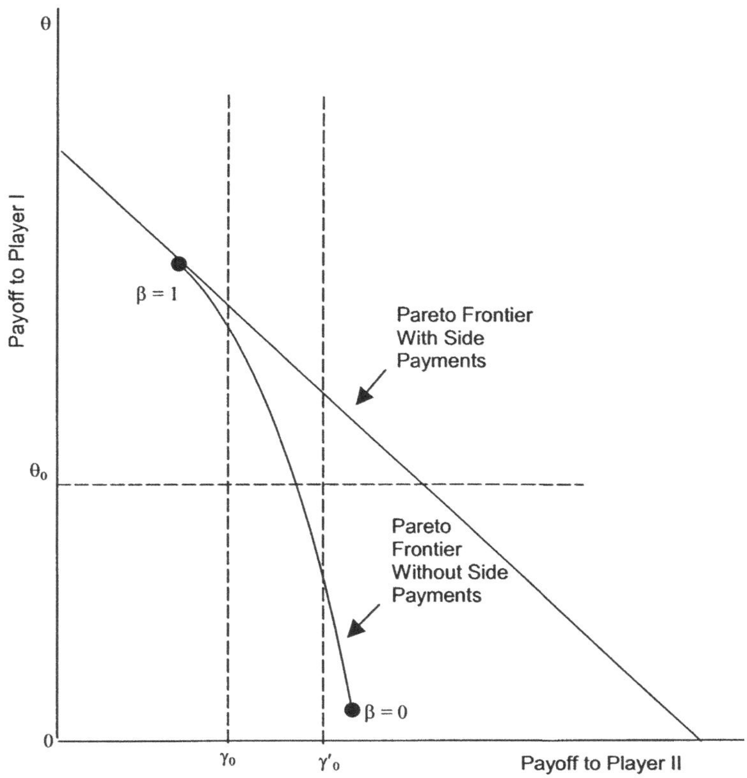

Fig. 1 Cooperative game with and without side payments.

---

[3] *The term comes from the name of an early 20th century Italian economist, Wilfredo Pareto, who argued that, in any dealings between individuals, an outcome was not optimal unless one could make one individual better off, only at the expense of another.*

payoff to Player I measures the stream of economic returns through time to Player I arising from a given resource management programme. Correspondingly, a given payoff to Player II measures the stream of economic returns to that player from a given resource management programme.

Consider the curve labelled Pareto Frontier Without Side Payments. It shows the payoffs from cooperative management regimes, in the absence of side payments, in which it is not possible to make Player I better off, except at the expense of II, and vice versa. If we commence at the top of the curve at $\beta = 1$, we would have a cooperative management programme that would maximize the benefits from the fishery to Player I. As we move down the curve, Player II would become successively better off, but only at the expense of Player I. By way of contrast, if we were at *any* point below the Pareto Frontier, both Players I and II could be made better off by adjusting the cooperative resource management programme.

The parameter β is, in fact, a bargaining parameter, $0 \leq \beta \leq 1$. If $\beta = 1$, then the management preferences of Player I are wholly dominant, whereas the management preferences of Player II count for nothing. If $\beta = 0$, the reverse is true.

Ignore, for the moment, the payoff γ'_0 on the horizontal axis. The payoffs θ_0 and γ_0 are those that Players I and II would enjoy respectively if there were no cooperation. They might be thought of as the payoffs associated with the solution to a non-cooperative game. Nash (1953) referred to this set of payoffs as the "Threat Point", because they represent the *minimum* payoffs that each of the two players must receive for the solution to a cooperative game to be stable. That part of the Pareto Frontier Without Side Payments segmented by the dashed lines emanating from the Threat Point payoffs represents the "core" of the game, in the absence of side payments. When side payments are feasible, the Pareto Frontier becomes a simple 45° line. The implication is that the players seek to maximize the global returns from the fishery without regard to differences in management objectives.

Side payments become truly significant when the management goals of the coastal states sharing the resource differ. It has been argued that, when there are differences in management goals, it is invariably the case that one player places a higher value on the fishery than does the other. When side payments are possible, then the optimal policy is one in which the management preferences of that player placing the highest value on the resource should be given full rein. That player should, in turn, then proceed to compensate its fellow player, or players, through the use of side payments, which can take any number of forms, monetary or non-monetary. This has been referred to as the *Compensation Principle* (Munro, 1987).

In our example, Player I is the player who places the highest value on the resource. If side payments are banned, and Player II's Threat Point payoff is γ_0, there will be cooperation, but with an inferior management regime, which will be to the detriment of both players. If, on the other hand, Player II's Threat Point payoff is γ'_0 rather than γ_0, the banning of side payments will ensure the collapse of cooperation. The "core" of the cooperative game would prove to be empty.

There are many other considerations to be taken into account in cooperative fisheries games. Certainly one of the most important is that of "time consistency". A "time consistent" outcome of, or solution to, the game, is one that can withstand the shocks of unexpected change, environmental change in particular. It must be stressed that an arrangement that is binding in a legal sense is not immune to "time inconsistency". An example is provided by the cooperative management of Pacific salmon by Canada and the United States. The cooperative management arrangement is contained within a formal, and hence binding, treaty, signed in 1985 (Treaty, 1985).

The Treaty appeared to work well during the first few years of its existence. It then broke down, with clear signs of the "players" reverting to competitive behaviour, with potentially grave consequences for the resource. While several factors led to the breakdown, unquestionably one very significant factor was an unpredicted, and unpredictable, climatic shift, which had a negative impact on salmon resources in the southern area covered by the Treaty, and a positive impact on such resources in the northern area. The Treaty proved, in the end, to lack the flexibility and robustness to withstand the stresses created by unexpected climatic shifts (Miller *et al.*, 2001).

To summarize, the analysis informs us first that achieving effective cooperation must involve reconciling possible conflicting management goals, as well as allocating the economic benefits from the fishery. The analysis also implies, management goals to one side, that one cannot safely assume that simple mechanical formulae for the allocation of benefits from the fishery will prove satisfactory. For good or ill, allocations will have to reflect the relative bargaining strength of the players, which will vary from case to case and which can, with respect to any individual case, be expected to shift through time. They will also have to reflect the players' perceptions of equity. Thus, for example, allocations based on the fractions of the resource found in each player's EEZ might seem to provide an eminently sensible basis on which to determine allocations. However, if the resultant formula leads to one player receiving a payoff less than its Threat Point payoff, now or in the future, then rigid application of the formula will lead ultimately to the certain collapse of the cooperative arrangement. We shall return to the question of allocation issues later.

ECONOMICS OF THE MANAGEMENT OF STRADDLING STOCKS

We turn now to straddling stocks, which under the terms of the UN Fish Stocks Agreement (United Nations, 1995) are to be managed through regional fisheries management organizations (RFMOs), having both coastal states and distant-water fishing nations/entities (DWFNs) as members. As noted earlier, economic analysis of the management of straddling fish stocks rests on a foundation provided by the economic analysis of transboundary fish stocks. The question to be asked is what modifications, if any, must be made to the economics of transboundary fish stock management in order to accommodate the particular characteristics of straddling fish stocks.

One part of this question can be answered quickly. Economic analysis of the non-cooperative management of straddling fish stocks differs not at all from economic analysis of the non-cooperative management of transboundary fish stocks. Except in unusual circumstances, non-cooperative management of straddling fish stocks will lead to the resources being mismanaged from society's point of view, and will do so for exactly the same reasons that non-cooperative management leads to the mismanagement of transboundary fish stocks – the "Prisoner's Dilemma" once again. The near destruction of pollock resources in the Bering Sea Doughnut Hole, and the Canada–Spain "fish war" on the Grand Bank of Newfoundland, are but two of many real world examples (Munro, 2000; Balton, 2001).

It is in cooperative management that distinctions between transboundary and straddling stock management appear. There are two distinctions, which are particularly striking.

Nature and number of participants through time

In a cooperative transboundary fishery management regime, the nature and the number of participants can be expected to remain unchanged, except in the most unusual circumstances. In the case of a RFMO, because some participants are DFWNs, new participants may appear, i.e. DWFNs that were not original, or "charter", members of the RFMO. Indeed, the UN Fish Stocks Agreement makes specific provisions for New Members (UN Fish Stocks Agreement, 1995, Article 11). It is this feature that probably most clearly distinguishes the cooperative management of straddling stocks from that of transboundary stocks.

Under the terms of the UN Fish Stocks Agreement, prospective New Members who are willing to abide by the terms of the RFMO management regime cannot be barred outright by "charter" members. Some legal experts contend that this implies that such New Members must be offered *just and reasonable* shares of the TAC(s) set by the RFMO (Orebech *et al.*, 1998). The interpretation of *just and reasonable* has implications for the stability of the RFMO cooperative regime.

Kaitala & Munro (1997) have demonstrated that if *just and reasonable* implies that New Members, on joining a RFMO, should receive, at no further cost, shares of the Total Allowable Catch, or the equivalent, on a *pro rata* basis, then, when planning is undertaken for the establishment of a RFMO, prospective "charter" members could well calculate that their expected payoffs from cooperation would fall below their respective Threat Point payoffs. Hence, the RFMO would be stillborn.

The aforementioned interpretation of *just and reasonable* poses the threat described, because it may give rise to a type of "free rider" problem. If the "charter" members commence with an overexploited resource which they undertake to rebuild, the New Members could then come to share in the fruits of the resource investment "free of charge", without having incurred any of the costs of the resource investment. Thus, they would be "free riders".

Exploitation of the resources by entities not party to the cooperative arrangement

In the case of a transboundary resource, any attempt by a non-member of the cooperative arrangement to exploit the resource(s) in the EEZ of a member of the arrangement without the express permission of that member would clearly be illegal. The member could take vigorous measures to repel the intruder (see FAO, 2001: International Plan of Action to Prevent, Deter and Eliminate Illegal, Unreported and Unregulated Fishing [IPOA–IUU, hereafter], para. 3.1.1). In the case of a straddling stock, a state or entity that is a non-member of the RFMO found fishing the stock on the high seas governed by the RFMO, in a manner inconsistent with the conservation and management measures of the RFMO, would be deemed to be engaging not in illegal fishing, but rather in *unregulated* fishing (United Nations, 2001, para. 3.3.1). Unregulated fishing is dealt with in the UN Fish Stocks Agreement in Article 17. The article calls on members of a RFMO to deter unregulated fishing in a manner consistent with international law. It was not, and is not, immediately obvious what powers international law confers on members of a RFMO (Munro, 2000). Therefore,the action which the members of the RFMO could take to deal with unregulated fishing is less clear than the action which they could take if confronted with illegal fishing. The IPOA–IUU attempts to address this problem.

Unregulated fishing is another form of "free riding". If it is uncontrolled, then it is easy to show, with the aid of game theory, that its existence or threatened existence can serve to undermine a RFMO. In a way, it is like a particularly virulent form of the New Member problem. One can also show, with game theory, that if the only way that unregulated fishing can be controlled is by persuading non-members to join the RFMO on a strictly voluntary basis, the stability of the RFMO will be in serious doubt. Recent work has shown the number of players that full cooperation can support under these circumstances is depressingly small (often no more than two). With a large number of players, defection, "free riding", becomes too easy and too attractive. If, on the other hand, effective punishment can be meted out to those who refuse to desist from unregulated fishing, then, not surprisingly, the likelihood of achieving a stable cooperative agreement with large numbers is greatly enhanced (see, for example, Lindroos, 2002; Pintassilgo, in press).

In the management of transboundary stocks, the number of participants (players) tends, with few exceptions, to be small. By way of contrast, in the management of straddling stocks, large numbers of participants can be expected to be the rule rather than the exception. The implications are obvious. If the RFMO regime is to be sustainable through time, effective implementation of the FAO IPOA–IUU is not merely desirable. It is mandatory.

ALLOCATION ISSUES: A COMMENT

Although we warned that rigid, mechanical rules of allocation of the benefits from a cooperatively managed fishery will not prove satisfactory, we did not mean to dismiss the issue. The issue of costs and their sharing is complex, to say the least, and does deserve further comment. First, allocation issues need to be considered in the context of the principal requirements of a stable cooperative management arrangement, specifically that (i) every country should be better off with cooperation than without it; (ii) the outcome of cooperation should be perceived as fair and equitable; and (iii) conditions (i) and (ii) need to hold over time (time-consistency).

Second, the requirement that every country participating (or wishing to participate) in the harvesting of a shared stock should be better off with cooperation than without it must be interpreted in terms of the *net* benefits attained. Therefore, allocation needs to address the sharing of (gross) benefits of cooperation. This poses several non-trivial problems, because the benefits of cooperation usually accrue to fishers or the fishing industry, whereas the costs of cooperation, largely management costs, are often borne by governments. The issue is further complicated by the need of government to further subdivide the country's allocation of catch quotas (or effort/capacity quotas) to individual fishers or fishing communities. Therefore, governments can expect to be under pressure on two accounts: first, to get the best deal for the country as a whole; second, to satisfy the competing interests of different groups of fishers.

Next, the structure of the fishing industry targeting the shared fish stock is likely to have a bearing on management costs and on a government's ability to enforce the cooperation agreement. A case in point is the resources of small pelagic fish shared between several West African countries. In some countries (e.g. Mauritania), small pelagic fish are exploited by a small number of large factory ships, whereas in other countries (e.g. Senegal), the same stocks are fished by thousands of small-scale fishers. The allocation of quotas and their enforcement is presumably more complex and costly where there are many participants in the fishery than where there are only few.

In the case of these small pelagic fisheries, it is not only the structures of the fishing industries that are different. The countries' management objectives can also be expected to differ. In Senegal, the priority objectives appear to include maintaining and enhancing domestic fish supply and providing a stable source of income (and employment) to rural people. In Mauritania, on the other hand, the emphasis appears to lie in hard currency earnings from licensing foreign fishing vessels. This raises the question of whether differences in the fishing industry's structure and in the objectives of fisheries development and management should be reflected in allocation criteria.

The International Commission for the Conservation of Atlantic Tuna (ICCAT) convened a Working Group on Allocation Criteria. The outcome (ICCAT, 2002) suggests a firm yes to the question. The ICCAT Working Group places allocation criteria into four categories:

(a) criteria relating to past/present fishing activity of qualifying participants;
(b) criteria relating to the status of the stock(s) to be allocated and the fisheries;
(c) criteria relating to the status of the qualifying participants;
(d) criteria relating to compliance/data submission/scientific research by qualifying participants.

Category (c) is of particular relevance to our question. The category lists, *inter alia*, the interests of artisanal, subsistence and small-scale coastal fisheries, the needs of coastal fishing communities and of coastal states' regions whose economies are overwhelmingly dependent on fisheries, and the contribution of the fisheries to national food security/ needs, domestic consumption, export income and employment.

The ICCAT negotiations of these criteria were influenced by, on one hand, the interest of DWFNs to maintain their fishing possibilities with reference to historical and present exploitation of the stocks in question, and on the other hand, by the interest of developing countries to allow future development of their fisheries. The latter group of countries argued their case, *inter alia*, with reference to the UN Fish Stocks Agreement, Article 24 (UN, 1995), that calls upon states to:

> *"... give full recognition to the special requirements of developing States in relation to conservation and management of straddling fish stocks and highly migratory fish stocks and development of fisheries for such stocks. .... , in particular: (a) the vulnerability of developing States which are dependent on the exploitation of living marine resources, including for meeting the nutritional requirements of their populations or parts thereof; (b) the need to avoid adverse impacts on, and ensure access to fisheries by, subsistence, small-scale and artisanal fishers and women fishworkers, as well as indigenous people in developing States, particularly small island developing States; and (c) the need to ensure that such measures do not result in transferring, directly or indirectly, a disproportionate burden of conservation action onto developing States."*

Arguably, the allocation criteria developed by ICCAT are far-ranging and not sufficiently specific to make them readily operational in allocating fishing quotas. This would likely be the case with any type of general principle or recommendation one might wish to establish for allocation criteria, be it at the level of RFMOs or within an international forum such as FAO's Committee on Fisheries. The precise allocation criteria (or allocation rules) would always be a matter of negotiation among the countries sharing a common resource. General sharing [equity] principles might, however, play a useful role in framing the expectations of the bargaining parties, so facilitating the process of achieving a satisfactory agreement.

SOME SELECTED CASE STUDIES

We turn now to a few case studies to illustrate some of the points made in earlier sections of this paper. One important, and significant, case study not discussed in this paper, is that of Caribbean spiny lobster. That case is discussed at length in an accompanying chapter (Cochrane *et al.*, 2003) in this volume.

PACIFIC ISLAND NATIONS TROPICAL TUNA FISHERIES

The Pacific Islands Region constitutes one of the richest tropical tuna grounds in the world. The tuna resources were, and are, of fundamental economic importance to the Islands. Consequently, it could be maintained that the Pacific Island Nations were, collectively, one of the big "winners" from the advent of the EEZ regime in 1982. Having said this, however, it was not at all clear at the time that the economic benefits that these countries would enjoy from the new regime would be other than ephemeral.

Collectively, the Pacific Island Nations (PINs) EEZs cover an area of 29 000 million km^2; their collective landmass is just 500 000 km^2. Most tuna within these EEZs, 80% or more, were harvested by DWFNs under access (to EEZs) agreements. Finally, the PINs were generally at low levels of development. Hence, they faced what appeared to be insurmountable monitoring and surveillance problems. The difficulties were compounded by other factors. First, the PINs initially faced one DFWN only (Japan), a major power in the Asia/Pacific region. As a provider of harvesting services, this powerful nation was in the position of a monopolist within the Pacific Islands Region. Second, the right of coastal states to assert management jurisdiction over tuna resources within their EEZs was bitterly contested at the close of the LOSC.

The PINs had an incentive to cooperate. Without cooperation, it was inevitable that the single DWFN, and later other DWFNs (e.g. the United States), would attempt to play one Island country off against the other, and would do so successfully (Munro, 1991). Thus, competitive behaviour among the PINs would indeed have led to a "Prisoner's Dilemma" type of outcome.

Achieving effective cooperation was, however, difficult. The independent PINs were (and are) members of the South Pacific Forum, so they attempted to cooperate through the Forum by establishing, in 1979, the South Pacific Forum Fisheries Agency (FFA). There were, however, 14 countries involved, which varied enormously in size, and which were spread over vast distances (Munro, 1991). This was the exception to the rule that the number of players in a transboundary fishery game tends to be small. It is commonplace in game theory that the larger the number of players in a cooperative game, the more difficult it is to achieve a stable solution. In the early 1980s, it appeared to most observers that the PIN 14-player cooperative tuna game was headed towards collapse (Munro, 1991). In cooperative games involving more than two players, there will always be the possibility of sub-coalitions emerging. In the case of the PIN tuna game, sub-coalitions did emerge, to the great benefit of the PINs, and, one might add, to the resource.

The tuna resources in the South Pacific are not evenly spread, tending to concentrate around the Equator. The consequence is that there are, in relative terms, "haves" and "have nots", among the Pacific Island Nations. Just 7 of the 14 countries could be regarded as "haves". Concerned about the lack of progress in the FFA, the seven met on the island of Nauru (one of the seven) and signed a formal agreement, the Nauru Agreement, and became known as the Nauru Group thereafter. The Nauru Group made it known, that, while the Group had no wish to see the FFA disintegrate, the Group would go it alone unless the others engaged in serious cooperation. The others decided that serious cooperation was indeed in their best interest. Therefore, two sub-coalitions were formed, the Nauru Group ("haves"), and the "have nots"'. It helped that two major Island nations, Papua New Guinea (PNG) and Fiji, were in different sub-coalitions. PNG was in the "haves" sub-coalition, and became its leader, whereas Fiji became the leader of the "have nots" sub-coalition. An intractable 14-player game had evolved into what amounted to a two-player game (Munro, 1991).

To the surprise of no-one, the management goals of the two sub-coalitions proved to be dissimilar. The Nauru Group was more concerned about the long-term stability of the resources than was the less well-off sub-coalition. Clearly, the Nauru Group placed higher value on the resource. Theory tells us that the optimal outcome would be for the management preferences of the sub-coalition placing the higher value on the resource to dominate, and for that sub-coalition to compensate its fellow sub-coalition.

The predictive power of the theory in this instance proved to be strong. The Nauru Group became the cutting edge in terms of formulating management policy. Various forms of side payments emerged, through which the "have not" sub-coalition was compensated (Aikman, 1987; Munro, 1991). These compensations continue to the present day. Moreover, the "have nots" sub-coalition has played an increasingly important role in cooperative management of the resource (David Doulman, FAO, pers. comm.), which attests to the growing strength of the cooperative resource management arrangement.

NORWEGIAN SPRING-SPAWNING HERRING FISHERY

The Norwegian spring-spawning herring is among the largest and biologically most productive fishery resources in the world. When healthy, the resource has a total biomass of 15–20 million tonnes and a spawning biomass averaging 10 million tonnes (Arnason *et al.*, 2000). The resource when healthy is both a transboundary and straddling stock, and it spawns, as the name suggests, in Norwegian seas. After spawning, the resource migrates from the Norwegian EEZ through the EEZs of the EU, the Faeroe Islands and Iceland. It also migrates through a large high seas enclave known by some as the "Ocean Loop", and by others as the "Herring Loop" (Arnason *et al.*, 2000). When depressed, the resource is confined to Norwegian waters (Bjørndal *et al.*, 1998; Arnason *et al.*, 2000).

The resource has the characteristic, typical of clupeoids, of being an intense schooling species, and is therefore highly vulnerable to overfishing. In the pre-LOSC era, the fishery was international and open access. The economics models of non-cooperative management of shared fish stocks proved, by the late 1960s, to have powerful predictive power. The resource collapsed as a result of overexploitation and came within a hair's breadth of extinction (Arnason *et al.*, 2000). An international moratorium was declared in 1969.

Through the good fortune of an exceptionally strong year class in the late 1980s and the continuing harvest moratorium, the resource recovered. Spawning biomass recovered to the healthy level of 10 million tonnes and the moratorium was lifted. Further, it was recognized that cooperative management of the resources was required if the health of the resource was to be maintained.

Five countries/entities exploiting the resource – originally Norway, Iceland, Russia, the Faeroe Islands, and later the EU – came together in the mid-1990s to establish a cooperative resource management arrangement. The initial attempt to establish cooperative resource management was disappointing (Bjørndal *et al.*, 1998). By 1996, however, the cooperative resource management regime achieved stability and has apparently remained successful up to the present time. It can be argued that, after 1995, the UN Fish Stocks Agreement provided the framework required for a successful cooperative resource management regime.

It can be noted in passing that the case of the Norwegian spring-spawning herring provides us with an example of the inadequacy of simple formulae for allocating benefits from the fishery among the players. There was some suggestion at an early stage of the negotiations that so-called "biological zonal attachment" (determined by the amount of the biomass in each EEZ and the high seas, and the amount of time spent by the biomass in each EEZ and the high seas) be given substantial weight in setting quota shares among the five. Heavy emphasis on "biological zonal attachment" would have given the EU a negligible quota share (Bjørndal *et al.*, 1998). However, the EU's bargaining strength was (is) such that its quota share was not, and is not, negligible.

In any event, the incentive to cooperate is strong. It is obvious to all players that the economic benefits from cooperation are large, particularly because a reversion to competitive behaviour would carry with it the distinct threat of extinction of the resource (Arnason *et al.*, 2000). There also seems to be some signs of "time consistency", although the cooperative arrangement has not been in place long enough to provide convincing evidence. Nonetheless, it is recognized that the migratory pattern of the resource is certain to vary over time, and that this will necessitate renegotiation of the quotas (Arnason *et al.*, 2000). Finally, the cooperative resource management regime has not been tested with respect to New Members, nor is there any likelihood that such testing will arise in the foreseeable future.

While there is no evidence of monetary side payments having been made in this instance, there are numerous "side arrangements" between players, for example allowing one player to take part of its quota in another player's zone, in exchange for various *quid pro quo*. If these arrangements do not meet a strict definition of side payments, they carry with them the distinct flavour of side payments.

PACIFIC SALMON – CANADA AND THE UNITED STATES

Wild Pacific salmon, as an anadromous species, are spawned in freshwater, spend most of their lives in the ocean, and then return to their freshwater origins to spawn and die. In Pacific North America, wild salmon are produced in rivers and streams from California through Oregon, Washington, British Columbia, to Alaska. Historically, the two single most important salmon river systems have been the Columbia, primarily (but not exclusively) in the United States, and the Fraser, wholly confined to Canada.

Some American-produced salmon are inevitably "intercepted," i.e. caught, by Canadian fishers; some Canadian-produced salmon are inevitably "intercepted" by American fishers. Hence, the resource is inescapably transboundary in nature. The fish are normally harvested as they approach river mouths on their way to the spawning grounds. They are easy to catch, and are therefore highly vulnerable to overexploitation. Hence, the consequences of non-cooperative management of the resource can be severe.

Canada–United States Pacific salmon negotiations, initially focused on the Fraser River, broadened in the early 1970s with the objective of covering all salmon produced from northern California to southern Alaska. The negotiations proved to be extraordinarily difficult. The negotiators were, however, spurred on by the threatened emergence of a "fish war," which both sides realized would be highly destructive, and by the blocking of enhancement projects on both sides of the border (Munro & Stokes, 1989).

In 1985, the Canada–United States Pacific Salmon Treaty came into being (Treaty, 1985). The division of returns from the fisheries was incorporated in the Treaty, in the so-called Equity Principle, in which each country was to receive economic benefits commensurate with the salmon produced in that country's rivers and streams. Achieving equity was to be through balancing interceptions alone. No thought was given to the possibility of side payments.

At the time the Treaty was signed, the Fraser and Columbia Rivers were seen as the heart of the cooperative resource management agreement. The Americans intercepted primarily Fraser River salmon; the Canadians intercepted primarily Columbia River salmon. Alaska was essentially a "side show." The interceptions appeared to be roughly balanced (Munro & Stokes, 1989). Initially, the cooperative agreement prospered. While no actual estimates were made, it was agreed that if the economic benefits of cooperation were to be measured, they would prove to be very large indeed.

However, two problems emerged over time. The first was that, while Canada could be viewed as a single player, the United States was in fact an unstable coalition, in which Washington/Oregon and Alaska were key players. The second problem arose from the fact that a climatic shift was under way in 1985, which was to prove detrimental to salmon stocks in Washington, Oregon and southern British Columbia, but highly beneficial to salmon stocks in Alaska (Miller *et al.*, 2001).

The impact of the climatic shifts became increasingly evident over time. Columbia River salmon showed signs of severe deterioration, which led in turn to declining Canadian interceptions. The booming Alaskan stocks, in their turn, resulted in increased Alaskan interceptions of Canadian-produced salmon, as Alaskan fishers sought to reap their bounty. Alaskan interception, initially minor from a Canadian perspective, achieved greater and greater importance (Miller *et al.*, 2001).

The rough interception balance of the early years of the Treaty was upset. The Alaskans were pressed to reduce their interceptions, which they insisted that they could do only by forgoing their bounty (Miller *et al.*, 2001). By 1993, the Treaty was in disarray. A very real threat of a "fish war," i.e. reversion to destructive competitive behaviour, loomed (Miller *et al.*, 2001). A fundamental condition for cooperative resource management had now been violated. The Individual Rationality Constraint had become binding. A major player, Alaska, was no longer better off with the Treaty than without.

The Pacific salmon case illustrates the paramount importance of flexibility and "time consistency" in cooperative fisheries management agreements. As noted earlier, the Canada–United States Pacific Salmon Treaty was as binding an agreement as one could hope to achieve. Yet, the Canada–USA Pacific salmon cooperative game had proved to lack the resilience needed to accommodate changing conditions. The harvest allocation mechanism set in place by the Treaty effectively broke down._One might add that a significant factor underlying the lack of resilience, and noted by an increasing number of observers, was the absence of the possibility of side payments (Miller *et al.*, 2001).

In 1999, Canada and the USA signed an Agreement in an attempt, initially successful, to restore cooperation (United States of America, 1999). One aspect of the Treaty, in its original form, which came to be particularly "time inconsistent", was the manner in which interceptions were dealt with. The Americans were faced with a ceiling on the number of

Fraser River salmon that they could catch; the Canadians were faced with a similar ceiling on the number of Columbia River salmon that they could harvest. The ceilings worked well when the stocks were increasing. The returns from Canadian (American) efforts to rebuild Fraser River (Columbia River) stocks went to the Canadians (Americans). The ceilings worked badly, however, when stocks were declining. Each side had an incentive to fish to its ceiling, regardless of how depleted the relevant stock, or stocks, might be. The Agreement calls for the ceilings to be replaced by "abundance based" management. Thus, for example, American harvest of Fraser River stocks will increase when the stocks are increasing and decrease when the stocks are declining. Hopefully, the new management regime will be more resilient than its predecessor in the face of environmental shocks.

The Treaty in its original form was also criticized for excluding the possibility of side payments. The Agreement sets a precedent by introducing side payments, although the term "side payment" is nowhere mentioned in the Agreement. Two Endowment Funds are to be established to strengthen and rebuild stocks on both sides of the border. Initially, at least, the funding will come from the United States alone, leading to an implicit side payment from the United States to Canada (Miller *et al.*, 2001). It is too early at this stage to pass judgement on the Agreement. It remains to be seen whether it will lead to a lasting peace, or whether it will prove to be no more than a temporary truce (Miller *et al.*, 2001).

CONCLUSIONS

The management of shared fish stocks continues to be a major issue in the management of world fisheries. Although there are as many as 1500 transboundary fish stocks and many more straddling fish stocks, only a limited number of these shared fishery resources are subject to effective cooperative management. The scope for improved management is, therefore, immense.

With few exceptions, cooperation in the management of shared fishery resources does matter. It is dangerous to assume that non-cooperative management of shared fishery resources will lead to resource management programmes that are adequate. Cooperative management at the secondary level, involving full joint management, is, admittedly difficult and costly, but there are some examples of effective cooperative resource management that can serve as examples for others.

Stability in cooperative resource management arrangements requires that certain requirements be met. Several of these are obvious. First, for a given arrangement to be stable, it must not be possible to find an alternative arrangement capable of making all "players" better off. Second, the so-called Individual Rationality constraint must be satisfied. Even if only one "player" or sub-coalition of "players" party to the arrangement concludes that it can do better by refusing to cooperate, the full cooperative arrangement will not hold.

In the case of straddling stocks, means must be found of accommodating New Members in accordance with the UN Fish Stocks Agreement that do not, at the same time, undermine the long-term viability of RFMOs. Means must also be found of dealing with the issue of *unregulated* fishing in waters governed by RFMOs. In this regard, it is of utmost importance that the FAO IPOA–IUU be fully adopted and implemented.

A less obvious, but highly important, requirement relevant to both transboundary and straddling fish stocks is that the cooperative management arrangement be "time consistent". The cooperative management arrangement must have the flexibility and robustness to withstand through time the shocks of unexpected and unpredictable changes.

REFERENCES

Aikman, C. C. (1987) Island Nations of the South Pacific and jurisdiction over highly migratory species. *Victoria University of Wellington Law Review*, **17**: 101–124.

Arnason, R., Magnusson, G. & Agnarsson, S. (2000) The Norwegian spring- spawning herring fishery: a stylized game. *Marine Resource Economics*, **15**: 293–320.

Balton, D. (2001) The Bering Sea Doughnut Hole Convention: regional solution, global implications. In: *Governing High Seas Fisheries: The Interplay of Global and Regional Regimes*. (Ed. by O. Stokke), pp. 143–178. Oxford University Press.

Bjørndal, T., Hole, A. D., Slinde, W. M. & Asche F. (1998) Norwegian spring spawning herring – some biological and economic issues. *Norwegian School of Economics and Business Administration, Centre for Fisheries Economics, Papers on Fisheries Economics*, **38**.

Caddy, J. (1997) Establishing a consultative mechanism or arrangement for managing shared stocks within the jurisdiction of contiguous states. In: *Taking Stock: Defining and Managing Shared Resources. Australian Society for Fish Biology and Aquatic Resource Management Association of Australasia Joint Workshop Proceedings, Darwin, NT, June 1997*. (Ed. by D. Hancock), pp. 81–123. Australian Society for Fish Biology, Sydney.

Cochrane, K. L., Chakalall, B. & Munro, G. (2003) The whole could be greater than the sum of the parts: the potential benefits of cooperative management of the Caribbean spiny lobster. (this volume)

Clark, C. (1980) Restricted access to a common property resource. In: *Dynamic Optimization and Mathematical Economics*. (Ed. by P. Liu), pp. 117–132. Plenum Press, New York.

Clark, C. (1990) *Mathematical Bioeconomics: The Optimal Management of Renewable Resources*, 2nd edn. Wiley, New York.

FAO. (1979) Interim Report of the ACMRR Working Party on the Scientific Basis of Determining Management Measures. *FAO Fisheries Circular*, **718**.

FAO. (1999) The International Plan of Action for the Management of Fishing Capacity. FAO, Rome.

FAO. (2001) International Plan of Action to Prevent, Deter and Eliminate Illegal, Unreported and Unregulated Fishing. FAO, Rome.

FAO. (2002) Report of the Norway–FAO Expert Consultation on the Management of Shared Fish Stocks, Bergen, Norway 7–10 October 2002. *FAO Fisheries Report*, **695**.

Gordon, H. S. (1954) The economic theory of a common property resource: the fishery. *Journal of Political Economy*, **62**: 124–142.

Gulland, J. A. (1980) Some problems of the management of shared stocks. *FAO Fisheries Technical Paper*, **206**.

ICCAT. (2002) Report for Biennial Period 2000–2001, Part 2 (2001), Vol. **1** (in press).

Kaitala, V. & Munro, G. (1993) The management of high seas fisheries. *Marine Resource Economics*, **8**: 313–329.

Kaitala, V. & Munro, G. (1997) The conservation and management of high seas fishery resources under the new Law of the Sea. *Natural Resource Modeling*, **10**: 87–108.

Lindroos, M. (2002) Coalition in Fisheries. Paper prepared for the 14th Annual Conference of the European Association of Fisheries Economists, Faro, Portugal.

McRae, D. & Munro, G. (1989) Coastal state "Rights" within the 200 mile Exclusive Economic Zone. In: *Rights Based Fishing*. (Ed. by P. Neher, R. Arnason and N. Mollet), pp. 97–112. Kluwer, Dordrecht.

Miller, K., Munro, G., McDorman, T., McKelvey, R. & Tydemers, P. (2001) The 1999 Pacific Salmon Agreement: a sustainable solution? *Occasional Papers: Canadian–American Public Policy*, **47**. Canadian–American Center, University of Maine, Orono.

Munro, G. (1979) The optimal management of transboundary renewable resources. *Canadian Journal of Economics*, **12**: 355–377.

Munro, G. (1987) The management of shared fishery resources under Extended Jurisdiction. *Marine Resource Economics*, **3**: 271–296.

Munro, G. (1991) The management of migratory fishery resources in the Pacific: tropical tuna and Pacific salmon. In: *Essays on the Economics of Migratory Fish Stocks*. (Ed. by R. Arnason and T. Bjørndal), pp. 85–106. Springer-Verlag, Berlin.

Munro, G. (2000) The UN Fish Stocks Agreement of 1995: history and problems of implementation. *Marine Resource Economics*, **15**: 265–280.

Munro, G. (2003) On the management of shared fish stocks. In: *Papers Presented at the Norway–FAO Expert Consultation on the Management of Shared Fish Stocks, Bergen, Norway, 7–10 October 2002. FAO Fisheries Report*, **695**(Suppl.): 2–29.

Munro, G. & Stokes, R. (1989) The Canada–United States Pacific Salmon Treaty. In: *Canadian Oceans Policy: National Strategies and the New Law of the Sea*. (Ed. by D. McRae and G. Munro), pp. 17–35. University of British Columbia Press, Vancouver.

Nash, J. (1951) Noncooperative Games. *Annals of Mathematics*, **54**: 289–295.

Nash, J. (1953) Two-person cooperative games. *Econometrica*, **21**: 128–140.

OECD. (1997) *Towards Sustainable Fisheries: Economic Aspects of the Management of Living Marine Resources*. Paris.

Orebech, P., Sigurjonsson, K. & McDorman, T. L. (1998) The 1995 United Nations Straddling and Highly Migratory Fish Stocks Agreement: management, enforcement and dispute settlement. *The International Journal of Marine and Coastal Law*, **15**: 361–378.

Pintassilgo, P. (in press) A coalition approach to the management of high seas fisheries in the presence of externalities. *Natural Resource Modeling*.

Treaty (1985) Treaty between the Government of Canada and the Government of the United States of America Concerning Pacific Salmon, March 1985. http://www.psc.org/Treaty/TREATY.HTM (last accessed 5 February 2003).

Tucker, A. W. (1950) *A Two-Person Dilemma*. Stanford University, unpublished.

United Nations. (1982) United Nations Convention on the Law of the Sea. *UN Document*, **A/Conf. 62/122**.

United Nations. (1995) United Nations Conference on Straddling Fish Stocks and Highly Migratory Fish Stocks. *Agreement for the Implementation of the United Nations Convention on the Law of the Sea of 10 December 1982 Relating to the Conservation and Management of Straddling Fish Stocks and Highly Migratory Fish Stocks*. UN Doc. A/Conf./164/37.

United States of America (1999) Diplomatic Note No. 0225 from Canada to the United States; reply attached Agreement, June 30. http://www.state.gov.

A review of Mediterranean shared stocks, assessment and management

Jordi Lleonart
Institut de Ciències del Mar (CSIC), Passeig Marítim 37, 08003 Barcelona, Spain

ABSTRACT: The continental shelf of the Mediterranean Sea is, with minor exception, narrow, so most of the fisheries for demersal and small pelagic fish are constrained to a thin band close to the coast. In addition, no Exclusive Economic Zone has been implemented. Therefore, although most countries have territorial waters out to 12 nautical miles, the rest of the sea lies in international waters, and the fisheries there are based almost exclusively on large pelagic fish. As a consequence, besides large pelagic fish, there are few shared stocks. However, there are wide shelves in the Adriatic Sea, the Gulf of Gabès, the northern Black Sea and to a lesser extent the Gulf of Lions, where fish stocks other than large pelagics are exploited by more than one country; examples are hake and anchovy in the Adriatic Sea and Gulf of Lions. Large pelagic fish in the Mediterranean include bluefin tuna, swordfish, albacore, dolphinfish and several sharks, and those species are fished by countries bordering the Mediterranean and others, including flag of convenience fleets. The recent development of tuna farming in many Mediterranean countries makes it more difficult to obtain reliable data for assessment purposes. These characteristics have led to most assessments being carried out within the framework of scientific projects, with little continuity over time, with the results rarely being incorporated in management. International management is not enforced rigorously, and few regular assessments are performed by the two international organizations operative in the Mediterranean: the International Commission for the Conservation of Atlantic Tunas, which deals solely with large pelagic fish, and the General Fisheries Commission for the Mediterranean, which covers all fisheries. The latter created a Scientific Advisory Committee in 1999, and that body is currently developing methods for assessment of fisheries internationally, focusing on shared stocks.

INTRODUCTION

The Mediterranean Sea, herein defined as including the Black Sea, is semi-enclosed and has a surface area of about 3 million km^2, some 0.8% of the world's total sea surface. It is generally considered to be oligotrophic (Margalef, 1985; Estrada, 1996; Stergiou *et al.*, 1997a). In terms of fisheries, its two fundamental features are the large variety of species harvested and the absence of large single-species stocks comparable to those inhabiting the coastal borders of open oceans and the subject of extensive fisheries. The continental shelf is generally narrow, the exceptions being the Adriatic Sea, the Gulf of Gabès, the northern Black Sea and, to a lesser extent, the Gulf of Lions (Figure 1).

In all, 21 countries border the Mediterranean, 26 including the Black Sea. Of these, the four EU members (Spain, France, Italy and Greece) constitute one-third of the coastline (Farrugio, 1996). No Exclusive Economic Zone (EEZ) has been implemented, except in the Black Sea where an EEZ has been declared by equidistant lines. Most countries have territorial waters out to 12 nautical miles from the coast (Ronzitti, 1999), the exceptions being Syria, to 35 nautical miles, Greece and Turkey, to 6 nautical miles, and the United Kingdom, to 3 nautical miles (Gibraltar and the Cypriot bases of Akrotiri and Dhekelia). Four countries have established Fishing Zones: Algeria (32 miles from the border with Morocco and 52 miles from the border with Tunisia), Libya (20 miles or a depth of 200 m), Malta (25 miles) and Spain (an equidistant border, with the exception of the Alboran Sea).

(a)

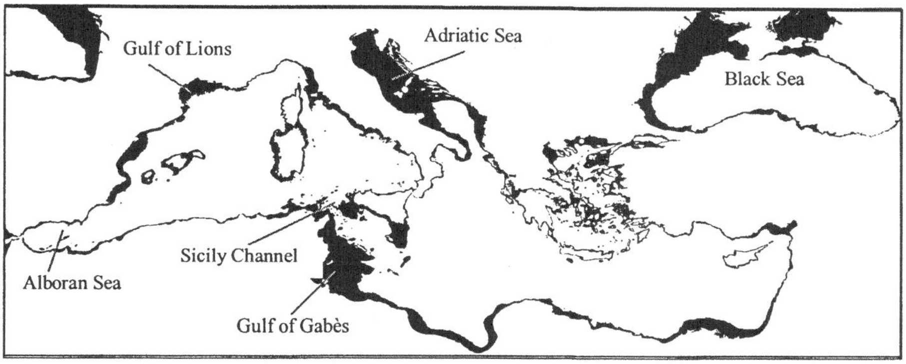

(b)

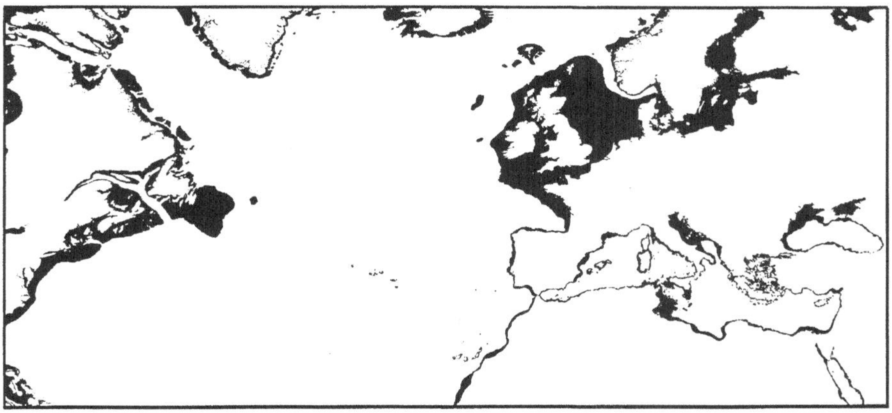

Fig. 1 (a) Mediterranean and (b) North Atlantic (including Mediterranean) continental shelves, defined here as depths out to 200 m.

Mediterranean fisheries have been prosecuted for thousands of years (classical Greek and Roman documents on fishing are abundant), so the current pattern is the result of a long history and not simply the outcome of a specific (relatively recent) management policy. Trawling, however, only commenced in the 18th century, and it burgeoned as a technique despite fierce opposition from other fishers, a pattern often repeated since the beginning of Mediterranean fisheries, to the extent that it is now the dominant gear. Nowadays, the Mediterranean trawl fishery is based on mesh sizes of about 40 mm, so most fisheries are for juvenile fish, with just a few fish escaping and reaching adulthood by becoming inaccessible to the fleets. This pattern, referred to by Caddy (1990) as the "spawning refugia paradigm", is in stark contrast to that (referred to as the "maximum size limit paradigm") implemented in the Atlantic and elsewhere, agreeing with Beverton & Holt's (1957) theory. However, the Mediterranean, with catch rates lower than in the Atlantic, but without "industrial" fleets in the Atlantic sense, has suffered few collapses and displays an apparent, though likely untrue, image of stability. Farrugio *et al.* (1993), Farrugio (1996), Anon. (2001), Bas (2002) and Lleonart & Maynou (2003), among others, have reviewed Mediterranean fisheries.

The Food and Agriculture Organization of the United Nations provides official bulletins and on-line databases of production statistics since 1970 by year, area, subarea and species (FAO Fishery Information Data and Statistics Unit, 2003). These data, updated regularly, are likely to be underestimates of actual catches, because under-reporting is widespread and discarding common. Therefore, official commercial catch statistics should be viewed with caution. Nevertheless, they are useful for indicating relative, as opposed to absolute, trends.

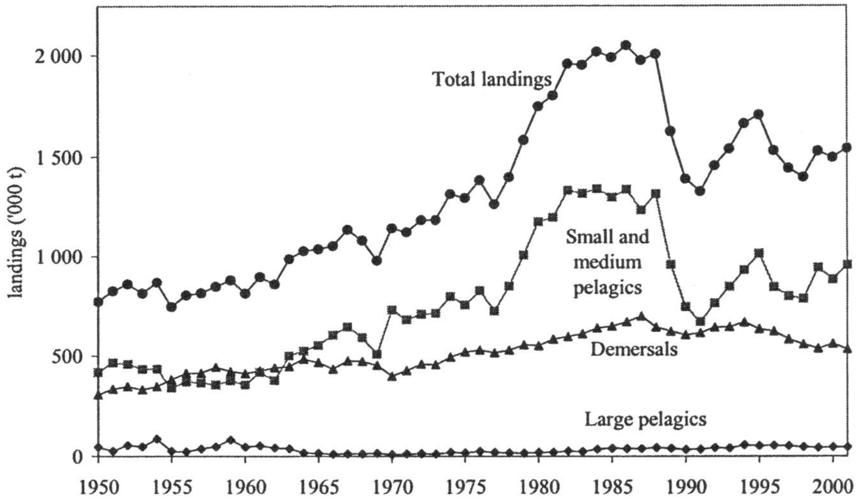

Fig. 2 Time-series of world fishery landings by species group, 1950–2001 (from FAO databases).

Average annual fish landings in the Mediterranean for the past decade have been of the order of 1.5 million tonnes (1.49 million in 2000, according to the FAO FISHSTAT database updated in 2003; Figure 2). Consequently, Mediterranean fisheries only provide a small portion of the world's production of some 100 million tonnes. However, the value of those landings (mainly sold fresh) is markedly above average world market prices. In the European Union alone, Mediterranean fisheries represent about 20% of the catch, but some 35% of the total value.

Table 1. Landings of the 30 main species or species groups (out of a total of 238) caught in 2001 in the Mediterranean and Black Seas (from the FAO's FISHBASE). The total catch was 1 540 767 t.

Species	Mass (t)	Percentage	Cumulative percentage
Engraulis encrasicolus	436 007	29.35	29.35
Sardina pilchardus	195 961	13.19	42.55
Sardinella spp.	69 490	4.68	47.22
Osteichthyes	63 933	4.30	51.53
Sprattus sprattus	63 125	4.25	55.78
Mytilus galloprovincialis	46 092	3.10	58.88
Chamelea gallina	42 657	2.87	61.75
Mugilidae	34 704	2.34	64.09
Micromesistius poutassou	27 933	1.88	65.97
Clupeonella cultriventris	27 777	1.87	67.84
Trachurus spp.	27 497	1.85	69.69
Merluccius merluccius	24 753	1.67	71.36
Thunnus thynnus thynnus	23 725	1.60	72.95
Boops boops	22 479	1.51	74.47
Trachurus mediterraneus	19 308	1.30	75.77
Sarda sarda	19 161	1.29	77.06
Mullus spp.	15 846	1.07	78.12
Mollusca	15 404	1.04	79.16
Octopus vulgaris	15 325	1.03	80.19
Pomatomus saltatrix	15 291	1.03	81.22
Xiphias gladius	14 988	1.01	82.23
Trachurus trachurus	11 602	0.78	83.01
Parapenaeus longirostris	11 093	0.75	83.76
Merlangius merlangus	11 045	0.74	84.50
Scomber japonicus	10 803	0.73	85.23
Sepia officinalis	10 698	0.72	85.95
Mullus surmuletus	9 598	0.65	86.60
Scomber spp.	9 148	0.62	87.21
Spicara spp.	8 928	0.60	87.81
Sparidae	8 805	0.59	88.41

In 2001, 21 species (or species groups) that each constituted >1% of the total reported landings (in weight), accounted for >82% of the total catch (Table 1). However, the order of "importance" of each species or species group varies in the different parts of the Mediterranean. It should also be noted that the "species" third in importance in Table 1 is a taxon not identified to species level.

FISHING FLEETS

Three types of fleet operate: artisanal, industrial and semi-industrial (Farrugio, 1996; Farrugio & Papaconstantinou, 1998).

The artisanal fleet comprises small, relatively cheap, boats, that usually belong to fishers personally, generally fish close inshore, and deploy a broad diversity of fishing gears. In number, Farrugio (1996) estimates 42 000 such boats for the four European countries alone and some 100 000 for the whole Mediterranean. Demersal species are the target species.

The industrial fleet comprises larger, more expensive, boats often purchased through investments made by companies or financial groups. In the Mediterranean, they operate mainly in the tuna fishery, and they include the fleets of Mediterranean countries as well as foreign fleets and flags of convenience.

The "semi-industrial" fleet is a group of boats intermediate between the other two classifications, but closest in characteristics to the artisanal fleet. It consists mainly of trawlers, purse-seiners and some longliners. Catches are landed usually daily or every other day, so catches are generally taken near the coast, on the shelf or upper slope. Trawl catches comprise mainly juvenile fish of many different species.

INTERNATIONAL ORGANIZATIONS

Two international organizations are mandated to assess and manage Mediterranean fisheries; the General Fisheries Commission for the Mediterranean (GFCM), of general nature, and the International Commission for the Conservation of Atlantic Tunas (ICCAT), which specializes on large pelagic fish. In addition, diplomatic conferences on Mediterranean fisheries were held in 1994 and 1996.

The GFCM was created in 1949 on the initiative of the FAO. The role of the GFCM is essentially advisory, its objectives being:

- to promote the development, conservation and management of living marine resources;
- to formulate and recommend conservation measures;
- to encourage training and cooperative projects.

Its area of responsibility is the Mediterranean and Black Seas. During the period 1999–2000, the GFCM created a Scientific Advisory Committee (SAC) and several subcommittees to coordinate fisheries research and assessment in the Mediterranean. The GFCM members are Albania, Algeria, Bulgaria, Croatia, Cyprus, Egypt, the European Community, France, Greece, Israel, Italy, Japan, Lebanon, Libya, Malta, Monaco, Morocco, Romania, Serbia and Montenegro, Slovenia, Spain, Syria, Tunisia and Turkey.

ICCAT was created in 1966 to ensure conservation of the resources of tuna and tuna-like fish of the Atlantic Ocean and adjacent seas. The Commission compiles fishery statistics from its members and from all entities fishing for such species in the Atlantic Ocean. In addition, it coordinates research, including stock assessment, on behalf of its members, develops scientifically based management advice, provides a mechanism for contracting parties to agree on management measures, and produces relevant publications. About 30 species are of direct concern to ICCAT; those specifically caught in the Mediterranean include bluefin tuna (*Thunnus thynnus thynnus*), albacore (*Thunnus alalunga*), swordfish (*Xiphias gladius*), dolphinfish (*Coryphaena hippurus*) and three species of shark, *Prionace glauca, Lamna nasus* and *Isurus oxyrhinchus*. Currently, ICCAT has 32 contracting parties, of which seven border the Mediterranean: Tunisia, Libya, Algeria, Morocco, Russia, Croatia and the EU (the four Mediterranean European countries, France, Greece, Italy and Spain, are a single contracting party).

Diplomatic Conferences on Fisheries Management in the Mediterranean were held in Iraklion, Crete, Greece, in 1994, and in Venice, Italy, in 1996. In Iraklion, a Solemn Declaration on the Conservation and Management of the Fishery Resources of the Mediterranean was made for the 15 Mediterranean countries and for the nine non-coastal states of the Mediterranean that have been linked traditionally to that sea. The Declaration pointed out for the first time the necessity to protect the fishery resources of the Mediterranean by ensuring regional cooperation covering the resources, the environment and the application of legal principles. The cooperation between different countries was agreed upon. At the second Diplomatic Conference in Venice, a new Solemn Declaration was made that declared new rules to protect fishery resources of the region (reduction of fishing effort, better statistics, more research), and new measures of cooperation through regional organizations were implemented.

LANDINGS

From an ecological and fisheries perspective, all the species of commercial importance caught in the Mediterranean can be assigned to four broad groups: small pelagics, medium-sized pelagics, large pelagics and demersals. The FAO's official catch statistics for 2001 assign 36% of the Mediterranean catch to the demersal group, 52% to small pelagics, 8% to medium-sized pelagics and 4% to large pelagics.

LARGE PELAGICS

The assessment and management problems of large pelagics are very different from those of the other groups. Their oceanic life and migratory behaviour means that all large pelagic species are shared. Bluefin tuna, swordfish and albacore are the most important of the large pelagic species in the Mediterranean. Although combined they constitute <4% of the total reported landings (Figure 2; Table 1), their economic importance is great. ICCAT considers bluefin tuna in the Eastern Atlantic and Mediterranean to constitute a single stock, the Mediterranean being its main spawning area. With an annual reported catch in 2000 and 2001 of some 24 000 t, and commanding a large price in the Japanese market, pressure on the resource is huge. Driftnets, longlines and seines are deployed to catch the species, often under flags of convenience, and vessels targeting it constitute the only purely industrial fleet in the Mediterranean. Swordfish are the second most important large pelagic species in terms of catch (about 15 000 t in 2000 and 2001). ICCAT considers there to be a single swordfish stock in the Mediterranean, fished by longline and driftnet. Albacore is the third most important species by catch weight (some 5 500 t in 2000 and 2001).

Driftnets for tuna and other large pelagic species are a major ecological problem in the Mediterranean because of their significant incidental catch of mammals, turtles and birds. The GFCM-recommended maximum length of driftnets in the Mediterranean is 2.5 km, but the regulation has never been strictly enforced; most nets currently deployed are therefore generally much longer. As from 1 January 2002, driftnetting by the four European Mediterranean countries was banned, but the gear is far from being extinct yet.

The minimum legal size established for swordfish, 120 cm according to EC regulation No. 1626/94, was overturned by EC regulation No. 973/2001. This last regulation set the minimum legal size for swordfish at 25 kg or 125 cm (lower jaw length), but only for the Atlantic, implicitly abolishing a minimum size for the Mediterranean, leaving the species with no size limit there, a decision that seems contrary to the precautionary principle.

Tuna farming (or fattening, or penning; the accepted description is still a matter of debate) is an activity that interferes with the assessment and management of bluefin tuna in the Mediterranean (Tudela, 2003). According to Anon. (2002) and the Glossary of the GFCM, tuna farming involves the collection of wild fish, both small and large, and rearing them in floating cages for periods of a few months to 1–2 years. Fish weight increments or a change in the fat content of the flesh are obtained through standard fish-farming practices. The confinement of captured tuna for periods of 2–6 months, aimed mostly at increasing the fat content of the flesh and strongly influencing the price of the product on the Japanese sashimi market, is what is generally described as “tuna fattening”.

However, notwithstanding the description above, two practices in the Mediterranean are respectively described as “tuna fattening” and “tuna farming”. The latter is currently carried out mainly in Croatia, the total weight of small tuna being increased substantially over periods of up to 2 years. The former, mainly carried out in Spain, Malta and Turkey (Oray & Karakulak, 2003), consists of larger tuna being caged for short periods to increase their fat content for the Japanese market. In the procedure, small and medium-sized tuna (10–200 kg) caught by purse-seine are put in cages 50 m in diameter and 20–30 m deep. There, they are fattened for some months before being exported to Japan for the sashimi market. This procedure started in 1996, and the export of such “farmed” bluefin tuna to Japan started in 1997. By 2001, more than 7 700 t of the bluefin tuna exported to Japan originated in cages, so some two-thirds of the Mediterranean bluefin tuna sent to Japan now come from that source.

Economic gains in bluefin tuna fattening have led the private sector to invest heavily in the process. Interest, as reflected by the increased number of fattening units established throughout the Mediterranean and new licence applications being submitted to the relevant national authorities, has risen dramatically. Between 1996 and 2001, the number of cages in the Mediterranean increased at least 20-fold. However, according to Anon. (2002), the main problems for the assessment and management of bluefin tuna generated by farming are (i) statistical, by making data collection difficult and reducing the accuracy of the data, (ii) biological, by diminishing the availability of biological samples, and (iii) administrative, because the activity makes it more difficult to assess and to control the fishery. Other potential problems could arise in the future: (iv) environmental, through the impact on wild marine populations of other species, mainly small pelagics, used as bait, and the alteration of local environments, (v) social and economic, generating conflicts among fishers, aquaculture and tourism, and (vi) managerial, by the probable increase in the market for small and medium-sized bluefin tuna.

OTHER SPECIES GROUPS

Most representatives of the other species groups, small pelagics, medium-sized pelagics and demersals, live on the shelf or slope. With the Mediterranean continental shelf being so narrow, most fishing grounds are close to the coast and most fishing for these species is by national fleets. Further, given the weakness of the GFCM until its creation of a SAC, few shared stocks have been identified and even fewer assessed, so none are currently managed on a shared basis. According to Christy (1997), a shared stock is defined as a stock of fish that migrates across the EEZ boundary of adjacent or opposite coastal states. However, as knowledge of fish migration patterns in the Mediterranean is limited, another definition of a shared stock is applied in the GFCM glossary: "a stock fished by two or more countries". Nowithstanding that description, stock definition and identification are areas that have not been developed.

A further shortcoming of management of these species is that, despite the generally extensive collaboration between scientists of the different Mediterranean countries, especially around the Adriatic Sea and the Gulf of Lions, no actual assessments were published until 1993. The first assessment carried out by collaborating scientists from different countries involved in the exploitation of a shared-stock species was for hake (*Merluccius merluccius*) in the Gulf of Lions (Aldebert *et al.*, 1993). In the SAC–GFCM meeting of 2002, up to 17 stocks of economic importance were recognized as shared by several countries, so assessment of these should become a priority (Table 2). Until now, however, only hake and anchovy (*Engraulis encrasicolus*) in the Gulf of Lions and anchovy and sardine (*Sardina pilchardus*) in the Adriatic Sea have been the subject of assessments presented to the SAC–GFCM.

Table 2. Mediterranean shared stocks identified by the SAC–GFCM in 2002.

Species	Area	Countries
Hake	Gulf of Lions	France, Spain
Hake	Adriatic	Albania, Croatia, Italy, Slovenia
Hake	Sicily Channel	Italy, Tunisia, Libya and Malta
Anchovy	Gulf of Lions	France, Spain
Anchovy	Adriatic Sea	Albania, Croatia, Italy, Slovenia
Sardine	Adriatic Sea	Albania, Croatia, Italy, Slovenia
Sprat	Adriatic Sea	Croatia, Italy, Slovenia
Red mullet	Adriatic Sea	Albania, Croatia, Italy, Slovenia
Blue whiting	Adriatic Sea	Albania, Croatia, Italy
Norway lobster	Adriatic Sea	Albania, Croatia, Italy, Slovenia
Bluefin tuna	Whole Mediterranean	All countries
Swordfish	Whole Mediterranean	All countries
Albacore	Whole Mediterranean	All countries
Prionace glauca	Whole Mediterranean	All countries
Lamna nasus	Whole Mediterranean	All countries
Isurus oxyrhinchus	Whole Mediterranean	All countries
Dolphinfish	Western Mediterranean	Italy, Malta, Spain, Tunisia

Small pelagics dominate Mediterranean catches (Figure 2; Table 1). Anchovy, sardine, sardinella and sprat (*Sprattus sprattus*) constitute >50% of the total landings by weight. Adding horse mackerels (*Trachurus* spp.), bogue (*Boops boops*) and bonito (*Sarda sarda*) raises the proportion to 59%. The collapse of small pelagics in the Black Sea in the late 1980s and early 1990s as a consequence of the introduction of the ctenophore *Mnemiopsis leidyi* can be clearly seen in the catch time-series. The gear used to catch small pelagics is mainly semi-industrial: purse-seine (with or without light) and pelagic trawl. The EU allows the use of pelagic trawls with a mesh size of 20 mm, but some countries (e.g. Spain) forbid the use of such gear.

More than 100 demersal species are exploited, but none contributes more than 3% to the total catch (Table 1). The artisanal fleet (with its many different gears) and bottom trawlers exploit demersal species, and catches are generally multispecific.

ASSESSMENT

The Mediterranean Sea as an area has a long-established tradition for oceanographic research. In a purely biological context, zoological endeavour dominates. Historically, the research has not been focused according to economic demand, so the species studied best were not necessarily those of most commercial importance (Farrugio *et al.*, 1993). Moreover, quantitative studies at a population level were almost totally neglected. Therefore, the passage from the realm of academic marine biology to applied fisheries research in the Mediterranean is comparatively recent (Farrugio *et al.*, 1993). The first attempts to apply mathematics in analysing fish population dynamics in the Mediterranean were based on the evolution of yields, the demographic composition of catches and the application of global production models (under equilibrium assumptions) during the 1970s (Oliver, 1983). There are two main problems in assessing Mediterranean stocks:

1. There is a general lack of shared stocks. With the exception of large pelagics, which are oceanic and fished in international waters by a number of fleets from different countries, not all Mediterranean, most other stocks, be they small or medium-sized pelagics or demersals, are fished by a single country, so there is little international pressure to assess them.
2. Appropriate datasets are generally unavailable. Although a large quantity of fisheries data around the Mediterranean can be found, they are not the consequence of homogeneous and long-term planned monitoring, but have been gathered by local initiatives or within the framework of ongoing short-term research projects, for example. Hence, data are highly heterogeneous with respect to reliability and time-series length.

Both these shortcomings are now being addressed as a result of the actions of the GFCM, in terms of all countries fishing in the Mediterranean, and the European Union, in respect of data collection by the four European Mediterranean countries. Nevertheless, as a consequence of such shortcomings, it is difficult to achieve the data needs of standard assessment methods such as Virtual Population Analysis (VPA). Therefore, few sound assessments that fulfil the standards required of international fisheries commissions have been presented as yet. Despite the paucity of data, Mediterranean scientists have tried to assess the fisheries with several tools, using available data. Methods developed for assessment and parameter estimation in tropical fisheries have been applied (Le Fur, 1998), not only because of the scarcity of suitable data, but also because of the complexity of the multispecies artisanal fisheries.

Another element contributing to the uncertainty of fisheries assessments is that, in most Mediterranean fisheries, catches are mainly recruits. As understanding of the recruitment process is more limited than than that of the adult life of fish species generally, the assessments are therefore also more uncertain. Studies on recruitment processes are scant (or even lacking) in the Mediterranean.

METHODS

A range of standard stock assessment procedures has been applied to Mediterranean stocks. They are listed below.

1 Indirect. These methods are based on fisheries data such as catch, effort and age structure of the catch and are usually known as population dynamics methods.

1.1 Analytical, or age-structured models.

1.1.1 VPA. This, one of the most widely applied methods in stock assessment, is not extensively used on Mediterranean stocks because data series tend to be too short. However, it has been applied to assess hake (Oliver, 1993; Aldebert & Recasens, 1996).

1.1.2 Length Cohort Analysis (LCA). This is a simplification of VPA, working on pseudocohorts and assuming steady state, so results are not as reliable as those of VPA. Nevertheless, it has often been applied to Mediterranean stocks and most of the assessments presented to the SAC have been obtained on the basis of this procedure.

1.1.3 Yield per recruit (Y/R). This method gives a general overview of the status of a fishery, and it has been used regularly in the Mediterranean together with LCA.

1.2 Global or production model. This method, based on analysis of catch per unit effort (cpue), requires a long series of reliable data to be applied effectively. That shortcoming, as well as the difficulties associated with allocating effort to the different species in a multispecies fishery, renders its use for Mediterranean fisheries tenuous.

2 Direct methods, or research surveys. These fishery-independent methods are generally used to avoid biases in commercial catch data. Traditionally, they are used to estimate abundance and demographic structure at sea, and to collect other biological information.

2.1 Bottom trawl surveys. These are used to evaluate the status of demersal fish and have been used in the Mediterranean for some time, particularly in Italy (Levi *et al.*, 1993, Ardizzone & Corsi, 1997). Since 1994, a large, EU-funded project named MEDITS (MEDiterranean International Trawl Survey) has been carried out by the four Mediterranean European countries (Bertrand & Relini, 1998; Bertrand *et al.*, 1998, Abella *et al.*, 1999) and recently has been extended to other countries (Croatia, Malta).

2.2 Acoustic surveys. Systematic acoustic surveys, which are appropriate for small pelagic fish in particular, have been carried out in some areas of the Mediterranean for many years (Abad *et al.*, 1996; Guennegan *et al.*, 2000; Patti *et al.*, 2000). The results seem to be consistent with those derived from other indirect assessment methods (e.g. DEPM; Pertierra & Lleonart, 1996).

2.3 Daily Egg Production Method (DEPM). Assessments based on this procedure, which again is suitable for small pelagics, have been carried out non-systematically in several areas of the Mediterranean. The results compare well with other assessment methods, so the method clearly has value in the Mediterranean (Chavance, 1980; Regner, 1990; Palomera & Pertierra, 1993; Garcia & Palomera, 1996; Somarakis & Tsimenides, 1997; Casavola *et al.*, 1998; Casavola, 1999; Quintanilla *et al.*, 2000), but it is expensive in manpower.

3 Other approaches. Regression analysis, generalized linear models (GLM) and time-series analysis have been used to analyse some fisheries, in particular series of catch and cpue (Stergiou & Christou, 1996; Stergiou *et al.*, 1997b; Daskalov, 1998, 1999; Goñi *et al.*, 1999; Lloret *et al.*, 2000, 2001).

STATE OF THE RESOURCES

Since the establishment of a Subcommittee on Stock Assessment (SCSA) of the SAC of the GFCM in 2000, several assessments have been accepted and some advice provided to the Commission. The task of the 2000 meeting was to gather the assessments published in the previous 10 years, but at subsequent meetings only new assessments specifically prepared for a specific meeting were presented, according to a specified format. Table 3 lists the assessments of target species presented to SCSA meetings to date. Assessments of some Black Sea resources have been published in other media (GFCM, 1994; Prodanov *et al.*, 1997; Öztürk & Karakulak, 2003).

Table 3. Assessments presented to SAC–GFCM from 2001 to 2003.

Species	2001	2002	2003	Comments
Merluccius merluccius	4	2	3	General growth-overfishing
Mullus barbatus	3	1	2	Overfished or fully fished
Pagellus erythrinus	3			Fully exploited
Diplodus annularis	3			Fully exploited
Nephrops norvegicus	1	1		Fully exploited
Aristeus antennatus	2	3	3	Overfished or fully fished
Aristeomorpha foliaciea		1		Fully fished
Parapenaeus longirostris			1	Underexploited
Sardina pilchardus	3	5	4	Overfished or fully fished
Engraulis encrasicolus	3	6	3	Risk of recruitment-overfishing
Trachurus trachurus	1		1	Fully exploited
Boops boops	1			Fully exploited
Total	24	19	17	

Hake stocks seem to be overexploited in all the assessments presented, but the status of other demersals is not so clear. Further, some anchovy stocks appear to be at risk of recruitment-overfishing. Nevertheless, it has been argued that, overall, the Mediterranean fisheries have reached a level of "sustainable overfishing", and given the available data, it is difficult to disagree. However, with technological improvements in capture methods and the continuous decline in the cpue of some demersal stocks, proponents of the sustainable overfishing theory seem to be losing support, for some species at least. Moreover, the recent introduction of new fishing practices (modern longliners operating in the northwestern Mediterranean) has eliminated the "spawning refugia" for species such as hake. As a consequence of that change in Mediterranean fisheries strategy, the pattern of overfishing is also changing.

Traditionally, two types of overexploitation have been described, according to their impact on the fishery system: growth-overfishing and recruitment-overfishing (Cushing, 1996). Recruitment-overfishing may be affecting the short-lived anchovy stock in the northwestern Mediterranean, because the pattern of exploitation has changed in recent years. The fishing season is now practically a full calendar year and a significant proportion of the population is caught before attaining maturity. In contrast, some demersal stocks have traditionally been subject to growth-overfishing, because effort was mainly directed at juveniles. On top of this, species such as hake may have started to suffer locally from recruitment-overfishing, as the parent stock became available to exploitation through longlining; at the same time, the level of effort directed at juveniles has increased. Therefore the risk of a Mediterranean hake stock collapse is increasing, and such a collapse could lead to crucial changes to the ecosystem in which hake is a major player. With managers and decision-makers nowadays regularly invoking the terms "ecosystem management" and management according to the "precautionary principle", such an outcome for hake can only be described as a potential disaster.

MANAGEMENT

Management of Mediterranean fisheries is based on effort control and technical measures such as minimum landing and mesh sizes, which are implemented but not always enforced. In most cases, size limits are lower in the Mediterranean than in the Atlantic. A Total Allowable Catch is applied only to bluefin tuna.

Most of the rules for managing demersal fisheries have been developed for the trawl fishery, not only because it is the gear that contributes most to demersal catches, but also because the selectivity pattern of trawlers (which catch large numbers of relatively small fish) compares poorly with that of most artisanal gears (e.g. nets and lines). However, relatively few Mediterranean countries have taken any real action to limit increases in their fishing effort, despite repeated recommendations coming out of the GFCM (Caddy, 1993). In reality, when a control rule to limit any increase in fishing effort has been implemented (e.g. limiting the number of boats operating or the time they spend at sea), any increase in fishing mortality attributable to technological progress or engine power increase has been overlooked. This means that, although effort limitation may have been enforced effectively, pressure on the stocks in the form of fishing mortality has actually increased as a consequence of technological improvement and increased engine horsepower. Some Mediterranean countries have closed areas to trawling, usually the three first nautical miles seawards from the coast or to a depth of 50 m. Overall, though, the low level of enforcement of legislation remains a major obstacle to effective fisheries management in the Mediterranean.

Almost all management pilot projects carried out in the Mediterranean Sea target trawlers. Three such projects seem to have achieved some success and the target stocks have seemingly recovered (in biomass and length structure): Spain's *Plan Castelló* (Suau, 1979; Lostado *et al.*, 1999), Cyprus fisheries management (Garcia & Demetropoulos, 1986), and the Sicilian Golfo de Castellammare management project (Pipitone *et al.*, 1996). All these control projects were in national waters, so the conclusion seems to be that implementation of a shared-stock management pilot project could be a very useful tool indeed.

In international waters, i.e. usually >12 nautical miles offshore, few rules apply. Therefore, the various industrial fleets (from both Mediterranean and non-Mediterranean countries, plus those under flags of convenience) are almost uncontrolled in terms of effort and gear. As an exception to this principle of free access, Spain extended its fisheries mandate in the Mediterranean in 1997 to an area limited by the equidistant line between countries (excluding the Alboran Sea) in order to be legally able to enforce international rules on fishing for tuna. Those waters can be described as the first truly European waters in the Mediterranean Sea, because EC provisions are now applicable there. However, national fisheries management in the Mediterranean is performed by the three organizations described below.

EUROPEAN UNION MANAGEMENT

Until now, the main resource management regulation issued by the EU for the Mediterranean is EC Regulation No. 1626/94. It covers species and environments that are endangered, characteristics of gear, minimum mesh sizes (trawl 40 mm, pelagic trawl 20 mm, purse-seine 14 mm) and minimum legal size of 17 species of fish, two species and a family of Crustacea and three species of bivalve (Table 4). However, the mesh and minimum landing sizes still allow immature fish of some species to be caught and landed.

Table 4. Minimum landing sizes for the Mediterranean established by EC Regulation No. 1626/94, and some comparative values for the Atlantic.

Species	Mediterranean	Atlantic	Unit	
Dicentrarchus labrax	23	36	cm	
Diplodus spp.	15		cm	
Engraulis encrasicolus	9	12	cm	
Epinephelus spp.	45		cm	
Lophius spp.	30	?	cm	
Merluccius merluccius	20	27	cm	
Mugil spp.	16		cm	
Mullus spp.	11	15	cm	
Pagellus spp.	12		cm	
Pagrus pagrus	18	15	cm	
Polyprion americanus	45		cm	
Scomber scombrus	18	20	cm	
Solea vulgaris	20	24	cm	
Sparus aurata	20	19	cm	
Thunnus thynnus thynnus	70	70	cm	or 6.4 kg
Trachurus spp.	12	15	cm	
Xiphias gladius	120 *	125	cm	
Hommarus gammarus	240		mm	
Nephrops norvegicus	70	70	mm	
Palinuridae	240		mm	
Pecten spp.	100		mm	
Venerupis spp.	25		mm	
Venus spp.	25		mm	

* *Overturned by subsequent regulation (973/2001 of 14 May 2001)*

NATIONAL MANAGEMENT

National management generally consists of technical and economic measures, such as limiting vessel power and tonnage, limiting the number of boats or licences, limiting the daily time at sea, declaring closed areas and, occasionally, implementing closed seasons.

However, governments still support the fishing sector through subsidies for modernization, infrastructure and fuel (often using EU funds). In general, economic measures are more effective than technical ones in managing Mediterranean fisheries.

FISHER ORGANIZATION MANAGEMENT

Some fisher organizations contribute to the regulation of local fisheries through gentlemen's agreements. In some cases and during a certain period, the associations have been able to implement a system of "self-regulation", based on specific rules to be followed by the whole fishing community, a behaviour that has been studied by some social scientists. However, they still have to follow government directives, which have higher priority than their own rules. A case study in progress is the clam fishery of the Adriatic Sea, for which consortia have been created legally (with agreement of the producers) to manage exploitation and to attempt seeding experiments. Quotas have been fixed annually on the basis of dredge surveys, and research input will continue to form the basis of management decisions by the consortia.

Community-based management is another interesting approach to Mediterranean fisheries management. It involves giving some authority to fishers concerning the regulations and in protecting both resource and local fishing activities. The approach implies a debate on global versus regional management.

Mediterranean fisheries are diverse. There are many cultures with different tastes for sea products. The small (artisanal) fishing is of paramount importance, covering many different fishing methods, some sharing areas and landing at many sites along the coast. It is difficult to find targets applicable to all areas and gears in the Mediterranean. Nevertheless, the market is global and unique. It is problematic when products obtained according to different rules (but often from the same unit fish stock) compete in the same market. Hence, two approaches, with advantages and disadvantages (Table 5), should be considered:

- a uniform management policy for the whole Mediterranean Sea;
- regional management – different areas apply different rules according to their requirements.

Table 5. Advantages and disadvantages of uniform and regional management policies for the Mediterranean Sea.

Management policy	Advantages	Disadvantages
Uniform	Management is simple and easy to control.	Regression in the management of the most advanced areas (e.g. the introduction of pelagic trawling into areas where it is banned)
Regional	Management allows harmonious development of regions according to their biological, economic and social potential. Management facilitates the implementation of pilot areas and projects.	Border effects (neighbours of a well-managed zone take advantage of success without any cost to themselves)

Rules may or may not be applicable at a regional level, any anomalies being based on the management units and the market. Rules affecting only the area where they are applied can be regional. For instance, different areas can be regulated by different effort limits (e.g. engine power, time fishing, number of boats) according to their own stocks, fleets, tradition or socio-economic requirements. Rules that cannot be applied in this way are those that affect shared stocks or markets. Minimum legal sizes of fish cannot then be applied regionally because the fish enter the same market. The minimum legal size of European hake is 20 cm in the European Mediterranean (EC regulation No. 1626/94) and 26 cm in the Atlantic, but the fish are sold into the same market. (Therefore, a hake 22 cm long found on a market is illegal if it comes from the Atlantic, but legal if it comes from the Mediterranean!) The case of anchovy is similar, the legal minimum size for the European Mediterranean being 9 cm and for the Atlantic 12 cm.

Clearly, effective fisheries management dictates that such anomalies be dealt with effectively around a table.

REFERENCES

Abad, R, Miquel, J. & Iglesias, M. (1996) Campañas de evaluacion por metodos acusticos de sardina, boqueron y ochavo en el Mediterraneo Occidental. *FAO Fisheries Report*, **537**: 191–193.

Abella, A., Belluscio, A., Bertrand, J., Carbonara, P. L., Giordano, D., Sbrana, M. & Zamboni, A. (1999) Use of MEDITS trawl survey data and commercial fleet information for the assessment of some Mediterranean demersal resources. *Aquatic Living Resources*, **12**: 155–166.

Aldebert, Y. & Recasens, L. (1996) Comparison of methods for stock assessment of European hake *Merluccius merluccius* in the Gulf of Lions (Northwestern Mediterranean). *Aquatic Living Resources*, **9**: 13–22.

Aldebert, Y., Recasens, L. & Lleonart, J. (1993) Analysis of gear interactions in a hake fishery: the case of the Gulf of Lions (NW Mediterranean). In: *Northwestern Mediterranean Fisheries*. (Ed. by J. Lleonart). *Scientia Marina*, **57**: 207–217.

Anon. (2001) Towards holistic fisheries management: a Mediterranean perspective. Report of a workshop held in Heraklion, Crete, March 2001, under the auspices of the European Union. Accompanying Measures Programme, Contract Q5AM–2000–00002. 22 pp.

Anon. (2002) Report of the sixth GFCM–ICCAT meeting on stocks of large pelagic fishes in the Mediterranean, Sliema, Malta, April 2002. 36 pp.

Ardizzone, G. D. & Corsi, F. (Eds) (1997) Atlas of Italian demersal fishery resources. Trawl surveys 1985–1987. *Biologia Marina Mediterranea*, **4**(2): 568 pp.

Bas. C. (2002). *El Mar Mediterráneo: Recursos Vivos y Explotación*. Ariel Ciencia, Barcelona. 518 pp.

Bertrand, J., Gil de Sola, L., Papaconstantinou, C., Relini, G. & Souplet, A. (1998) An international bottom trawl survey in the Mediterranean: the MEDITS programme. In: *Demersal Resources in the Mediterranean*. (Coordinated by J. A. Bertrand and G. Relini). *Actes de Colloques IFREMER*, **26**: 76–93.

Bertrand, J. & Relini, G. (Coordinators) (1998) Demersal Resources in the Mediterranean. *Actes de Colloques IFREMER*, **26**.

Beverton, R. J. H. & Holt, S. J. (1957) On the dynamics of exploited fish populations. *Fisheries Investigation Series 2*, **19**. Her Majesty's Stationery Office, London.

Caddy, J. F. (1990) Options for the regulation of Mediterranean demersal fisheries. *Natural Resource Modeling*, **4**: 427–475.

Caddy, J. F. (1993) Some future perspectives for assessment and management of Mediterranean fisheries. In: *Northwestern Mediterranean Fisheries*. (Ed. by J. Lleonart). *Scientia Marina*, **57**: 121–130.

Casavola N. (1999) Valutazione della biomassa di alici mediante la stima della produzione giornaliera di uova lungo le coste adriatiche pugliesi nel 1995. *Biologia Marina Mediterranea*, **6**(1): 553–555.

Casavola, N., De Ruggieri, P., Rizzi, E. & Lo Caputo, S. (1998) Daily egg production method for spawning biomass estimates of sardine in the south-western Adriatic Sea. *Rapports et Procès-verbaux Commission Internationale Exploration Scientifique Mer Méditeranée*, **35**(2): p. 396.

Christy, F. T. (1997) The development and management of marine fisheries in Latin America and the Caribbean. Policy Research Paper, Inter-American Development Bank, Washington, D.C., **ENV–110**.

Chavance, P. (1980) Production des aires de ponte, survie larvaire et biomasse adulte de la sardine et de l'anchois dans l'est du Golfe du Lion (Méditerranée occidentale). *Téthys*, **9**: 399–413.

Cushing, D. H. (1996) Towards a science of recruitment in fish populations. *Excellence in Ecology*, **7**: 175 pp.

Daskalov, G. (1998) Pêcheries et changement environnemental à long terme en Mer Noire. Thèse de doctorat, Université de la Méditerranée (Aix–Marseille II).

Daskalov, G. (1999) Relating fish recruitment to stock biomass and physical environment in the Black Sea using generalized additive models. *Fisheries Research*, **41**: 1–23.

Estrada, M. (1996) Primary production in the northwestern Mediterranean. In: *The European Anchovy and its Environment*. (Ed. by I. Palomera and P. Rubiés). *Scientia Marina*, **60**(Suppl. 2): 55–64.

FAO Fishery Information Data and Statistics Unit. (2003) Capture production 1950–2001. Available in FISHSTAT Plus (Universal software for fishery statistical time series, Version 2.30) at http://www.fao.org/fi/statist/fisoft/fishplus.asp and CD-rom of March 2003. FAO, Rome.

Farrugio, H. (1996) Mediterranean fisheries status and management. Evolution of the research and improvement of regional cooperation. Diplomatic Conference on Fisheries Management in the Mediterranean, Venice, November 1996. 18 pp.

Farrugio, H., Oliver, P. & Biagi, F. (1993) An overview of the history, knowledge, recent and future research trends in Mediterranean fisheries. In: *Northwestern Mediterranean Fisheries*. (Ed. by J. Lleonart). *Scientia Marina*, **57**: 105–119.

Farrugio, H. & Papaconstantinou, C. (1998) The status of fisheries resources in the Mediterranean. In: *Workshop on Gaps in Fishery Science*. (Moderated by R. Robles and Ph. Ferlin). *CIESM Workshop Series*, **5**: 13–24.

Garcia, A. & Palomera, I. (1996) Anchovy early life history and its relation to its surrounding environment in the western Mediterranean basin. In: *The European Anchovy and its Environment*. (Ed. by I. Palomera and P. Rubiés). *Scientia Marina*, **60**(Suppl. 2): 155–166.

Garcia, S. & Demetropoulos, P. (1986) Management of Cyprus fisheries. *FAO Fisheries Technical Paper*, **250**. 43 pp.

GFCM. (1994) Report of the Second Technical Consultation on Stock Assessment in the Black Sea, Ankara, Turkey, February 1993. *FAO Fisheries Report*, **495**. 199 pp.

Goñi, R., Álvarez, F. & Adlerstein, S. (1999) Application of generalized linear modelling to catch rate analysis of western Mediterranean fisheries: the Castellón trawl fleet as a case study. *Fisheries Research*, **42**: 291–302.

Guennegan, Y., Liorzou, B. & Bigot, J. L. (2000) Exploitation des petits pelagiques dans le Golf du Lion et suivi de l'evolution des stocks par echo-integration de 1999 a 2000. Paper presented at the Working Group on Small Pelagics, Fuengirola, Spain, March 2000: 27 pp.

Le Fur, J. (1998) Adaptabilité potentielle de modèles de pêche artisanale tropicale (complexe) aux exploitations halieutiques méditerranées. In: *Workshop on Gaps in Fishery Science.* (Moderated by R. Robles and Ph. Ferlin). *CIESM Workshop Series*, **5**: 31–39.

Levi, D., Andreoli, M. G. & Giusto, G. B. (1993) An analysis based on trawl-survey data of the state of the "Italian" stock of *Mullus barbatus* in the Sicilian Channel, including management advice. *Fisheries Research*, **17**: 333–341.

Lleonart, J. & Maynou, F. (2003) Fish stock assessments in the Mediterranean: state of the art. *Scientia Marina*, **67**(Suppl. 1): 37–49.

Lloret, J., Lleonart, J. & Solé, I. (2000) Time series modelling of landings in northwest Mediterranean Sea. *ICES Journal of Marine Science*, **57**: 171–184.

Lloret, J., Lleonart, J., Solé, I. & Fromentin, J-M. (2001) Fluctuations of landings and environmental conditions in Northwest Mediterranean Sea. *Fisheries Oceanography*, **10**: 33–50.

Lostado, R., del Rio Orduña, V. & Vivas, D. (1999) The Castellón trawling project (1961–1966): teachings for a sustainable fisheries management. *Informes y Estudios COPEMED*, **2**.

Margalef, R. (1985) Introduction to the Mediterranean. In: *Western Mediterranean*, pp. 1–16. (Ed. by R. Margalef). Key Environments Series, Pergamon Press. 363 pp.

Oliver, P. (1983) Los recursos pesqueros del Mediterráneo. Primera parte: Mediterráneo occidental. *Studies and Reviews of the General Fisheries Commission for the Mediterranean*, **59**. 139 pp.

Oliver, P. (1993) Analysis of fluctuations observed in the trawl fleet landings of the Balearic Islands. In: *Northwestern Mediterranean Fisheries.* (Ed. by J. Lleonart). *Scientia Marina*, **57**: 219–227.

Oray, I. & Karakulak, F. S. (Eds) (2003) Workshop on Farming Management and Conservation of Bluefin Tuna. (5–7 April, 2003). Published by the Turkish Marine Research Foundation, Istanbul, Turkey. Publication Number **13**. 145 pp.

Öztürk, B. & Karakulak, F. S. (Eds) (2003) Workshop on Demersal Resources in the Black Sea and Azov Sea (15–17 April, 2003). Published by the Turkish Marine Research Foundation, Istanbul, Turkey. Publication Number **14**. 130 pp.

Palomera, I. & Pertierra, J. P. (1993) Anchovy biomass estimate by the daily egg production method in 1990 in the western Mediterranean. In: *Northwestern Mediterranean Fisheries.* (Ed. by J. Lleonart). *Scientia Marina*, **57**: 243–251.

Patti, B., Mazzola, S., Calise, L., Bonanno, A., Buscaino, G. & Cosimi, G. (2000) Echosurveys estimates and distribution of small pelagic fish concentrations in the Strait of Sicily during June 1998. Paper presented at the Working Group on Small Pelagics, Fuengirola, Spain, March 2000. 11 pp.

Pertierra, J. P. & Lleonart, J. (1996) NW Mediterranean anchovy fisheries. In: *The European Anchovy and its Environment.* (Ed. by I. Palomera and P. Rubiés). *Scientia Marina*, **60**(Suppl. 2): 257–267.

Pipitone, C., Badalamenti, F., D'Anna, G. & Patti, B. (1996) Divieto di pesca a strascico nel Golfo di Castellammare (Sicilia nord-occidentale): alcune considerazioni. *Biologia Marina Mediterranea*, **3**: 200–204.

Prodanov, K., Mikhailov, K., Daskalov, G., Maxim, C., Chashchin, A., Arkhipov, A., Shlyakhov, V. & Ozdamar, E. (1997) Environmental management of fish resources in the Black Sea and their rational exploitation. *Studies and Reviews of the General Fisheries Commission for the Mediterranean*, **68**. 178 pp.

Quintanilla L. F., Garcia, A., Giraldez, A. & Cuttitta, A. (2000) Daily egg production estimate of the spawning biomass of the Sicilian Channel anchovy during July 1998. Paper presented at the Working Group on Small Pelagics, Fuengirola, Spain, March 2000. 43 pp.

Regner, S. (1990) Stock assessment of the Adriatic sardine and anchovy using egg surveys. Atti di seminario "Reproductive biology of small pelagics and stock assessment through ichthyoplanktonic methods. *ICRAP Quaderno Pesca*, **4**: 17–31.

Ronzitti, N. (1999) Le zone di pesca nel Mediterraneo e la tutela degli interessi italiani. *Supplemento alla Rivista Marittima*, **6**. 96 pp.

Somarakis, S. & Tsimenides, N. (1997) A daily egg production method biomass estimate of the northern Aegean Sea anchovy stock. *Oceanography*, **2**: 133–148.

Stergiou, K. I. & Christou, E. (1996) Modelling and forecasting annual fisheries catches: comparison of regression, univariate and multivariate time series methods. *Fisheries Research*, **25**: 103–138.

Stergiou, K. I., Christou, E., Georgopoulos, D., Zenetos, A. & Souvermezoglou, C. (1997a) The Hellenic seas: physics, chemistry, biology and fisheries. *Oceanography and Marine Biology. An Annual Review*, **35**: 415–538.

Stergiou, K. I., Christou, E. & Petrakis, G. (1997b) Modelling and forecasting monthly fisheries catches: comparison of regression, univariate and multivariate time series methods. *Fisheries Research*, **29**: 55–95.

Suau, P. (1979) Un ejemplo de regulación en pesquerías. *Investigación Pesqueras, Barcelona*, **43**: 21–29.

Tudela, S. (2003) Tuna farming in the Mediterranean: the "coup de grace" to a dwindling population. In: *Workshop on Farming Management and Conservation of Bluefin Tuna*. (Ed. by I. Oray and F. S. Karakulak). Published by the Turkish Marine Research Foundation, Istanbul, Turkey. Publication Number **13**: 53–67.

The experience of Antarctic whaling

Sidney Holt*
4 Upper House Farm, Crickhowell, NP8 1BZ, Powys, UK

ABSTRACT: From 1929 the few nations engaged in commercial whaling and associated trade sought to limit production and to agree on national and company shares. They had three contrasting objectives: to conserve stocks, to control prices of commodities and to impede participation by other nations. After an introduction to scientific matters, three phases of the negotiations are mentioned: the 1930s, under various international agreements; the three decades after the International Whaling Commission (IWC) began to operate in 1949; and the years after 1979, when only two countries – Japan and the USSR – participated in Antarctic pelagic whaling, until the USSR ceased operations in 1987. Levels of scientific input differed greatly between the three phases, and will differ again in a fourth phase if commercial whaling – in principle suspended since the "moratorium" decision of 1982 – was again to be legitimized, with catch limits for baleen whales being set on the basis of the IWC's *Revised Management Procedure* (RMP), embedded in a *Revised Management System* (RMS). Although an International Observer Scheme was negotiated by the IWC in the 1960s, and is being revived now, shares of catches must be agreed by separate negotiation outside the IWC, which is constitutionally prohibited from making national or fleet allocations. Thus, there is a complex interaction between the processes of settling total exploitation levels and agreeing on shares. Antarctic waters remain designated as High Seas, so national jurisdictions over sea areas have not played a significant part in these negotiations, but national jurisdictions over flags, sub-Antarctic islands and operatives and crews have been important. Two special considerations now arise, especially from the longevity of whales and their charisma. One concerns the economic and cultural values increasingly being attached to live whales as well as to dead ones. The other is the matter of sharing resource use between present and future human generations. Brief consideration is given to other aspects of management than by setting catch limits, such as by establishing protected areas, and the implications of environmental change. An Appendix reviews recent examination of the roots of population models that have been applied to the management of fisheries, including whaling.

INTRODUCTION

This paper is largely a synthesis of three published studies (Holt, 1998, 2000, 2002b). To these I have added some further ideas and new information.

---

* *The author is currently scientific adviser to the Third Millenium Foundation, Paciano, Italy.*

THE SCIENCE

In a recent paper, the Canadian fisheries scientist Chris Corkett (2002) argued that there is a fundamental difference between predictions of the consequences of regulations such as setting catch limits by calculation from population models of the class that Ray Beverton and I investigated about 50 years ago, and the "dynamic pool" model associated mainly with Michael Graham and Benny Schaefer (although in fact that model was first devised and applied to Antarctic blue whales in 1933 by the Norwegian scientists Johan Hjort, G. Jarn and Per Ottestad – Hjort *et al.*, 1933), to which is added an economic function put forward by H. Scott Gordon in 1953. Corkett calls this latter the "falsifiable Gordon–Schaefer model"; in it costs are assumed linearly related to fishing effort, and catch values similarly to physical sustainable yield, from which simplistic assumptions net economic benefits may be calculated. Corkett did not notice that Beverton and I did exactly the same.

Further, as Colin Clarke showed a few years later, we were all wrong in omitting consideration of discount rates (e.g. Clark, 1976). In fact Clark showed how the financial interests of whalers might best be served by exterminating the resource within a suitable time-frame and investing the proceeds in some other profitable activity. This was very nearly what both the Norwegian and Scottish-based Antarctic whaling industries did, with the Norwegians building the city of Sandefjord and financing the shipbuilding and shipping industry, and Christian Salvesen of Leith, once the world's biggest whaling company, eventually mutating into one of Europe's biggest trucking companies, using capital accumulated from the profits of Antarctic whaling. That is "sustainable development"! David Lavigne (2002) has given other examples of such pseudo-sustainability.

Corkett sought to show in his essay that calculations and predictions that cannot in practice be invalidated by subsequent events are unscientific, in terms of Karl Popper's well known definition of "science". I regard this as a simplistic view of practical science-based management issues, but this is not the time and place to discuss such a philosophical matter in depth. However, Corbett's preoccupation with population and economic models does bring me to my subject here because the Scientific Committee of the International Whaling Commission (S.C. of IWC) has got very close to rejecting all such models in its efforts to devise a procedure for regulating commercial whaling. It has also rejected the idea of targeting a Maximum Sustainable Yield (MSY), despite that being now enshrined, in a qualified way, in the UN Convention on the Law of the Sea (UNCLOS).

Although the IWC began work in 1949, it was only in 1960 that it got around to agreeing that it should set catch limits so that whaling might be biologically sustainable. Ten years later the states of the major species–stocks in the Antarctic and the North Pacific were judged to be so bad that the United Nations called upon the IWC to enact a ten-year pause in all commercial whaling (a "moratorium"). Whaling countries rejected that idea but did enact, in 1974/75, a proposal by the Australian Government (then still engaged in whaling from shore stations) for what it originally called "a modified moratorium", but which became known as the New Management Procedure (NMP).

The essence of the NMP was the classification of each "management stock" of whales as a Protection Stock (PS) that had been "depleted" by whaling to less than about half of its original number and would have a zero catch limit; or as a Sustained Management Stock (SMS) judged to be between about half and four-fifths of its original number – for which the catch limit would be set at 90% of an estimate of MSY; or as an Initial Management Stock (IMS) which had apparently been little affected by whaling, and for which catch limits would be set rather more generously. Later, an additional criterion was introduced: whaling must not commence on a hitherto unexploited IMS stock until a satisfactory estimate of stock size, in terms of number of recruited whales, had been obtained.

By 1978 it was already clear that this Procedure was unsatisfactory. The *coup-de-grace* was given to it by an Australian scientist, William de la Mare, who demonstrated in a seminal study that even if the population model used was a true representation of biological reality, and the estimates of parameters in it precise, the NMP was a recipe for further depletions of stocks (de la Mare, 1986). This failure led to renewed demands for a general complete pause in commercial whaling, of unlimited duration, which was eventually voted in 1982, to come into effect in 1986. Meanwhile the IWC had decided, in 1980, to prohibit all pelagic whaling using factory ships (including the factory–catchers developed and used in the North Atlantic by Norway, and later in the North Pacific by Japan) except for the catching of minke whales. This selective prohibition arose from the perception that pelagic whaling was much more difficult to regulate effectively than whaling from land-stations, but it was secured politically by countries that were still operating land-stations voting with the non-whaling countries against the few remaining pelagic operators. The IWC had also decided, in 1981, to prohibit all catching of sperm whales, by any means and for any purposes. This was because the NMP–MSY approach was evidently not appropriate for this sexually dimorphic species with a "harem" social structure. These three "moratoria" are still in effect, but in practice the general one is flouted by Norway and Japan, the only countries now engaged in commercial whaling; Norway can legally do this because it has an "objection" both to the moratorium and to the designation of the Northeast Atlantic minke whale stock as a Protection Stock under the NMP, while Japan has taken advantage – in a manner that most Member states of IWC regard as unethical – of an article in the ICRW that allows a Member to award itself as many permits as it likes for the taking of whales for scientific purposes. (Whaling under objections to IWC decisions, and under special scientific permits but with the products being marketed, are both equally "commercial", a term not otherwise specifically defined in the ICRW 1946.)

The 1982 decision for a pause was predicated on the need to go back to basics and to try to develop new rules for regulating any future whaling, as well as to allow time for recovery of the most depleted stocks. This gave the S.C. an unprecedented opportunity to pause in its efforts to calculate sustainable catch limits every year for every stock of whales, as the NMP required. After nearly ten years of effort it agreed on a new procedure, called the Revised Management Procedure (RMP) That "procedure" allows for calculation of sustainable and precautionary catch limits for baleen whales that are not dependent on any particular population model. The RMP has been accepted by the Commission but has not yet been implemented. The reason for this is that the Commission is at the same time trying to negotiate an International Observer Scheme and other arrangements to prevent future fraud, especially an international database of DNA profiles of legally caught whales, to be incorporated with the RMP in a Revised Management Scheme (RMS). In this connection it should be noted that recent investigations have revealed that in the past some whaling countries, especially the USSR, but also Japan with respect to its coastal whaling operations, were guilty of large-scale falsification of catch statistics provided to the IWC, and that this continued even after appointment in the 1960s and 1970s of IWC-accredited Observers (see e.g. Danilov–Danilyan & Yablokov, 1996; Tormosov *et al.*, 1998; Yablokov, 2000; Kasuya & Brownell, 2001).

The RMP was developed through a competitive process, involving at various times five groups of scientists. Intensive computer tests were made to determine which of several "candidate" procedures best fulfilled management criteria established by the IWC. There were three such criteria: first, there should be a very low probability of any stock ever being inadvertently reduced below some intermediate level, around half the original number; second, there should be, within the constraint of the first criterion, the highest possible cumulative catch over a long period (this was taken to be one century, this being determined principally by available computing power for testing the robustness of any candidate algorithm); third, the catch limit should not vary from one year to the next more than experience showed to be absolutely necessary. This last was, of course, purely for the convenience of the industry.

The method adopted, by scientific consensus, was to generate, by computer, "catch" and population size "data" using what had previously been accepted as a working population model for baleen whales. This was essentially the Schaefer–Graham dynamic pool model, but incorporating a non-linear relationship between stock size and the net reproductive rate. The parameter values were set so that MSY would theoretically be obtained when the stock was at about 60% of its initial value.

No restraints were placed on the structure of the algorithm that would eventually be used to calculate catch limits. As it happened, the algorithm finally accepted, because it had, overall, the best performance in terms of meeting all three management criteria, was that developed by Justin Cooke, a British scientist working in Germany (Cooke, 1995). Cooke's algorithm did include a function looking suspiciously like the starting population model, but this is not in fact critical; other candidates, that had almost as good performance (one by de la Mare, for example) did not incorporate such a function.

After the first round of testing, new data were generated from the original population model with drastically changed parameter values (such that, for instance, the MSY would occur at much higher or lower population values, and with a very wide range of hypothetical values for the intrinsic mortality and recruitment rates), and then using different types of model. There was in fact a free-for-all in attempting to devise models the "data" from which would break the algorithm. Questions were also raised about the consequences of large environmental changes – beneficial and adverse – during the 100-year simulation, and about the effects of falsified statistics, erroneous historical catch data, and biases in the sightings surveys on which periodic stock estimates would depend. Simulations of these factors led to relatively minor adjustments in the algorithms and the modified procedures passed all the tests of robustness.

Japanese scientists hypothesized in the 1980s concerning strong competitive interdependencies of baleen whale species feeding on the same resource – for example blue and minke whales feeding on Antarctic krill. There is still no evidence for this but the hypothesis seemed plausible, until one reckons that some fish, squid, birds and seals as well as other cetaceans also feed on krill. It may be modelled as a special case of the effects of environmental change, and also multispecies data-generation models were proposed. However, agreement was finally reached that such multispecific interactive modelling was unnecessary, essentially because the periodic surveys, with estimated statistical errors, would in due time pick up directly the effects of any significant interactions.

The following features of the RMP should especially be noted:

- Notwithstanding the aforementioned reference to MSY in UNCLOS, no effort is made to seek MSY or anything like it, essentially because this concept is now widely regarded by scientists as spurious. (This took me, as a participant in this work, back to my instructions from Director Michael Graham, when I started work at Lowestoft in 1947, that we were looking for regulations to correct or prevent over-ishing that would lead to *improvement* in the situation, not its *optimization*, as the US authorities tended to prefer.)

- The S.C. rejected the use of catch-based estimates of stock size (such as catch per unit effort, and marking experiments) in favour of direct visual survey. (There is no reason in principle why this should not be replaced eventually by other suitable direct methods, such as acoustic survey, especially if it were to be applied to sperm and other toothed, deep-diving cetaceans).
- Neither single- nor multispecies population models are used in the calculation of catch limits, so these do not depend at all on the validity of any particular model.
- The Procedure seeks to allow natural recovery of all stocks to relatively high levels, and sustainability, but the latter is defined in a flexible way.
- The Procedure is applied in such a way as to minimize the possibility that, given great uncertainties about the existence and distributions of separate subpopulations of a species within a region, a single catch limit will inadvertently be applied to two or more distinct populations. This is, of course, yet another precautionary measure, and a very important one.

LAW, POLITICS, DIPLOMACY

All the baleen whales, the sperm whale, and most of the other toothed whales (but not dolphins and porpoises) are by legal definition "shared stocks" because they are Highly Migratory Species under UNCLOS, the conservation and exploitation of which are to be managed through appropriate inter-governmental organizations. IWC was not specifically named as such in the text of UNCLOS (though that was certainly intended, as the records of the negotiations show), but it was so named later in the consensus document *Agenda 21* that emerged from the UN Conference on Environment and development, held in 1992 in Rio de Janeiro.

Agenda 21 is important especially because it formalized, in "soft law", the *precautionary principle* or a "precautionary approach" to management. This approach was embodied in the RMP by setting very low probabilities for inadvertent further depletions, and in other ways, as indicated above. In particular, all catch limits are zero *by default* until designated otherwise.

Marine mammals, including especially the cetaceans, are in UNCLOS set apart from fish, molluscs and crustaceans in specific ways. In particular, their exploitation is exempted from the requirement to seek some conditional MSY through authority being given to national authorities and the competent international organizations to take more conservative protective measures than would be mandated for other species such as fish and shellfish. This provision does not in principle exclude the possibility of complete, permanent protection. Some nations, and many non-governmental organizations, would like that to happen, for diverse reasons. However, insofar as the UNCLOS provisions are to be executed through the IWC, it has to be recognized that the IWC has no mandate to enforce any such ban, except in the unlikely event of a consensus of all Parties to the 1946 International Convention for the Regulation of Whaling (ICRW 1946) from which the IWC derives its authority. Majority votes for a ban would be formally objected to by those states that did not agree and they would therefore have no purpose other than the "flagging" of a majority view.

Whales differ from other marine species in another way, in terms of economic use. This is the worldwide growth of commercial whale watching within the burgeoning ecotourism industry. Hence, there are elements of "sharing" not envisaged in most fisheries, and questions arise such as: Are whale watching and whaling compatible? Do whale watching operations adversely affect the individual whales and their populations and therefore need regulating? Answers to these and related questions call for special kinds of research and possibly for regulatory practices similar to those increasingly used to regulate the various and often competing uses of maritime space – fishing, shipping, waste disposal, mining, etc.

In present circumstances the old problem of how permitted catches should be shared between countries, fleets and nationalities hardly arises while pelagic whaling is prohibited except for minke whales, and the Southern Ocean is closed to all commercial whaling, as a Sanctuary (in ICRW terms). Most whaling that may be envisaged for the future will be in the northern hemisphere and mainly in national Exclusive Economic Zones. Any (unlikely) sharing of permitted catches within each of those Zones with other nationals would be subject to negotiation with the coastal state, but subject to the overall limits. It should be noted, too, that the RMP for baleen whales on their feeding grounds is applied to so-called "small areas". These are largely arbitrary, set so as to avoid the possibility of more than one stock being subject to an overall catch limit, and may easily be defined with boundaries such that different catch limits are set within and outside distinct jurisdictions.

In my study for the FAO (Holt, 2002a) I described in detail the negotiations between whaling states through the 1930s to the mid-1980s, when Soviet pelagic whaling in the Antarctic ceased, leaving the Japanese fleet to operate there alone. The ICRW prohibits the IWC from setting national quotas, so these have always been negotiated directly by the countries and/or the companies involved; that is, when they were not in mutually destructive competition. There were four periods of such negotiation, and all were determined primarily by the nature of the commodity markets – for oil in earlier days, and for meat in later years. Before World War II, Norway and the UK were competing vigorously with each other but also cooperating in seeking to exclude other nations from Antarctic whaling[1]. In the IWC era of the 1960s the competition was mainly among five "pelagic whaling nations" – Norway, UK, Netherlands, Japan, USSR – but also with the secondary objective of excluding newcomers (which may or may not be members of the IWC). Baleen whale oil was the determining commodity.

The arguments about national quotas and the setting of IWC overall catch limits were inextricably entangled. The pretence was that the overall limits would first be set in accordance with scientific advice about stocks and then shared by negotiation. The reality was that the nations would not vote for overall catch limits such as would lead to one or more of them not securing a high enough catch to keep their operations profitable. The first loser in this was the Netherlands, having only one factory ship and so not being able to adjust its whaling effort very flexibly (all could, within operational limits, vary the number of catcher boats working with a factory ship, and the USSR especially did this extensively – but even that was complicated by the fact that the Russians were training a new generation of gunners, to replace the Norwegians who previously had a

[1] *Some efforts to impede other nations entering continued after World War II, while Japan, under US military occupation, was positively encouraged to resume whaling in the Antarctic as a means of countering food shortages, Germany, in part under UK occupation, was prevented from so doing. (Chambers of Industry and Commerce, 1947, Whaling Committee).*

quasi-monopoly of this skill). As time passed, and some countries were deciding to get out of the business, matters were complicated by sales of factory ships and catchers to remaining countries, and the touchy issue of whether such transferred vessels carried with them the negotiated share of the catch limit, whatever that might be. Eventually this was tolerated, and led in some cases to factories being purchased but then never employed for whaling; some were converted to other use, some broken up[2].

By the 1970s only Japan and USSR were left in the pelagic industry, and the primary commodity had become whale meat for human consumption. As the market for meat was almost entirely located in Japan, the negotiation of shares was eventually between just the two countries, with Japan dominating the negotiations; the USSR interest was in obtaining hard currency. At that time, however, there was still a mass market for whale meat in Japan, and it was relatively cheap, so the traders had an interest in keeping substantial supplies moving, so encouraging the Soviet operations. The situation evolved such that the overall catch limit was shared roughly equally (with a small allocation to Brazil, which was still engaged in coastal whaling for southern hemisphere minke whales, but serving only the Japanese market from a joint venture with Japan). Sharing was modified to a degree by the fact that the USSR was much more interested in industrial oil from sperm whales than was Japan

Subsequent sharing problems have little general interest, because they simply involved trade in whale catches, of the same or similar species but from different stocks, for example in the southern and northern hemispheres. There are at present severe restrictions on international trade in products from whales, under CITES regulations, but both Japan and Norway have reservations to these. Trade may soon open up again, but now, with relatively small catches in view, and very high prices in the Japanese market, but much lower ones in the limited Norwegian domestic market, the interests of the Japanese whaling industry and of Japanese meat traders are not in harmony; the traders would like to import from Norway (and perhaps eventually from Iceland and South Korea), but the whalers appear to be cool about increasing imports which could depress prices at the point of production. The trade issue has become complicated by the fact that there is a market for blubber in Japan, but not in Norway. Although this and other commodities as by-products are lower in value than the prime meat, their sale can make the difference between profit and loss[3].

The question of sharing the vast Antarctic whale resources will only arise in future if the Southern Ocean Sanctuary is abolished, as Japan will seek to ensure when its status is reviewed in 2004 by the IWC. Meanwhile, Japan has raised two new questions of some significance concerning future sharing. One is the claim that whales are consuming so much of the marine living resources of interest to humans that whales should be "culled" to prevent their further recovery and even to reduce populations. This of course

[2] *Recently (The Japan Times, 24 December 2002), the existence of a secret pact, signed in 1962, between Japan, Netherlands, Norway and UK was revealed. The pact called on the four countries not to seek, for three years, any changes in quotas amounting to 80% of the total (the USSR receiving the other 20% of the total baleen whale catch limit). This would apparently have nullified a 1960 commitment made within the IWC to reduce catches to sustainable levels in accordance with scientific advice, but became void when the UK pulled out of whaling in 1963 followed by the Netherlands in 1964.*

[3] *Japan, with Norway, failed, at the 2002 Conference of parties to CITES, to obtain the two-thirds majority vote needed to re-legitimise international trade in products from minke and Bryde's whales in the northern hemisphere. This now obstructs the intention of some elements among the authorities and industrialists of Iceland to renew commercial whaling there.*

represents a fundamental departure from the present policy of sustainability and recovery. It is a hypothesis that seems plausible to many lay observers, but is not in fact supported by scientific evidence[4]. It is also the precise converse of a provision in UNCLOS, which is replicated in the Convention for the Conservation of Antarctic Marine Living Resources (CCAMLR), which requires fisheries on prey species to be regulated in such a way that biological productivity of predatory species is not adversely affected.

The Government of Japan also now contends that the present moratorium on commercial whaling should be lifted in order to permit contribution of whale meat to relief of world hunger and to global food security generally. This seems to be an absurd proposition in the present state of most whale stocks. As the moratoria were declared partly to allow recovery of depleted stocks, and the time-scale of recovery is measured in terms of decades or perhaps even centuries, it would seem more reasonable to retain the moratorium for as long as may be needed, and leave it to future generations of humans to decide, in circumstances by then prevailing, whether resumption of whaling is called for. They will also be in a better position to confront the moral issue of whether to allow what many even now regard as an inhumane process and an ethically controversial industry to resume. However, the combination of the "global food security" argument, and the claim that whales are destroying fisheries, could together perhaps stampede nations into lifting the moratorium prematurely. Presumably that is why these arguments are being pushed simultaneously. It is, in this context, worth noting that the negotiators of the ICRW in 1946 (which did not include Japan, then under US occupation) were forward-looking in recognizing specifically, in the first paragraph of the Preamble to the new convention, "... the interest of the nations of the world in safeguarding *for future generations* the great natural resources represented by the whale stocks." (my emphasis).

There is a deeper question to which little professional consideration has been given so far. It is this: could commercial, *biologically* sustainable whaling, under precautionary catch limits set by a procedure like the RMP, be *economically* sustainable and, if so, under what circumstances? Certainly biologically *unsustainable* whaling has been immensely profitable. It seems likely that biologically sustainable levels of whaling on some recovered stocks could be economically sustainable in a trivial sense – that is profitable in that the value of commodities is greater than the cost of the whaling operation. (In fact, although conservation propaganda was made from the 1930s through to 1960, justifying efforts to place some restraints on Antarctic operation, in reality the motivation was to keep the price of whale oil high. This was eventually undone by a number of factors, one being the availability of substitutes for whale oil.) However, the question of which operations might be sufficiently profitable to allow financial resources to accumulate sufficiently for re-investment in new vessels and equipment and so on, making plausible assumptions about future costs and prices, has not been publicly examined in depth, although the problem has been recognized by, for example, Conrad & Bjørndal (1993) and Amundsen *et al.* (1995). Unfortunately, these authors used, in the biological part of their bio-economic models, the old and discredited "NMP" approach, as well as management objectives the IWC long ago rejected and, further, made rather dubious assumptions about the values of parameters in it. Furthermore, they considered the day-to-day costs of whaling operations and the price structure and gross income (including price elasticity), but not the costs of vessel and infrastructure

[4] *This controversy is the subject of a burgeoning literature. An introduction to it may be obtained through references in the Appendix.*

renewal over time. In the particular case studied it is also relevant that minke whaling provides only a part of the incomes of the crews and operators, and occupies only a part of the time of the vessels engaged; this makes any realistic analysis of the particular industry they chose for study very much more complicated. Finally, the overt and also covert subsidies to these industries from government have to be appropriately discounted if economic sustainability is to be claimed.

Presumably the executives of the bigger whaling industries, such as those exploiting the Antarctic resources, quietly make their own more realistic calculations!

To round off this section I return to the matter of models of population dynamics. One whale population, that of the northeast Pacific grey whale, has recovered under protection from near extermination in the 19th century. It has continued to increase in recent years despite substantial authorized catching allegedly for subsistence purpose, in the Russian Arctic. It has now reached numbers that appear to be far above (2–3 times) the numbers existing before commercial American whaling began, and that is despite the presumed reduction in carrying capacity attributable, among other things, to the reduction in the number and size of breeding lagoons in Mexico and California. The phenomenon has been described in the S.C. as "overshoot", and could perhaps be expected if there is a longer lag time between birth and first reproduction than present knowledge suggests. The IWC is still wrestling with this anomaly, and there is even now a suggestion that this phenomenon has appeared in the recovering humpback whale population of the western North Atlantic.

Concerning this phenomenon, Lars Witting (2001a), a Danish mathematical geneticist working in Greenland, has written:

"Assuming that pre-historic populations are in population dynamic equilibrium (which is expected for most density regulated models) this method *(back calculation using catch histories and current population estimates, as well as estimates or guesses of density-dependent net reproductive rates – SJH)* gives us an estimate of the equilibrium abundance. But when this method is applied to the Eastern North Pacific grey whale the estimated trajectories do not show the strong increase that has occurred in the population since the 1960s. The discrepancy between the data and the density-regulated model is not explained by moderate under-reporting of the commercial catch, nor by the inclusion of aboriginal catches in the catch history."

Before he began to be interested in the grey whale data, Witting (1997, 2000) had developed a class of new population models which, following the author, I here refer to as *models with inertial dynamics*. In these, on the basis of evolutionary considerations, the intrinsic rate of population growth (that is the notional rate in a vanishingly small population) is not Malthusian exponential (or, in the discrete case, geometric) but rather is *hyper-geometric*. The driver of these models is the selection pressure of density-dependent competitive interaction (i.e. intraspecifically, between the individuals in the population, which may or may not be in the same cohort). Such models exhibit intrinsic oscillatory (cyclic) behaviour, and have been applied successfully to data for animal species showing such behaviour.

Subsequently Witting (2001a, b) modified the version of the discrete logistic used by the IWC S.C., programmed as BALEEN II (Punt, 1999), to inertial dynamic form, and applied it to the grey whale data. His modification included the substitution of the Pella–Tomlinson stock–recruitment function, describing the density-dependence of the reproductive rate, by two alternative formulations (an exponential form advanced in 1954 by W. E. Ricker, and a power density function) which do not possess, as does the Pella–

Tomlinson function, the unacceptable feature of giving zero or (an impossible) negative reproduction when the population is at, or is larger than, an equilibrium abundance.

As expected, the inertial model displays oscillatory behaviour. For the grey whale the equilibrium abundance is predicted to be between 8000 and 10 000 animals, with the population being far above the equilibrium during the past two or three decades. It further predicts that the population is about to experience a major decline. The model, according to the author, *also predicts that constant future annual catches of 50 or 170 whales* (the present average annual "aboriginal subsistence" catch is about 180 whales) *may cause the extinction of the population during the predicted phase of abundance decline.* This contrasts starkly with the recent management advice by the IWC S.C., based on BALEEN II, that "a take of up to 482 whales per year is sustainable".

The inertial model indicates that the carrying capacity in 2000 was, with 95% probability, between 280 and 370% of the capacity in 1846, but not implying any corresponding environmental change. The calculated trajectories fit the data well for the 30-year period of surveys. The period of oscillation is of the order of two centuries[5].

In these exercises Witting has, I think, taken us back to the very roots of the science, assumptions and procedures of population dynamics. In consideration of the alternative models used to generate "data" for the development and testing of the RMP, it seems to me to be desirable to examine further the properties of Witting's model. In the Appendix I give a relevant extract from a forthcoming essay (Holt, 2003) in which I set out some of the relevant history.

MARINE PROTECTED AREAS

In recent years there has been renewed interest in closing substantial areas to fishing, temporarily or for the long-term, as an element in conservation and management regimes (see, for example, Roberts & Polunin, 1991, and Pearce, 2002). In 1946 the IWC was empowered to declare "closed areas, including sanctuaries" on the high seas as well as in waters under national jurisdictions. This resulted from prior consultations among governments, extending back many years, regarding their own powers to participate in such decisions.

In 1979 the Government of the Republic of Seychelles, which had just joined the IWC, proposed the designation of the entire Indian Ocean as a sanctuary, from the coasts of Africa, Middle East (South Asia) and the Indian subcontinent in the north, to the Antarctic ice-edge. This action was supported by all Indian Ocean coastal states (both IWC Members and others) and by the countries having island possessions in the Antarctic sector. The main intent was to protect the whales that breed in the temperate and tropical waters of the region and migrate through it, and also the southern feeding area of most of them. These whale populations were thought to be distinct from others in the southern hemisphere, although it was known that some baleen whales breeding in, for instance, the tropical Atlantic, moved into the Indian Ocean Antarctic sector to feed, and vice versa. A second objective was to put an end to the casual killing of whales by pelagic whaling fleets in transit to and from the Antarctic. In particular the Soviet fleets were the culprits because they carried out target practice on these voyages, as well as catching – legally – large numbers of sperm whales.

[5] *Since I presented these ideas to the Lowestoft symposium the grey whale population has been reported as having crashed by 20% last year; increasing numbers of dead animals – mostly calves – have been stranded.*

The Seychelles proposal required a three-fourths majority vote of states participating in the 1979 meeting, excluding abstentions, and it was evident that it would be blocked by a minority of Member states that were still whaling. A compromise was reached by accepting, for the time being, the truncation of the sanctuary at 55°S. In following years, the Indian Ocean states made collective representations to the IWC to revert to the original proposal, but this did not happen until 1994 when the entire Southern Ocean was declared as a sanctuary. Since then proposals have come forward for the declaration also of the South Pacific and the South Atlantic as sanctuaries. If these were to be accepted, practically the entire southern hemisphere would be closed indefinitely to commercial whaling regardless of what happens to the "moratorium" decisions of 1980, 1981 and 1982. It seems that this is what is desired by southern hemisphere coastal states. Adoption of these proposals has, however, been blocked by Japan, with the assistance of Norway and a group of "client" governments, mostly small island states that receive substantial economic assistance from Japan in the field of fisheries.

The IWC's authority to declare waters closed to whaling fleets is unusual, perhaps unique, in international law. Closure of the southern hemisphere (while perhaps allowing regulated whaling in the northern hemisphere), would meet the requirements for MPAs defined as covering the entire breeding, migratory and breeding distributions of migratory species (IUCN, 1999). The whale populations of the two hemispheres are quite separate, although the "equator" for the whales is not exactly the one shown on maps by humans, especially in the Indian Ocean sector, but it is close. At present, there is consensus only that the powers of the IWC extend to baleen whales, the sperm whale, the killer whale (*Orca*) and the North Atlantic bottlenose whale, although a majority of Parties to the 1946 Convention consider that "whales" may be taken to mean all marine cetaceans or, at least, to include the animals commonly known as "whales" – pilot whales, beluga, narwhal, all beaked whales and bottlenose whales, all of which are Highly Migratory Species in international law.

The IWC does not itself have power to take other actions to protect whales from adverse human action, although it can make recommendations to governments and to other international organizations having appropriate powers. The IWC can do that with considerable authority because its S.C. has long been the principal focus in the world for consideration of scientific findings regarding all cetaceans. Governments of many coastal states have enacted additional protective measures in their EEZs (affecting, for example, ship-routing and some types of fishing) and it has been suggested that complementary actions could be taken in Antarctic waters under the Antarctic Treaty of 1957 (whose maritime jurisdiction extends northwards to 60°S) and its Environmental Protocol. Such complementary actions could relate to other human activities than whaling, or to critical areas of habitat for whales, or both.

There have been a few proposals for the declaration of sanctuaries also in the northern hemisphere: some years ago by Jamaica regarding the northwest Atlantic and by the UK regarding the northeast Atlantic. These were, however never put to the vote and were withdrawn. More seriously, the coastal states of the Mediterranean (where no EEZs have been established, so national jurisdictions extend only out to 12 miles, in the territorial seas) have agreed, through the Barcelona Convention, that the Mediterranean Sea should be a sanctuary for cetaceans. This convention does not itself have powers to enact this, but the Parties to it suggested that the IWC should take such action. This, however, would call for one or more of the Mediterranean Members of the IWC (Spain, Monaco, France, Italy at present) to make a formal proposal, but they have not yet done so. However, three of them – France, Monaco, Italy – have made, as a first step, a trilateral declaration of the Ligurian Sea as a sanctuary, because substantial numbers of large whales, especially fin and sperm whales, are known to frequent the region. Legal advisers to these countries also argue that their trilateral action confers some powers over activities in the offshore area, other than through their control of their flag vessels and of port facilities (Scovazzi, 2001).

The closure of entire ranges of distinct populations of whale species is, of course, different from the current discussions concerning closure of partial ranges of certain fish stocks, although this was the effect of the *de facto* Indian Ocean sanctuary from 1979 to 1994. There is nothing to stop the IWC from engaging in partial closures in future, and in fact it has traditionally acted in this manner to a certain degree throughout its existence. For example, the waters of the southern hemisphere north of 40°S have always been closed to factory ships with respect to the pelagic killing of baleen whales, migrating and on their breeding grounds. However, although such measures were long regarded as having a conservation effect, in reality they were based mainly on the fact that baleen whales in temperate and tropical waters are less fatty than when on the feeding grounds; the restrictions were made for commercial reasons, to keep oil yields per whale high.

RESTRAINTS OTHER THAN CATCH LIMITS AND QUOTAS

Various conditions related to sharing have applied from the very beginning of Antarctic whaling. Initially the whaling was based on sub-Antarctic islands, mostly under UK jurisdiction, especially South Georgia and also the South Shetlands. The UK/Falklands authorities granted licences that, among other things, limited the whaling effort[6], and also, by virtue of the particular geography of the Antarctic, impeded whalers from encircling the globe However, this led to Norwegian operators developing floating factories that could anchor in most longitudes, and also escape British licensing rules and penalties. With the invention of the stern ramp these became the "pelagic" factory ships we know today – and which themselves evolved directly, as whaling declined, into the large factory-trawlers that were used especially by Soviet fishing fleets, but derived from the British *Fairtry*, which was a commercial failure.

In the early days Norway sought to impede expansion by other countries through domestic legislation prohibiting Norwegian gunners from working on foreign ships. This led to many gunners moving to other countries' vessels and incidentally to their getting higher wages. The response by Japan and the USSR was to train their own gunners. In the pre-World War II period, the UK and Norway sought to retain monopoly by control of the global oil market. Eventually this failed too.

In the post-War/IWC period actions have been taken to impede trade in whaling equipment and ships – restricting such sales or transfers at least to IWC Member states – and to impede trade in products from whales. The final stage of this process was the prohibition of international trade in whale products through listing of all species of large whales on Appendix I of the CITES convention. This action was taken in lock-step with the IWC declaration setting all catch limits to zero in 1986. However, countries

---

[6] *A duty on whale oil was used to fund scientific research on whales in the Antarctic, through the Discovery Committee (See Marsden, 1999).*

may lodge reservations to CITES decisions just as they can object to IWC decisions, and Japan and Norway have duly done so. Norway recently announced its intention to reopen trade in whale products with Iceland. Iceland withdrew from the IWC after a period of "scientific whaling" following the 1986 implementation of the 1982 moratorium decision. It has recently sought to re-join the Commission, but with an objection to the 1982 decision. As such an act is, strictly, contrary to the provisions of the ICRW, it has been contested by most Members. The status of Iceland's re-adherence is therefore, at the time of writing, unclear. Norway and Japan have meanwhile been seeking changes in the CITES listings in order to legitimize trade, but they have so far failed in this endeavour. However, if any non-zero catch limits are set in future, the rules of this diplomatic game will change.

ENVIRONMENTAL CHANGE

As described above, the RMP for baleen whales is demonstrably robust to substantial changes in environmental conditions. Successful trials included assumptions of up to 50% change in either direction, either instantaneously or spread over a long period. However, the possibility that such testing may have been insufficient now needs examination. Evidence is accumulating of the contraction of the Antarctic ice-edge throughout the past century. Interestingly, some of that evidence comes from studies of the locations of whaling vessels that, especially in the early days of Antarctic whaling, then mainly for blue and fin whales, operated close to the ice (de la Mare, 1997). It now seems that phytoplankters are trapped at the edge of the ice in winter. As spring and summer arrive these algae are released, multiply and provide food for euphausids, which in turn are consumed by whales and other carnivores (Brierley *et al.*, 2002).

It therefore seems possible that over this period the productivity of food for whales, and so of the whales themselves, has greatly diminished, so we perhaps should not expect the depleted whale stocks to recover towards their former abundances. To add to speculation about this, the results of recent counts of minke whales provide very much lower numbers (less than 50%) than obtained 10–20 years ago by essentially the same visual sighting methods. The reason for this is still a controversial matter, but some decline in minke whale abundance cannot be ruled out. Another possibility is that the intrinsic statistical error of surveys is much greater than had been thought – but if that is true it has important implications for application of the Revised Management Procedure. However this may be, the clear changes in Antarctic conditions will certainly have an impact on public perception of whether and when and under what circumstances and rules Antarctic whaling may ever be resumed, even on recovered stocks.

ACKNOWLEDGEMENTS

I acknowledge the support given me in this study by the International Fund for Animal Welfare (IFAW). I have benefited from conversations with Vassili Papastavrou, and the help of Olga Nève in extracting, compiling and analysing information and in other ways. I also appreciate the comments made by a referee, Daniel V. Gordon, on the draft of the oral presentation of this paper.

REFERENCES

Amundsen, R. S., Bjørndal, T. & Conrad, J. M. (1995) Open access harvesting of the Northeast Atlantic minke whale. *Environmental and Resource Economics*, **5**, 1–19.

Brierley, A. S., Fernandes, P. G., Brandon, M. A., Armstrong, F., Millard, N. W., McPhail, S. D., Pebody, M., Perrett, J., Squires, M., Bone, D. G. & Griffiths, G. (2002) Antarctic krill under sea ice: elevated abundance in the narrow band just south of the ice edge. *Science*, **295**, 1890–1892.

Chambers of Industry and Commerce of German Seaports (1947) German Whaling: Opinion of the Whaling Committee. 9 pp. [English translation of Handelskammern deutscher Seestadte: Bremen, Bremerhaven, Emden, Hamburg, Kiel, Lubeck.]

Clark, C. W. (1976) *Mathematical Bioeconomics: the Optimal Management of Renewable Resources*. Wiley-Interscience, New York.

Conrad, J. M. & Bjørndal, T. (1993) On the resumption of commercial whaling. *Arctic*, **46**, 164–171.

Cooke, J. G. (1995) The International Whaling Commission's Revised Management Procedure as an example of a new approach to fishery management. In: *Developments in Marine Biology*. **4**. *Whales, Seals, Fish and Man*. Proceedings of the International Symposium on the Biology of Marine Mammals in the North East Atlantic, Tromsø, Norway, November/December, 1994. (Ed. by A. S. Blix, L. Walløe and Ø. Ulltang). pp. 647–657. Elsevier Science, Amsterdam. 720 pp.

Corkett, C. J. (2002) Fish stock assessment as a non-falsifiable science: replacing an inductive and instrumental view with a critical rational one. *Fisheries Research*, **56**, 117–123.

Danilov-Danilyan, V .I. & Yablokov, A. V. (1996) Foreword, pp. 4–6. In: *Soviet Antarctic Whaling Data (1947–1972)*. Centre for Russian Environmental Policy, Moscow. 335 pp. [1996 bilingual English and Russian edition of 1995 original in Russian only].

De la Mare, W. K. (1986) Simulation studies on management procedures. *Reports of the International Whaling Commission*, **36**, 429–450.

De la Mare, W. K. (1997). Abrupt mid-twentieth-century decline in Antarctic sea-ice extent from whaling records. *Nature*, **389**, 5–60.

Holt, S. J. (1998). Fifty years on. *Reviews in Fish Biology and Fisheries*, **8**, 357–366.

Holt, S. J. (2000). Chapter 112. Whales and whaling. In: *Seas at the Millennium: an Environmental Evaluation*. (Ed. by C. R. C. Sheppard). *Global Issues and Processes*, **3**, 73-88.

Holt, S. J. (2002a) Sharing the catches of whales in the southern hemisphere. In: *Case Studies on the Allocation of Transferable Quota Rights in Fisheries* (Ed. by R. Shotton). *FAO Fisheries Technical Paper*, **141**, 322–373.

Holt, S. J. (2002b) Viewpoint: the whaling controversy. *Fisheries Research*, **54**, 145–151.

Holt, S. J. (2003) Reflections on sustainability and precaution. MS 17 pp. (in press).

Hjort, J., Jarn, G. & Ottestad, P. (1933) The optimum catch: essays on population. *Hvalraad. Skrifter*, **7**, 92–127.

IUCN (1999) *Guidelines for Marine Protected Areas*. Best Practice Protected Area Guidelines Series **3**. (Ed. and coordinated by G. Kelleher). IUCN World Commission on Protected Areas, Gland, Switzerland. 107 pp.

Kasuya, T. & Brownell, R. L. (2001) Illegal Japanese coastal whaling and other manipulation of catch records. *IWC Document,* **SC/53/RMP 24**.

Lavigne, D. (2002) Ecological footprints, doublespeak, and the evolution of the Machiavellian mind. In: *Sustainable Development: Mandate or Mantra*. (Ed. by W Chesworth, M. R. Moss and V. G. Thomas), pp. 63–92. The Kenneth Hammond Lectures on Environment, Energy and Resources, 2001 Series, University of Guelph, Ontario, Canada.

Marsden, R. R. G. (1999) The Discovery Committee – motivation, means and achievements. *Scottish Naturalist,* **111**, 69–92.

Pearce, J. (2002) The future of fisheries – marine protected areas – a new way forward or another management glitch? *Marine Pollution Bulletin*, **44**, 89–91.

Punt, A. E. (1999). A full description of the standard BALEEN II model and some variants thereof. *Journal of Cetacean Research and Management*, **1**(Suppl.), 267–276.

Roberts, C. M. & Polunin, V. C. (1991) Are marine reserves effective in management of reef fisheries? *Reviews of Fish Biology and Fisheries,* **1**, 65–91.

Scovazzi, T. (2001) The Mediterranean sanctuary for marine mammals. In: *Report of the Closing Workshop to Review various Aspects of Whale Watching*, Tuscany, Italy, February, 2000, pp. 79–84. IFAW, London [also issued as *IWC Document,* **IWC/53/22**].

Tormosov, D. D., Mikhaliev, Y. A., Best, P. B., Zemsky, V. A., Sekiguchi, K. & Brownell, R. L. (1998) Soviet catches of southern right whales *Eubalaena australis*, 1951–71. Biological data and conservation implications. *Biological Conserv*ation, **86**, 185–197.

Verhulst, P. F. (1838) Notice sur la loi que le population suit dans son accroissement. *Correspondance Mathématique et Physique,* **10**, 113–121.

Witting, L. (1997*) A General Theory of Evolution, by Means of Selection by Density Dependent Competitive Interactions*. Peregrine, Århus, Denmark. 332 pp.

Witting, L. (2000) Population cycles caused by selection by density dependent competitive interactions. *Bulletin of Mathematical Biology,* **62**, 1109–1136.

Witting, L. (2001a) On inertia dynamics in whale populations. The case of the eastern North Pacific Gray Whale. *IWC Document*, **SC/53/AWMP6**.

Witting, L., (2001b) Evolutionary dynamics of exploited and unexploited populations selected by density dependent competitive interactions. *IWC Document*, **SC/D2K/AWMP6** (Rev.).

Yablokov, A. V. (2000) Consequences and perspectives of whaling, pp. 6–10. In: *Soviet Whaling Data (1947–1972).* Centre for Russian Environmental Policy, Marine Mammal Council, Moscow, 408 pp. [Bilingual English and Russian edition of original, 2000, in Russian only.]

APPENDIX

EXTRACT FROM: *REFLECTIONS ON SUSTAINABILITY AND PRECAUTION (HOLT, 2003)*

Prologue: the sustainability paradigm

For nearly 70 years, efforts by scientists to provide advice on how to manage intensive – usually commercial – exploitation of wild animals, especially of fish and marine mammals, have been dominated by ideas first formulated by a Norwegian scientist, Per Ottestad. In a seminal paper (Hjort *et al*, 1933) he and his co-authors applied the "law" of logistic population growth expressed by Raymond Pearl (1920, 1925) to available data for the blue whales of the southern hemisphere. Pearl's own ideas derived from the original of P. F. Verhulst (1838). (Smith, 1994, gives some further historical details and connections, and Clark, 1976, Chapter 1, reviews the simple mathematical basis for these and subsequent developments.)

Pearl's own description of how he thought populations grew is worth recalling:
In a spatially limited universe the amount of increase in any particular unit of time, at any point of the single cycle of growth, is proportional to two things, viz:

(a) the absolute size already attained at the beginning of the unit interval under consideration, and

(b) the amount still unused or unexpended in the given universe (or area) of actual and potential resources for the support of growth.

The first part of this is, of course, a form of statement of the exponential law of unrestrained population growth (geometric increase, if regarded in terms of discrete increments) formulated by Thomas Malthus in 1798 and 1830 (see Flew, 1985). Malthus envisaged this "law", applied to human populations, as having catastrophic consequences (implying possible social revolution) when the limits of external resources became manifest. The developers of the logistic model appeared to justify hopes that equilibria would eventually be reached peacefully by virtue of the continuous operation of density-dependent processes.

The basic forms of the many versions and derivatives of the logistic model all predict the eventual attainment of the initial equilibrium after a single perturbation, or of different equilibria during the exercise of a continuous and steady external pressure, such as from exploitation by humans. If, however, the logistic model is expressed, by difference equations, as showing discrete growth, the population behaviour alters if a time-delay is introduced to the system, as by an inter-generational period dependent, for example, on the age of first reproduction. If such delay is relatively short, and non-linearities in the particular model not too severe, a stable equilibrium point is sustained. If, however, the delay is rather longer, the population oscillates between two locally stable fixed points, with period of two generations. More prolonged delays lead to increase in the number of stable points about which the population oscillates, and eventually to a chaotic regime in which there is an infinite number of interlocking periodic orbits. Cycling is also predicted when it is assumed that biological and environmental parameters vary over time. It has, further, long been known that oscillations may occur if consideration is given to interactions between two or three populations (as between a predator and its prey, or competition between two predators for a common resource – Lotka, 1925;

Volterra, 1926), with chaos soon appearing as the number of interacting species is increased. (see, for example, May, 1981, 1984, for further exposition of these phenomena, and Yodzis, 1994, 2001a, b, and Boyd, 2001, for discussion of the behaviour of complex multispecies systems).

Fishery biologists, especially those working on pelagic species such as sardines and herring, were for many years preoccupied with the problem of fluctuations in catches, usually reflecting year-to-year fluctuations in recruitment. Those fluctuations had significant economic implications and much effort went into trying to predict them. However, the fluctuations were usually regarded as overlaying, and perhaps obscuring, the longer-term patterns of change. In these circumstances, even those working on fluctuating fisheries were generally content to assume that, as far as the effects of fishing were concerned, these fluctuations were about an equilibrium, unless events showed otherwise (see Torensen & Jakobsson, 2002, for an account of one such history). In effect Occam's razor was in play, in the sense that simple models that displayed equilibria were preferred to more complex models displaying simple or multiple oscillatory periods or chaos.

Ottestad had remarked that many other mathematical curves than the logistic would fit the population data presented by Pearl – a view reiterated by Feller (1940) – and therefore that the goodness of fit to a data set did not prove that the data had been generated by the processes assumed in the derivation of the particular equation. Some subsequent experimental studies appeared to display a contrary phenomenon: that data were not fitted well by any of the single-species logistic forms (see above, regarding the grey whale). Nevertheless, essentially all further development of the population models applied to fisheries and wildlife data concentrated on the modification of the basic logistic model, even when it was incorporated in multispecies models of one type or another. This was true not only of the so-called "surplus production" models associated especially with Hjort and Ottestad, Oscar Sette (Schaefer *et al.*, 1951), Michael Graham and M. B. Schaefer, but also of the age-structured models associated particularly with Baranov, Beverton & Holt (1957), and W. E. Ricker. The Beverton & Holt "yield per recruit" equations implied such an intense dependence of natural mortality on density or abundance in the early stages of life that recruitment was, for practical purposes, independent of the abundance of parent animals, at least over a wide range of parental stock numbers; while their "self-regenerating model" simply introduced a graduated density-dependence in that early phase, such that the undisturbed population would be sustained at its "carrying capacity", and a perturbed population would tend towards a new equilibrium.

The modifications of the basic logistic model that have been extensively applied to fisheries concern mainly the middle levels and upper bounds of population abundance or density. They produce, for example, sigmoid population growth curves in which the inflexions (which indicate levels for "maximum sustainable yield") occur at higher or lower relative population sizes than the 50% of carrying capacity implied by the original simple logistic. In addition, the initially fixed parameters may be modified to stochastic variables, and the function describing density-dependence close to carrying capacity may be modified to prohibit the biologically impossible feature of the birthrate becoming negative if the population should, in recovering from a perturbation, temporarily "overshoot" the theoretical equilibrium carrying capacity.

Attention to the lower levels of population has mainly focused on the phenomenon of *depensation*, giving rise to the so-called "Allee effect" (Stephens & Sutherland, 1999; Courchamp *et al.*, 1999), whereby a minimum viable population level may exist, below which the population will irreversibly decline to extinction even if the external pressure (as by exploitation) is relaxed. Where such a level does exist we speak of *critical* depensation. It has proven very difficult to quantify depensation in exploited populations or even to demonstrate its existence or otherwise. It has therefore become the practice to try to manage fisheries such that the stock is never reduced, inadvertently or deliberately, below levels that are guessed to be well above any plausible level of critical depensation. This policy has been followed explicitly in the IWC's management procedure for aboriginal subsistence whaling (MPASW), as well as in the RMP. Although precise definition is usually lacking it has become customary to refer to populations below or close to such critical levels as *depleted, seriously depleted* or *critically depleted.* In the IWC context, however, depletion has also been the term used when referring to a population that is protected from whaling because it has been reduced to somewhat below its approximately guessed level for maximum sustainable yield (MSY).

The concept of population equilibria from which sustainable yields may be secured by following correct management procedures has proven to be a very powerful one. That wildlife should be used sustainably – if they are to be "used" at all in the sense of lethal exploitation – is now enshrined in innumerable declarations, resolutions, treaties and laws. It now finds its most universal expression, as far as marine life is concerned, in the UN Convention on the Law of the Sea (UNCLOS), although various economic and social conditions are specified therein to modulate it. In addition UNCLOS provides that, where a predatory animal is the prime target, users of marine resources should refrain from exploitation of its prey in such a way as to affect adversely its reproductive capability, and hence disturb equilibria and reduce carrying capacity.

Some persons and institutions having special economic interests, and swimming against the mainstream, have sought to justify *unsustainable* use, increasingly so in recent years. A current example is the pressure by whaling interests for legalization and resumption of commercial whaling at unsustainable levels (or such as to impede recovery of depleted populations) because whales are accused of consuming too much of "our" fish (see Yodzis, 1994, 2001a, b). Arguments for increased sealing have a similar history and seek rationalization of "culling". Nevertheless, such arguments are, by default, generally cast in the language of departures from the norm of equilibrium. The prevalence of such demands has led one group of specialists in population and ecosystem dynamics to prepare a protocol against which proposals for culling marine mammals should be scientifically evaluated (UNEP, 1999).

Similarly, arguments have been made for intensifying the exploitation of minke whales in the southern hemisphere, for example because they might be impeding the recovery of the more-or-less "protected" blue whales from the brink of extinction, by competing for their preferred food, Antarctic krill (*Euphausia superba*), copepods and amphipods. In this context recourse is commonly made to the need "to restore the balance of nature", regardless of the fact that, as long ago as 1927, Charles Elton persuasively challenged the very idea that any such "balance" existed. More recently, Witting (1997) insisted that the idea of balance "presents a paradox for classical population dynamics" because "the fundamental theorem of natural selection predicts a steady increase in the population equilibrium, and because this leads to a continuous deterioration of the resources (used by that population)."

The newer idea of "a precautionary approach" to the exploitation of wildlife is also closely associated with the concept of desirable and attainable steady states. The general approach is that if, as is practically always the case, we are uncertain about the values of parameters to use in population models, or about the structure of those models, we should "aim high", that is seek to avoid inadvertent depletions, or reductions below optimal population levels by, as it is sometimes popularly expressed, "giving the benefit of the doubt" to the fish, whales, seals or whatever. This approach, sometimes elevated to a "principle", is also focused on likely deviations or perturbations from a perceived or assumed equilibrium. It contrasts starkly with advice provided in several circumstances by scientific groups, based on "best estimates" (or even on thinly disguised best *guesses* – sometimes called "guestimates") of parameters.

In all the modifications and elaborations of the logistic population model, one feature has been left virtually untouched. That is the notion of a fixed "intrinsic (and maximal) rate of natural increase" in a vanishingly small population, as supposed by Thomas Malthus and the many who followed him. In some expositions of evolutionary theory this rate is referred to as a measure of *fitness* which, through natural selection, may be expected generally to increase on an evolutionary time-scale. Therefore, the fundamental theorem of natural selection, as elaborated by Fisher (1930), tells us that, if competitive interactions and density dependence are absent, as they should be at the vanishing limit, then the intrinsic growth rate is expected to increase at a rate given by the genetic variance in that growth rate (Witting, 1997). Witting demonstrates that a population within which there is genetic variation would be expected to increase at a rate higher than expected from the Malthusian argument, i.e. that the rate would be hyper-exponential, the logarithm of the number being non-linear with time but curvilinear and concave upwards. That finding has serious implications for the established ideas about population dynamics and for theories of managing human uses of living resources.

SELECTED REFERENCES *(OTHER REFERENCES IN MAIN TEXT)*

Beverton, R. J. H. & Holt, S. J. (1957) On the dynamics of exploited fish populations. *Fisheries Investigation Series* 2, **19**. HMSO, London.

Boyd, I. (2001) Culling predators to protect fisheries: a case of accumulating uncertainties. *Trends in Ecology and Evolution*, **16**, 281–282.

Courchamp, F., Clutton-Brock, T. & Grenfell, B. (1999) Inverse density dependence and the Allee effect. *Trends in Ecology and Evolution,* **14**, 405–410.

Elton, C. S. (1927) *Animal Ecology*. Sidgwick & Jackson, London.

Feller, W. (1940) On the logistic law of growth and its empirical verification in biology. *Acta Biotheoretica*, **5**, 51–66.

Fisher, R. A. (1930) *The Genetical Theory of Natural Selection*. Dover, New York.

Flew, A. (Ed.) (1985) Introduction to an Essay on the Principle of Population and a Summary View of the Principle of Population. Reprint of 1970 original, Penguin Classics, London.

Lotka, A. J. (1925). *Elements of Physical Biology*. Williams and Wilkins, Baltimore.

May, R. (Ed.) (1981) *Theoretical Ecology: Principles and Applications*. 2nd edn. Blackwell, Oxford.

May, R. M. (Ed.) (1984) *Exploitation of Marine Communities*. Report of the Dahlem Workshop on Exploitation of Marine Communities, Springer, Berlin.

Pearl, R. (1920) On the rate of growth of the population of the United States since 1750 and its mathematical representation. *Proceedings of the National Academy of Science USA,* **6**, 275–288.

Pearl, R. (1925) *The Biology of Population Growth*. Knopf, New York, 260 pp.
Schaefer, M. B., Sette, O. E. & Marr, J. C. (1951) Growth of the Pacific coast pilchard fishery to 1942. *US Fish and Wildlife Service Research Report,* **29**, 31 pp.
Smith, T. D. (1994) *Scaling Fisheries: the Science of Measuring the Effects of Fishing, 1855–1955*. Cambridge University Press.
Stephens, P. A. & Sutherland, W. J. (1999) Consequences of the Allee effect for behaviour, ecology and conservation. *Trends in Ecology and Evolution,* **14**, 401–405.
Torensen, R. & Jakobsson, J. (2002) Exploitation and management of Norwegian spring-spawning herring in the 20th century. *ICES Marine Symposia,* **215**, 558–571. [Report of a Symposium "100 Years of Science under ICES" held in Helsinki, August 2000].
UNEP (1999) Protocol for the scientific evaluation of proposals to cull marine mammals. *Report of the Scientific Advisory Committee of the Marine Mammals Action Plan.* A study supported by Greenpeace, World Wide Fund for Nature, International Fund for Animal Welfare and the United Nations Environment Programme. [Available from IMMA, Guelph, Ontario, Canada, and IFAW, London and Bristol, UK].
Volterra, V. (1926) Variazioni e fluttuazioni del numero d'individui in specie animali conviventi. *Memorie Academia Lincei*, **2**, 31–113.
Yodzis, P. (1994) Predator-prey theory and management of multispecies fisheries. *Ecological Applic*ations, **4**, 51–58.
Yodzis, P. (2001a) Must top predators be culled for the sake of fisheries? *Trends in Ecology and Evolution,* **16**, 78–84.
Yodzis, P. (2001b) Culling predators to protect fisheries: a case of accumulating uncertainties. *Trends in Ecology and Evolution,* **16**, 282–283.

Transbouundary issues in the purse-seine, trawl and crustacean fisheries of the Southeast Atlantic

Moses Maurihungirire
National Marine Information and Research Centre, Ministry of Fisheries and Marine Resources, P.O. Box 912, Swakopmund, Namibia

ABSTRACT: A short description of the habitat and spatial distribution of commercially important pelagic (sardine and anchovy), trawl (hake and horse mackerel) and crustacean (red crab) species in the Southeast Atlantic (FAO area 47) is given. Focus is on stock partitioning. The stocks move temporally across geopolitical boundaries, behaviour that warrants the development of cohesive management and research among the Southeast Atlantic coastal states. Possible connections among geopolitically segregated stocks of hake, horse mackerel and red crab are illustrated along with their life histories. Pelagic sardine and anchovy spawn mainly close to the coast in summer and autumn, shallow-water Cape hake spawn all year round, with an intense phase in midwinter, Cape horse mackerel spawn intensely in warmer water west of the shelf break in the central and southern Southeast Atlantic, and red crab spawn year-round over most of their distributional range. Broad national policies, legislation and structures for managing hake, horse mackerel and crab are outlined. Areas where a lack of knowledge of the resources could lead to mismanagement are also discussed. There is limited understanding of stock definition in respect of the cross-boundary migratory species addressed. Studies of the age structures are inadequate, and information on total population size is lacking. In solving all these aspects, South Africa seems to be more advanced than Namibia and Angola, and the situation of the latter has been worsened by the prolonged civil war in that country.

INTRODUCTION

Changes in the legal regime of the oceans, especially the introduction of the Law of the Sea, have resulted in a decline in regional and international collaboration in fisheries management and fishing. The declaration of exclusive economic zones (EEZs) has brought the long tradition of international and regional cooperation virtually to a halt, indicating that there has frequently been erratic interpretation of the Law of the Sea in this respect. Article 63 indicates the need for regional cooperation between coastal states with non-migratory shared stocks (transboundary; Paragraph 1) or stocks occurring within an EEZ and beyond and adjacent to it (straddling; Paragraph 2) in devising a regime that ensures their conservation (United Nations, 1983). Gulland (1980) classifies stocks on the basis of the extent of their movement across geopolitical boundaries as follows: regular seasonal migrations, movement attributable to development of individual fish, and dispersions and movement as a consequence of climatic changes. Following the Law of the Sea, the establishment of international and regional fisheries organizations was necessitated in some cases by fishing vessels congregating in areas of great fish abundance on the high seas, order of some kind being needed to ensure safe and efficient fishing. Declines in the stocks of various commercially important fish species in various parts of the world also added impetus to the need to establish international and regional fisheries organizations (Cushing, 1972).

This paper addresses the distribution and movements of some commercially important pelagic, midwater and demersal species within the Southeast Atlantic. The current level of collaboration in research and management is evaluated, and suggestions regarding appropriate measures for encouraging that collaboration are proposed.

THE AREA OF STUDY

The intent of this paper is not to elucidate the marine environment of the Southeast Atlantic, but rather to describe how the shared fisheries resources make use of it for their survival and propagation. The Benguela Current lies between about 17 and 35°S (Figure 1), but its boundaries are not fixed. It is important to note the direction of flow of water masses, and the positions of the upwelling cells and the waters that fall under its influence. The Benguela system is unusual because it is the only upwelling system in the world bounded (north and south) by warm-water frontal jet currents (Shannon, 1985). It can be subdivided into three distinct areas: the northern region stretches from about 17 to 25°S, the central region from around 25 to 31°S, and the southern region from around 31 to 35°S, 20°E (Shannon, 1985).

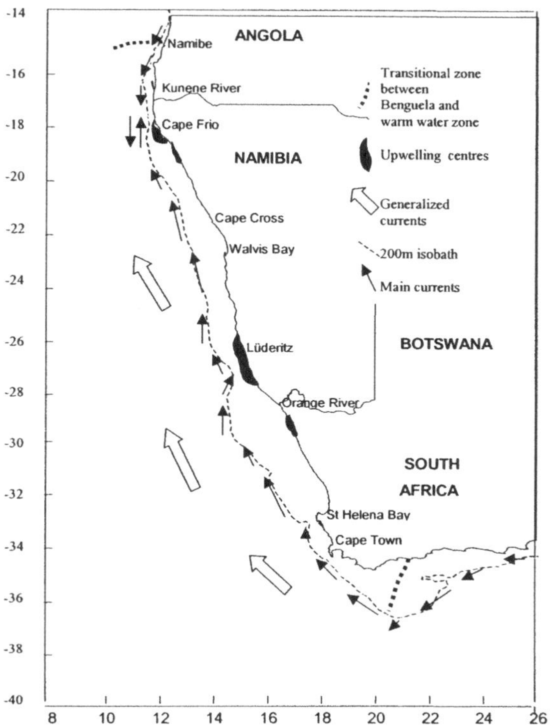

Fig. 1 Some physical oceanographic features of the Benguela upwelling system (after Shannon, 1989).

DISTRIBUTION AND MOVEMENTS

PELAGIC SARDINE (SARDINOPS SAGAX) AND ANCHOVY (ENGRAULIS CAPENSIS)

Virtual population analysis (VPA) estimates of the biomass of sardine in Namibian waters between 1952 and 1988 reveal a marked decline during the late 1960s (Figure 2), when catches peaked, and a further drop to below a million tons in the mid-1970s (Butterworth, 1983; Boyer, 1996). Namibian sardine spawn mainly within 60 km of the coast between 21°S and south of the confluence area of the Benguela and Angola Currents (O'Toole, 1977). Spawning occurs near the 200 m isobath in late summer/autumn in water temperature of 19–21°C.

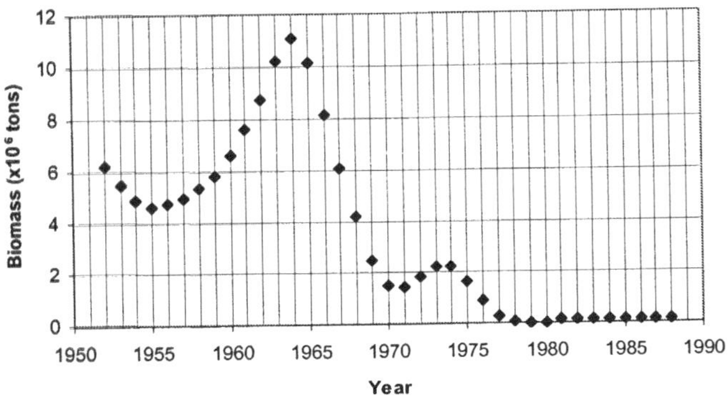

Fig. 2 VPA estimates of sardine biomass off Namibia, 1952–1988.

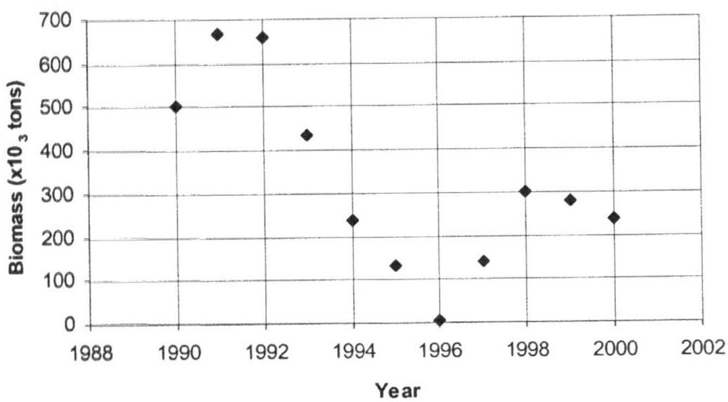

Fig. 3 Acosutic biomass estimates of sardine off Namibia and southern Angola, 1990–2000.

During the 1970s, the distribution of Namibian commercial catches suggested a northward shift in their core distribution (south of Walvis Bay) after the collapse of the fishery, perhaps resulting from depletion of the southern spawning population (north of Cape Town) and the discontinuation of associated migrations (Schwartzlose *et al.*, 1999). Acoustic survey estimates of sardine on the Namibian/southern Angolan shelf during the 1990s (Boyer and Hampton, 2001) indicated that the adult biomass was still well below pre-decline levels (Figure 3). The surveys also indicated a shift in distribution even farther north in 1994, when much of the biomass was off southern Angola (Barange *et al.*, 1999).

Anchovy off Namibia behave and are distributed essentially similarly to sardine. Larvae drift south close to the coast, recruiting as 0-year-olds north of 21°S. Subsequently, young adults return to the northern confluence area (between the Benguela and Angola Currents), where they spawn for the first time. Older fish return to the same area to spawn at around 21°S.

The principal upwelling cell in the Benguela system is close to Lüderitz, at around 27°S (Shannon, 1985). This is downstream of where sardine and anchovy recruit and spawn, as is the case with the California, Peru and north-west African upwelling systems (Bakun, 1996). In the southern Benguela off South Africa, spawning of pelagic fish used to be localized in two areas, in the relatively cool water north and west of 32.3°S, 18°E (main upwelling area), and south and east of 34.2°S, 18.25°E (main upwelling area) inshore in the Agulhas Current (Hutchings, 1996). The latter was the main spawning ground for sardine after the decline of the South African fishery for that species in the 1960s. Anchovy spawn offshore in the same area. Eggs and larvae are transported from the spawning grounds on South Africa's south coast to the west coast entrained in a perennial equatorward jet current between the 200 and the 500 m isobaths (Fowler & Boyd, 1998) along the offshore boundaries of the major upwelling cells.

South African pelagic fish recruitment is therefore also downstream of the upwelling cells. Some sardine and anchovy larvae spawned off South Africa recruit as far north as the Orange River (28.3°S), and the limit of distribution of these larvae during normal years can extend as far north as the southern boundary of the Lüderitz upwelling cell (26.5°S; Lutjeharms & Meeuwis, 1987). At that boundary, the larvae are either carried far offshore and lost to the stock or drawn inshore by compensatory flow and returned southwards by the inshore south-flowing counter-current. It has also been suggested that, in years when the Lüderitz upwelling cell is abnormally weak, eggs and larvae spawned in the southern Benguela may pass the upwelling barrier and recruit into the southern Namibian fishery, so linking northern and southern Benguela fisheries. Such a scenario may well have been prevalent in 1987, a year of peak anchovy catches in the southern Benguela. The peak was associated with a tenfold increase in anchovy catches off southern Namibia, and south of Walvis Bay in particular.

PELAGIC AND MIDWATER HORSE MACKEREL (TRACHURUS SPP.)

There are two species of horse mackerel off southern Africa, *Trachurus trecae*, which is found virtually only off Angola, and *T. trachurus capensis*, which extends north into Angolan waters but is the basis for a commercial fishery off Namibia that has yielded an average annual catch of almost half a million tons for more than 15 years. Of the latter species, at least two stocks exist, one off northern Namibia and southern Angola, and the other in the southern Benguela (Draganik, 1977). The two stocks are sustained by separate spawning populations (Babayan *et al.*, 1983). Spawning in both the south and the north takes place in warmer waters west of the shelf break, and the nursery areas are adjacent to the spawning areas but closer to shore. Substantial longshore and cross-shelf migrations of both juveniles and adults have been documented. For both stocks, juveniles are found closer inshore and adults more offshore. Biomass estimates of horse mackerel in Angolan waters between 1985 and 1997 were higher off central and southern Angola than in the north of that country, between Luanda and Cabinda (Hampton *et al.*, 1999).

DEMERSAL CAPE HAKES (MERLUCCIUS SPP.)

Gordoa *et al.* (1995) and Burmeister (2000) are of the opinion that the Namibian stock of deep-water Cape hake *M. paradoxus* is the sink of the South African spawning stock. This hypothesis can also be supported by the absence of spawning areas for *M. paradoxus* in Namibian waters (Gordoa *et al.*, 1995) and by the fact that the recruitment areas are only found in the south of Namibia. Off Namibia and South Africa, certain regional genetic variations for shallow-water Cape hake *M. capensis* have been observed, whereas there was no such distinction for *M. paradoxus* (Grant *et al.*, 1987). However, Payne (1989) disputed that there was much horizontal migration of hake apart from a gradual movement offshore into deeper water as the fish grow. Deep-water hake dominate the southern Benguela off western South Africa, whereas shallow-water Cape hake *M. capensis* have traditionally dominated the landings off Namibia (especially catches off central Namibia) and off South Africa's south coast (Payne, 1989; Figure 4). Shallow-water hake are found on the entire shelf of the Southeast Atlantic down to the Agulhas Bank (Konchina, 1987). The biomass of deep-water hake off South Africa increased through the 1990s and it has been suggested by several authors that the core of their distribution has recently spread northwards into Namibia. Certainly the biomass of the species in Namibian waters was low in 1992 (Hampton *et al.*, 1999), but it has since escalated and spread right up into the northern Benguela (Strømme, 1996). The catch mix of the two species of Cape hake off Namibia has changed markedly in recent years (Hampton *et al.*, 1999).

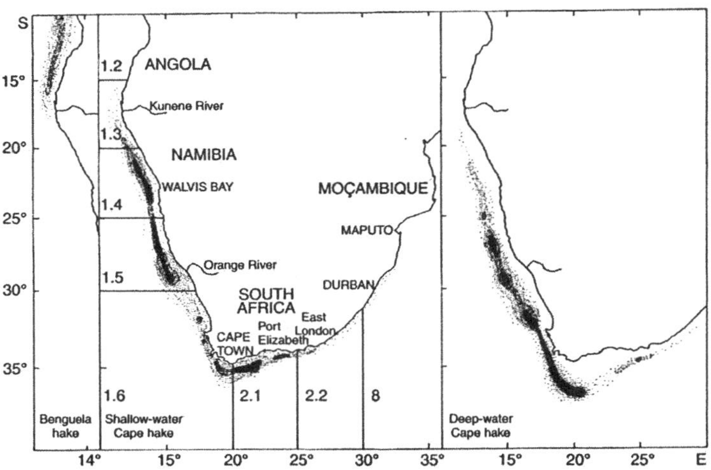

Fig. 4 Distribution of the three hake species in the Benguela ecosystem (after Payne, 1989)

There is no evidence of Angola and Namibia sharing any of the hake species except for Benguela hake (*M. polli*), which overlaps in distribution with the shallow-water Cape hake in the north of Namibia (Payne 1989). Benguela hake contribute little to Namibian catches, because the bulk of Namibian fishing effort is concentrated on the central fishing grounds. Although shallow-water Cape hake dominate hake catches in Namibian waters and off the south coast of South Africa, there is no apparent link between the two stocks. Payne (1989) suggests that the differences in local abundance of the two hake species are a function of the width of the continental shelf and the steepness of the adjacent continental slope. Deep-water Cape hake seem to dominate where the continental slope is less steep and the shelf is narrow, whereas shallow-water Cape hake are abundant on wider continental shelves.

BENTHIC DEEP-SEA RED CRAB (CHACEON MARITAE)

These crustaceans occur on the continental slope from approximately 27°S northwards into Angolan waters (Dias & Seita Machado, 1974), at depths of 300–900 m off both southern Angola and northern Namibia (Melville-Smith, 1983). Adult females migrate from Namibia into Angola to spawn, and this suggests a single stock of the species (Le Roux, 1997). Stock assessment has revealed that the biomass of the Namibian stock decreased from 40 000 tons in the 1980s to approximately 10 000 tons in the 1990s (Hampton *et al.*, 1999). Neto (1997) reported estimated biomasses off Angola of 47 000, 91 000 and 18 000 tons for 1977, 1982 and 1996 respectively. Clearly, the stock is being reduced on both sides of the international border.

SHARED FISHERY STATUS

SARDINE

Annual catches of sardine off Namibia rose from low levels in the early 1950s (when they started) to a peak of 1.4 million tons in 1968, after which they declined (Boyer and Hampton, 2001). Since the 1980s, annual catches have rarely exceeded 100 000 tons, only 2000 tons being caught in 1996 and a "zero *TAC*" effected in 2002.

Sardine catches off South Africa rose from 100 000 tons in the early 1950s to peak at 400 000 tons in the early 1960s, after which they plummeted as sardine was "replaced" by anchovy. By the 1980s, the South African sardine stock was very small and a conservative strategy was recommended by scientists and used by decision-makers to rebuild it (Payne & Bannister, 2003). Catches and *TAC*s have since recovered to upwards of 200 000 tons. Industrial catches of sardine in Angolan waters peaked at 146 000 tons in 1957. Catches by the Angolan fleet were generally determined by the extent to which the Namibian stock extended into southern Angola. However, since 1994, when the availability of sardine off northern Namibia decreased, the Namibian fleet has also been active off southern Angola, and the largest catch there, 47 000 tons, was made in 1995.

ANCHOVY

In contrast to the fortunes of sardine off both Namibia and South Africa, anchovy catches off both countries escalated rapidly upwards from their onset in 1964 and 1966 respectively. Catches varied around 200 000 and 300 000 tons in Namibia and South Africa, but all-time high catches were taken in 1988 and 1987 respectively. A remarkable recruitment in South Africa was responsible for the huge catches there, and the Namibian peak was thought to be a consequence of an abnormal influx of recruits from the expanded South African stock (Hampton *et al.*, 1999). Contemporary surveys have established that the anchovy stock along the Namibian coast is extremely depleted, virtually no biomass being found after the 1990s (Boyer and Hampton, 2001). Little or no anchovy is caught off Angola.

HAKE

Catches of shallow-water Cape hake and Benguela hake in Angolan waters have amounted to less than 1000 tons in recent years. In contrast, the fishery for hake off Namibia started in the late 1950s, rapidly attracted distant water fleets from Europe and Asia, and peaked at more than 800 000 tons in 1972. The total hake catch from the SE Atlantic that year exceeded 1.1 million tons. Largely as a consequence of this massive level of exploitation, mainly on newly recruited fish from an abundant year-class, the availability of hake declined. After then catches started to decline in the Southeast Atlantic and some foreign vessels were withdrawn (Alheit & Pitcher, 1995). The International Commission for the Southeast Atlantic Fisheries (ICSEAF) imposed a 110 mm minimum mesh size control measure for the hake fishery after the decline in hake catches (Alheit and Pitcher, 1995). South Africa soon declared an EEZ, in 1977 (Payne, 1989), and catches there were down to 120 000 tons by 1982 before aggressive management aided a recovery to allow catches of the present levels of some 160 000 tons.

As foreign fleets were displaced from South Africa, they focused their activities on Namibian waters, and annual catches off Namibia alone attained almost half a million tons. Catches then declined as Namibia moved towards Independence and declaration of its own EEZ in 1990. Stringent control was then introduced (Oelofsen, 1999), and an annual *TAC* of 50 000 tons in 1990 rose to some 120 000 tons over the period 1996–1998.

HORSE MACKEREL

Adult horse mackerel are the main target for midwater trawlers operating off Namibia and Angola, the Namibian fishery for that species being nationally its biggest fishery by volume for at least the final decade and a half of the 20th century. Annual trawl catches of the two target species rose from less than 50 000 tons in the early 1960s to between 600 000 and 700 000 tins from 1982 to 1984. Annual catches of mainly *T. trachurus capensis* in Namibia since Independence in 1990 initially fluctuated around 350 000 tons, but declined to between 200 000 and 250 000 tons in recent years. Catches of the same species in Angolan waters in the 1990s were insignificant.

The fishery for Cape horse mackerel in South Africa was initially a purse-seine operation for predominantly juveniles on its West Coast in the 1950s and early 1960s but switching to a bottom-trawl fishery for adults on the South Coast thereafter, caught by both local and foreign (mainly Japanese) trawlers. Annual catches rose from about 10 000 tons in 1964 to peak at more than 97 000 tons in 1978 (Hampton *et al.*, 1999), but then declined to average just 30 000 tons in the 1980s and 1990s.

RED CRAB

The red crab fishery in Namibia started in 1973. From small beginnings, catches peaked at some 10 000 tons in 1983, but after an initial mining out, they declined steadily to just 2 676 tons in 1991, and have remained stable at around this level ever since. Off Angola, a few hundred tons of red crab are taken by trawl each year as a by-catch of the deep-water prawn fishery. Since 1986 there has also been a directed crab fishery by a single Japanese vessel, which has caught a few thousand tons annually.

CURRENT MANAGEMENT STRATEGIES

Marine resources in Angola, Namibia and South Africa are managed sustainably according to national policies and legal frameworks. In all cases, current management strategies are based on the assumption of discrete national stocks. In Angola, acoustic methods are used to determine the status of sardine and other pelagic stocks, and the resources are managed through a *TAC*. Scientific personnel in the Ministry of Fisheries and Marine Resources in Namibia base their recommendations on acoustic/midwater trawl surveys. The *TAC* recommended for sardine for a season is generally 18% of the survey biomass estimate at the end of the previous fishing season.

South African pelagic fish (sardine, anchovy and round herring *Etrumeus whiteheadi*) have been surveyed acoustically since the 1980s (Hampton, 1992). In addition, the annual egg production method was applied for anchovy between 1984 and 1993, and the results of this method proved similar to those from acoustic surveying as well as providing an important means for estimating the bias of the acoustic surveys. Scientific information on estimates of population structure from commercial catches, plus survey estimates of recruitment and spawning biomass are utilized in the assessment process for anchovy and sardine through an Operational Management Procedure (Cochrane *et al.*, 1997). Round herring are not currently subject to a *TAC*.

In Angola, the statistics on horse mackerel catches are deemed unreliable for the purpose of stock assessment. Consequently, the results of fishery-independent surveys are used to determine an appropriate *TAC* for the two *Trachurus* species combined (Hampton *et al.*, 1999). Cape horse mackerel dominated Namibian catches before 1990, so ICSEAF made no distinction between the two species in their assessments of stock status (Boyer & Hampton, 2001). Schaefer and Fox surplus production models were used in these assessments. After Independence, scientists of the Ministry of Fisheries and Marine Resources based their recommendations on trends in acoustic survey estimates, in combination with age- and length-based VPA estimates obtained using commercial catch data (Boyer & Hampton, 2001). Since 1993, South African scientists have used a Beddington and Cook-type yield-per-recruit model from bottom-trawl survey estimates in concert with acoustic techniques to set a precautionary limit for Cape horse mackerel.

The two species of hake found off Angola (*M. polli* and *M. capensis*) are currently assessed on the basis of fishery-independent survey data. However, more use of catch rate data is envisaged in the future owing to recent improvements in the quality and availability of fishery-dependent data. The two hake species are assessed separately and a *TAC* set for each. In addition, minimum size, and effort and area limitation, are applied to manage Cape hake and other demersal species. In Namibia since the declaration of the Namibian EEZ in 1990, hake (*M. paradoxus* and *M. capensis*) *TAC* recommendations have been based on trends in abundance during fishery-independent bottom trawl surveys while making an adjustment to allow for fish off the bottom at night, as determined acoustically. Traditionally, the *TAC* was recommended as 20% of the estimated biomass of mature hake. Then, from 1998, an Interim Management Procedure (IMP) was implemented. Within the IMP, the *TAC* was adjusted according to the mean change in the survey and catch rate indices for the previous five years (Geromont *et al.*, 1999; Butterworth & Geromont, 2001). That IMP paved the way for a fully functional OMP in 2002.

From the year of declaration of its EEZ (1977) until 1983, South African *TAC* recommendations by ICSEAF and its own scientists were based on an equilibrium model in concert with an effort-averaging procedure in association with Fox's formulation of the surplus production function (Andrew, 1986). From 1984, South Africa resorted to a policy geared toward maintenance of annual sustainable yield of the hake resource in their waters. In this case an $f_{0.2}$-type harvesting strategy was implemented to allow stock recovery without excessive industry restriction (Payne & Punt, 1995). After dissolution of ICSEAF in 1990, South Africa initially continued with the $f_{0.2}$ harvesting strategy for their hake TAC recommendations, using the Butterworth–Andrew observation error estimator (Payne & Punt, 1995). Today, however, South African hake is assessed by a rigorous procedure that uses an age-aggregated production model in concert with commercial catch rate and fisheries-independent survey data to compute *TAC*s (Geromont *et al.*, 1999; Payne & Bannister, 2003). The two species of Cape hake off South Africa are assessed together, but with separation of the stocks on the west and south coasts, similar to the situation during and post-ICSEAF management.

Management of deep-sea red crab in Namibia is based on length-based cohort analysis and predictive models, adapted to fit the growth dynamics of the species (Le Roux, 1997). *TAC*s are recommended on the basis of a projection of future biomass of the stock as a function of the catch. Fishers are prohibited from operating in water shallower than 400 m and a minimum size limit is applied. Off Angola, assessment is based on trends in catch rates and estimates of maximum sustainable yield. Catches there are controlled by *TAC*, limitation of effort, prohibition of fishing shallower than 500 m, minimum size and through limitation of the crab by-catch in the prawn fishery.

THE WAY FORWARD FOR SUITABLE GOVERNANCE OF SHARED RESOURCES

Having illustrated the distribution and management regimes of some selected commercially important fish stocks in the Southeast Atlantic, it is necessary to investigate aspects of this knowledge that can ensure their sustainable utilization. Sardine and anchovy could possibly be classified as a transboundary resource (non-migratory shared) between South Africa and Namibia, and sardine alone similarly between northern Namibia and Angola. Cape horse mackerel are clearly transboundary and are shared between Namibia and southern Angola, but there does not seem to be any mixing of the Namibian and South African stocks. Deep-water Cape hake cross the international boundary between South Africa and Namibia, and red crab is a shared resource of Namibia and Angola.

CONCERTED EFFORTS IN RESEARCH

Because of the shared nature of some of the resources, it is crucial that the coastal states pool their knowledge and experience in researching the biological characteristics of the transboundary stocks, their management and exploitation. For example, the dynamics of red crab in Namibian waters will be better understood if the component of the stock resident off Angola is researched and managed as a unit with the Namibian component. Should Namibia disregard the dynamics of the Angolan component of this stock, it would be easy for the status of the resource off Namibia to be misinterpreted. The same situation applies in the case of deep-water Cape hake off South Africa and Namibia.

Cooperative and collaborative research by the three countries is therefore a matter of urgency as far as shared stocks are concerned. Historical politics made such an objective difficult, but huge strides have been made in recent years. Scientists from the three countries started cooperating meaningfully in the late 1990s, with the common goal of generating collaborative sustainable management of shared resources. The main collaborative research initiative is the Benguela Environment, Fisheries, Interaction and Training (BENEFIT) programme. From a small start in late 1995, BENEFIT has developed into a dynamic regional research initiative, underpinned by financial support from mainly Europe along with contributions from the three countries themselves. A lot of the work conducted thus far has been on the stocks that are shared, but to date collaborative research leading directly and specifically to improved management of the shared stocks across the common boundaries has not been accomplished. The reason is that BENEFIT has necessarily targeted its resources at the training of scientists mainly, but not only, in Angola and Namibia, the two countries where professional experience was lowest at the outset of the programme. Clearly, should the situation arise where quotas of these shared stocks need to be allocated to each country or where some other joint management strategy is implemented, it is crucial that each country be advised and represented by experienced suitably qualified scientists and that decisions be based on sound scientific advice.

A REGIONAL MANAGEMENT STRATEGY FOR SHARED RESOURCES

Due to discrepancies in the socio-economic and political milieu of the coastal states of the Southeast Atlantic, problems may well arise in setting a unified policy framework for the management of a shared stock. For the past 12 years Namibia has been in the process of building its fishery along with the process of Namibianization of the industry itself (Oelofsen, 1999). Angola was engaged in a 27-year-long civil war that countered any realistic efforts at generating sustainable management of its fisheries resources. In contrast, South Africa's marine resources are comparatively healthy and well managed in world terms (Cochrane *et al.* 1997; Payne & Bannister, 2003), but the industry was initially built on the foundation of minority participation (through apartheid). At present its policy is skewed towards permitting new entrants from the previously disadvantaged members of its society. With these differences in structure and current priorities among the national fisheries of the region, it becomes a daunting task to develop a unified and cooperative structure of governance of shared stocks.

Having alluded above to the promising move towards collaborative research, largely through donor funding, it still needs to be stressed that, for this process to be cost-effective, action on cooperative management needs to be implemented. Indeed, the establishment of national EEZs has created a sense of ownership of resources by each coastal state. Against this "ownership", collaborative management could be seen by some as a loss of authority over its national resources.

Namibia v. Angola

Sardine and deep-sea red crab off northern Namibia and southern Angola are shared stocks that exhibit regular cyclical migrations at different stages in their development. As a consequence, Angolan fishers capture more female deep-sea red crab during the spawning season of that resource. Similarly, Namibian fishers capitalize on adult spawning sardine, to which resource Angola only has access during its juvenile phase, when the fish are too small to be retained by the mesh size of the net in use. The same applies to anchovy, when it crosses the border northwards. Optimal management of such shared stocks is always going to be difficult. Harvesting too small or too old fish is suboptimal, and for all resources there is an age and size at which harvesting is optimal. In the case of anchovy and sardine this size is obviously found in Namibian waters, so joint management of anchovy and sardine will require Angola to protect juveniles for the benefit of Namibian fishers. For this arrangement to be agreed and effective, Namibia might have to allow Angolan fishers access to a quota for these stocks in the Namibian EEZ. Alternatively, other trade-offs in other political arenas would need to be offered to Angola in return for their protection of the juveniles.

In contrast, in the case of deep-sea red crab, Angola needs to protect its spawning and nursery areas to ensure sustainable harvesting of the adult component of the stock on both sides of the border.

Namibia v. South Africa

For management of shared stocks to succeed off Namibia and South Africa, there would need to be agreement on harvesting levels of sardine and anchovy and deep-water Cape hake. In the case of these resources, Namibia is considered to be the sink and South Africa the source. Anchovy and sardine juveniles are carried passively and then migrate to northern South African and southern Namibian fishing grounds and return as adults to spawn on the west and south coast of South Africa. Therefore, Namibia would have to protect those juveniles that reach its waters so that South African fishers can benefit. Again, quota or political trade-offs would need to be found.

Deep-water Cape hake constitute a transboundary stock, with spawning taking place in the South African waters and both countries benefiting from the fruits of a healthy stock. The South African macro-economy exceeds that of Namibia in magnitude, but both countries have similar socio-economic conditions. Their objectives of managing fisheries are also similar, notably sustainability, job creation and economic development. It should therefore be relatively simple for deep-water Cape hake to be managed jointly as a transboundary stock by the two countries. However, a first hurdle to be overcome would be to separate the combined management of the two Cape hake species, in the first instance by formulating separate assessments. However, because there is currently no means of employing operational selective harvesting for the two species, and because the time-series for Cape hake in each country is not differentiated, separate assessment is currently not viable, even if desirable.

CONCLUSION

I hope to have shown here that southern Africa has several commercially important stocks that can be categorized as "straddling", according to the definition given by the

FAO. However, despite the region comprising countries that have a long way to go to match the high-tech proficiency of Europe, the Americas, Australasia and the Far East, local politicians and decision-makers are distinctly aware of the challenges they face in optimal, regional fisheries management. That they respond positively to research initiatives and advice, and find the financial means to support it (with and without donor support), surely bodes well for the region. It is gratifying too to be able to report, as Payne & Bannister (2003) wrote, that many of the resources of the region are healthy and growing, perhaps for reasons not solely related to wise management and fishing practice in the past. Indeed, it is the current reasonable status of many of the resources that makes it imperative that regional management is taken seriously. Given the parlous state of many of the world's fisheries resources (FAO, 2000), positive governance initiatives in the southern African region are, in my opinion, a target for other regions to aspire to.

ACKNOWLEDGEMENTS

I sincerely thank the Norwegian *Nansen* programme for supporting my attendance of the symposium at which this paper was read, and Kevern Cochrane, Andy Payne and an anonymous reviewer for their valuable comments on earlier drafts that both focused my thoughts and hopefully made reading the finished product much easier.

REFERENCES

Alheit, J. & Pitcher, T. J. (1995) *Hake: Biology, Fisheries and Markets*. Chapman & Hall, London. 478 pp.

Andrew, P. A. (1986) Dynamic catch–effort models for the southern African hake populations. *Report of the Benguela Ecology Programme of Southern Africa*, **10**, 248 pp.

Babayan, V., Kolarov, P., Prodanov, K., Komarov, Yu. A., Vaske, B. & Wysokinski, A. (1983) Stock assessment and catch projections for Cape horse mackerel in ICSEAF Divisions 1.3+1.4+1.5. *Collection of Scientific Papers of the International Commission for the Southeast Atlantic Fisheries*, **10**(1), 55–62.

Bakun, A. (1996) *Patterns in the Ocean. Ocean Processes and Marine Population Dynamics*. Published jointly by Centro de Investigationes Biológicas del Noroeste, La Paz, Mexico, and University of California Sea Grant, San Diego, USA. 323 pp.

Barange, M., Hampton, I. & Roel, B. A. (1999) Trends in the abundance and distribution of anchovy and sardine on the South African continental shelf in the 1990s, deduced from acoustic surveys. *South African Journal of Marine Science*, **21**, 367–391.

Boyer, D. C. (1996) Stock dynamics and ecology of pilchard in the northern Benguela. In: *The Benguela Current and Comparable Eastern Boundary Upwelling Ecosystems* (Ed. by M. J. O'Toole), pp. 79–82. Deutsche Gesellschaft für Technische Zusammenarbeit (GTZ) Gmbh, Eschborn, Germany.

Boyer, D. C. & Hampton, I. (2001) An overview of the living marine resources of Namibia. *South African Journal of Marine Science*, **23**, 5–35.

Burmeister, L-M. (2000) Survey based assessment of the stock identity of *Merluccius paradoxus* (Franca) in the Benguela. MPhil thesis, University of Bergen: 93 pp.

Butterworth, D. S. (1983) Assessment and management of pelagic stocks in the southern Benguela region. In: *Proceedings of the Expert Consultation to Examine Changes in Abundance and Species Composition of Neritic Fish Resources, San José, Costa Rica, April 1983*. (Ed. by G. D. Sharp and J. Csirke). *FAO Fisheries Report*, **291**(2), 329–405.

Butterworth, D. S. & Geromont, H. F. (2001) Evaluation of a class of possible simple interim management procedures for the Namibian hake fishery. *South African Journal of Marine Science*, **23**, 357–374.

Cochrane, K. L., Butterworth, D. S. & Payne, A. I. L. (1997) South Africa's offshore living marine resources: the scientific basis for management of the fisheries. *Transactions of the Royal Society of South Africa*, **52**, 149–176.

Cushing, D. H. (1972) A history of the international fisheries commissions. *Proceedings of the Royal Society of Edinburgh B*, **73**, 361–390.

Dias, C. C. & Seita Machado, J. F. (1974) Preliminary report on the distribution and relative abundance of deep-sea red crab (*Geryon* sp.) off Angola. *Collection of Scientific Papers of the International Commission for the Southeast Atlantic Fisheries*, **1**, 258–270.

Draganik, B. (1977) Horse mackerel (*Trachurus trachurus*, L) fished in the southeast Atlantic. *Collection of Scientific Papers of the International Commission for the Southeast Atlantic Fisheries*, **4**, 25–65.

FAO. (2000) *World Review of Fisheries and Aquaculture. The State of World Fisheries and Aquaculture* **1**. FAO, Rome. 142 pp.

Fowler, J. L. & Boyd, A. J. (1998) Transport of anchovy and sardine eggs and larvae from the western Agulhas Bank to the West Coast during the 1993/4 and 1994/94 spawning season. *South African Journal of Marine Science*, **19**, 181–195.

Geromont, H. F., De Oliveira, J. A. A., Johnston, S. J. & Cunningham, C. L. (1999) Development and application of management procedures for fisheries in southern Africa. *ICES Journal of Marine Science*, **56**, 952–966.

Gordoa, A., Macpherson, E. & Olivar, M-P. (1995) Biology and fisheries of Namibian hakes (*M. paradoxus and M. capensis*). In: *Hake: Biology, Fisheries and Markets*. (Ed. by J. Alheit and T. J. Pitcher), pp. 49–88. Chapman & Hall, London.

Grant, W. S., Leslie, R. W. & Becker, I. I. (1987) Genetic stock structure of the southern African hakes *Merluccius capensis* and *Merluccius paradoxus*. *Marine Ecology Progress Series*, **41**, 9–20.

Gulland, J. A. (1980) Some problems of the management of shared stocks. *FAO Fisheries Technical Paper*, **206**, 22 pp.

Hampton, I. (1992) The role of acoustic surveys in the assessment of pelagic fish resources on the South African continental shelf. *South African Journal of Marine Science*, **12**, 1031–1050.

Hampton, I., Boyer, D. C., Penney, A. J., Pereira, A. F. & Sardinha, M. (1999) Integrated overview of fisheries of the Benguela Current region. *Thematic Report for the Benguela Current Large Marine Ecosystem Programme*. UNDP, Windhoek, Namibia. 92 pp.

Hutchings, L. (1996) Stock dynamics and ecology of pelagic fish and plankton in the Benguela Current region. In: *The Benguela Current and Comparable Eastern Boundary Upwelling Ecosystems* (Ed. by M. J. O'Toole), pp. 83–96. Deutsche Gesellschaft für Technische Zusammenarbeit (GTZ) Gmbh, Eschborn, Germany. Ministry of Fisheries and Marine Resources, and GTZ, Eschborn.

Konchina, J. V. (1987) Effect of predation on *Merluccius capensis* in the Benguela upwelling area. *Collection of Scientific Papers of the International Commission for the Southeast Atlantic Fish*eries, **14**(1), 249–261.

Le Roux, L. (1997) Stock assessment and population dynamics of the deep-sea red crab *Chaceon maritae* (Brachyura, Geryonidae) off the Namibian coast. MSc thesis, University of Iceland, Reykjavík. 88 pp.

Lutjeharms, J. R. E. & Meeuwis, J. M. (1987) The extent and variability of South-East Atlantic upwelling. *South African Journal of Marine Science*, **5**, 51–62.

Melville-Smith, R. (1983) Abundance of deep-sea red crab *Geryon maritae* in South West African waters from photography. *South African Journal of Marine Science*, **1**, 123–131.

Neto, V. (1997) Fisheries resources of Angola. In: *The Report on Symposium on Science in Africa. American Association for the Advancement of Science Annual Meeting, February 1997, Seattle, USA*, pp. 63–67.

Oelofsen, B. W. (1999) Fisheries management: the Namibian approach. *ICES Journal of Marine Science*, **56**, 999–1004.

O'Toole, M. J. (1977) Investigation into some important fish larvae in the South East Atlantic. PhD thesis, University of Cape Town. 299 pp.

Payne, A. I. L. (1989) Cape hakes. In *Oceans of Life off Southern Africa*. (Ed. by A. I. L. Payne and R. J. M. Crawford (Eds), pp. 136–147. Vlaeberg, Cape Town.

Payne, A. I. L. & Bannister, R. C. A. (2003) Science and fisheries management in southern Africa and Europe. *African Journal of Marine Science*, **24**, 1-23.

Payne, A. I. L. & Punt, A. E. (1995) Biology and fisheries of South African Cape hakes (*M. capensis* and *M. paradoxus*). In: *Hake: Biology, Fisheries and Markets*. (Ed. by J. Alheit and T. J. Pitcher), pp. 15–47. Chapman & Hall, London.

Strømme, T. (1996) An overview of hake research undertaken in Namibian waters by the Dr Fridtjof Nansen Programme. In: *The Benguela Current and Comparable Eastern Boundary Upwelling Ecosystems* (Ed. by M. J. O'Toole), pp. 100–107. Deutsche Gesellschaft für Technische Zusammenarbeit (GTZ) Gmbh, Eschborn, Germany.

Schwartzlose, R. A., Alheit, J., Bakun, A., Baumgartner, T. R., Cloete, R., Crawford, R. J. M., Fletcher, W. J., Green-Ruiz, Y., Hagen, E., Kawasaki, T., Lluch-Belda, D., Lluch-Cota, S. E., MacCall, A. D., Matsuura, Y., Nevárez-Martínez, M. O., Parrish, R. H., Roy, C., Serra, R., Shust, K. V., Ward, M. N. & Zuzunaga, J. Z. (1999) Worldwide large-scale fluctuations of sardine and anchovy populations. *South African Journal of Marine Science*, **21**, 289–347.

Shannon, L. V. (1985) The Benguela Current ecosystem 1. Evolution of the Benguela, physical features and processes. In *Oceanography and Marine Biology. An Annual Review*, **23** (Ed. by M. Barnes), pp. 105–182. University Press, Aberdeen.

Shannon, L. V. (1989) The physical environment. In: *Oceans of Life off Southern Africa* (Ed. by A. I. L. Payne and R. J. M. Crawford), pp. 12–27. Vlaeberg, Cape Town.

United Nations (1983) *The Law of the Sea*. United Nations, New York. 224 pp.

Allocation in high seas fisheries: avoiding meltdown

Douglas S. Butterworth
MARAM (Marine Resources Assessment and Management Group), Department of Mathematics and Applied Mathematics, University of Cape Town, Rondebosch 7701, South Africa
Andrew J. Penney
Pisces Environmental Services (Pty) Ltd, PO Box 31228, Tokai 7966, South Africa

ABSTRACT: The traditional basis for allocating Total Allowable Catches (TACs) for high seas fisheries has been in proportion to performance in terms of past catches. This basis is coming under increasing pressure in Regional Fisheries Management Organizations (RFMOs) as new members, particularly developing states, without records of substantial past performance demand shares, citing acknowledgement of their rights to such in legal instruments such as the 1995 UN Fish Stocks Agreement. Recent difficulties in reaching agreement on allocation in two tuna RFMOs, the International Commission for the Conservation of Atlantic Tunas (ICCAT) and the Commission for the Conservation of Southern Bluefin Tuna (CCSBT) are reviewed, together with an analogous domestic situation involving the redistribution of fishing rights to historically disadvantaged persons in South Africa. Unless rapid accommodation is reached on an operational basis for (re-)allocation, serious depletion of resources threatens if nations all catch what they claim, thereby exceeding sustainable levels. The mechanism of "attrition", whereby each year every rights holder returns a small proportion of their current right to a central authority such as an RFMO for re-distribution, is motivated as a framework to facilitate such an accommodation. Allocation problems can be further exacerbated when scientific disagreements arise over appropriate levels of TAC, and the appointment of arbitration panels of independent scientists (as in the case of CCSBT) is advocated as an appropriate approach to resolve such problems.

INTRODUCTION

The primary broad aims of fisheries management, and how to achieve them in principle, are well known. First, to avoid wastage of resource production potential that results from the tragedy of the commons (Hardin, 1968) requires the imposition of some overall limitation on a fishery, such as an annual Total Allowable Catch (TAC) or cap on total effort. Then there is the need to avoid the race-to-fish until this limitation is reached, and hence the resulting over-capitalization in the absence of entry limitations. This requires imposing individual allocations such as individual quotas.

However, any "imposition" requires an associated authority empowered to impose. National Exclusive Economic Zones (or EEZs), introduced in 1977, essentially solved this problem for coastal fisheries. Coastal states effectively became "owners", and hence allocation authorities, out to 200 nautical miles. However, the UN Law of the Sea Convention (LOSC) did not resolve the matter for resources on the high seas, for which there are no analogous "owners".

The 1995 UN Fish Stocks Agreement (Anon., 1995) is the first legal instrument with worldwide scope that has started to address the issue. Regional Fisheries Management Organizations (RFMOs) are to be the institutions to agree participation rights. Articles 10 and 11 of that Agreement (shown in Appendix 1) set out some general guidelines for the process of determining such rights. In particular, the interests of new members and of developing states are to be considered. However, and importantly, the Agreement goes no further than the broad statements of these Articles; in particular *no* specific formulae for the allocation process are offered.

Now the European Union's objectives for a future Common Fisheries Policy (Commission of the European Communities, 2001, p. 21) include "to promote the responsible and rational exploitation of fishery resources in international waters", an aim with which any other nation, if asked, would no doubt associate itself. With such appearances of good intentions, one might be led to infer that international fisheries allocation issues would be quickly and readily solved, and that the broad guidelines offered by the 1995 UN Fish Stocks Agreement are as far as there is any need to go. However, do such inferences reconcile with reality?

In the past, the primary basis for allocation in RFMOs has been to share in proportion to performance in terms of past catches (see, for example, frequent references to limiting catches to a proportion of "recent" past catches in the International Commission for the Conservation of Atlantic Tunas (ICCAT) recommendations for bluefin tuna, bigeye tuna and swordfish, and initial proportional ICCAT quota allocations for bigeye tuna and swordfish; ICCAT, 2002a). Even that simple formulation was not without its problems: arguments have ensued as to the reference period over which such past catches should be considered. More recently, however, the impacts of two other much more difficult issues have been growing rapidly. The first is increasing catches, often over and above the TAC, by countries that are not members of the relevant RFMO. Second, if new members do join RFMOs, under the historic performance criterion they find themselves offered hardly any share in the TAC. As they would choose to describe matters, the "historic performance" approach to allocation rewards nations in proportion to their contribution to past overfishing!

This paper addresses the dangers of fishing nations failing to reach a rapid accommodation on this issue, *viz.* the meltdown situation of each nation then proceeding to catch what it claims as its share irrespective of overall limits required for sustainability, with consequent overexploitation and depletion of resources. The material presented concentrates on two international tuna Commissions: ICCAT and the Commission for the Conservation of Southern Bluefin Tuna (CCSBT). There are two reasons for this. The first is the high value (bluefin tuna can command a price in excess of US$200 per kg) and consequent importance of fisheries for these highly migratory animals (in addition to the two RFMOs named above, the international tuna commissions also include the South Pacific Forum Fisheries Agency (FFA), the Inter-American Tropical Tuna Commission (IATTC) and the Indian Ocean Tuna Commission (IOTC)). Second, both authors are able to draw on personal experiences through their participation in activities of both ICCAT and CCSBT over a number of years. The latter reason underlies the inclusion as well of a brief discussion of a recent domestic fishing rights re-allocation process in South Africa. For reasons to be explained, the circumstances associated with this process mirror the very same problems with which RFMOs currently have to deal. It is therefore of interest to examine whether experiences gained in attempting to address these problems at a local level in South Africa offer any guidance for how best to deal with the high seas situation. The paper concludes by advocating a mechanism – "attrition", whereby each year each rights-holder returns a small proportion of their current right to a central authority (e.g. RFMO) for redistribution – to facilitate reaching agreement on rights allocation in high seas fisheries.

The authors stress that this paper makes no claim to constitute an exhaustive evaluation of the issue of allocation in high seas fisheries; rather, it is a reflection of personal views and suggestions arising from their own experiences in that area.

THE SOUTH AFRICAN EXPERIENCE

Following the first open elections in South Africa in 1994, pressures arose for greater participation in fisheries (at the level of "ownership") by historically disadvantaged persons – the fishing sector was seen to need such "transformation". Thus, South Africa faced essentially the same issues as arise with re-allocation of rights on the high seas: how to incorporate new entrants, given generally fully utilized resources, each with their own mature associated industry.

The first attempt at this exercise proved anything but successful. A new Act, the Marine Living Resources Act of 1998, followed by a call for quota applications, found the Marine & Coastal Management (MCM) section of the responsible Government Department of Environment Affairs & Tourism quite unable to deal effectively with the flood of applications it received. There were inadequate checks of information on applications, lack of a clear basis for adjudication and accusations of undue influence. MCM became overwhelmed in litigation, primarily in relation to reductions of past rights. Furthermore, many persons with newly awarded quotas considered their own immediate economic interests best served by hiring the existing industry to catch on their behalf, rather than making investments to develop the capacity to fish themselves. In the meantime, the existing industry was plagued by the uncertainties associated with an unpredictable annual re-allocation process.

The appointment of some new senior staff at MCM brought a new initiative in 2000. The two key emphases were to restore some stability to the industry and to minimize the potential for litigation. To these ends, rights were to be issued for four years, and stress was laid on careful specification of the processes to be followed so as to ensure legal defensibility (Kleinschmidt *et al.*, 2003).

This new initiative sought a number of features in applications submitted. These included evidence of transformation (a move towards greater involvement of historically disadvantaged persons) within companies at the level of management staff and ownership, and a track record of past performance and investment indicative of long-term future commitment to the industry. Particular stress was laid upon an assessed low likelihood of creating a "paper quota" – allocating a right that the holders would then use simply to pay existing industry to catch for them – because this was seen to be contrary to the objective of genuine transformation, and the "free lunch" connation made a poor impression on the public and particularly those engaged in the actual fishing. Legal teams checked applications, and point scores against the criteria were summed. For some fisheries, limited allowance was made for new entrants, although it should be noted that the process over the previous few years had already introduced some such historically disadvantaged persons to the industry.

These various aims were not without the difficulties of internal contradictions. How could a potential new entrant show investment intent in circumstances where there was pressure to peg vessel numbers in fully subscribed fisheries? Restrictions on transfers and related endeavours to avoid paper quotas served to constrain the flexibility for exchanges between companies that is needed in mixed species fisheries regulated by TACs, to lessen problems such as discarding catches of species no longer wanted when quota limits are about to be reached.

A few features seem, on initial consideration, to characterize the outcome of this process, which was completed early in 2002 for the major South African fisheries (for trawled hake, sardine and anchovy, and rock lobster). The primary function that the points-scoring system appears to have served was the exclusion of most paper quota-holders and of companies that showed little evidence of transformation. Otherwise, changes were of the order of a few percentage points only, both between existing holders and in terms of proportions allocated to new entrants.

LESSONS LEARNED

Important lessons that emerge from the South African process would seem to be:

(1) The system applied has to be imposed from the top; lengthy prior discussions including current and potential new rights-holders produced no workable compromise. This is hardly surprising, because their interests proved to be in total conflict: the former did not want to relinquish any of their existing holdings, whereas the latter laid claim to very large allocations.

(2) Redistribution between companies, though not necessarily transformation within companies, when viewed over the past few years, has been gradual, even though that was not always the intent. Anything else would probably not have been feasible anyway, because the greater the loss to an existing rights-holder through redistribution, the greater the chance of that holder initiating lengthy litigation.

(3) The overall process of redistribution also created undue expectations in the wider community. The inadequacy of the resource base to meet those expectations led to problems of lack of acceptance, with resultant compliance problems.

The case of abalone provides a dramatic example of this last point. Spiralling levels of poaching (see Figure 1), which have now exceeded the legal commercial TAC (Anon., 2003), coincided with the period of the rights re-allocation process (though the increasing involvement of a serious criminal element in this poaching, with links to drug-trafficking, also played a role in the increase; Hauck & Sweijd, 1999).

Do these lessons from the South African experience translate to the high seas? There is no equivalent high seas authority to impose detailed rules "from the top". However, it certainly does seem essential that any re-allocation process incorporates gradual change if it is to be acceptable to existing rights-holders, whose resistance is motivated not only by self-interest, but also because time is needed to be able to accommodate change. There is also a danger of creating undue expectations in high seas fisheries, which can lead, in turn, to lack of acceptance, compliance problems and, consequently, overexploitation.

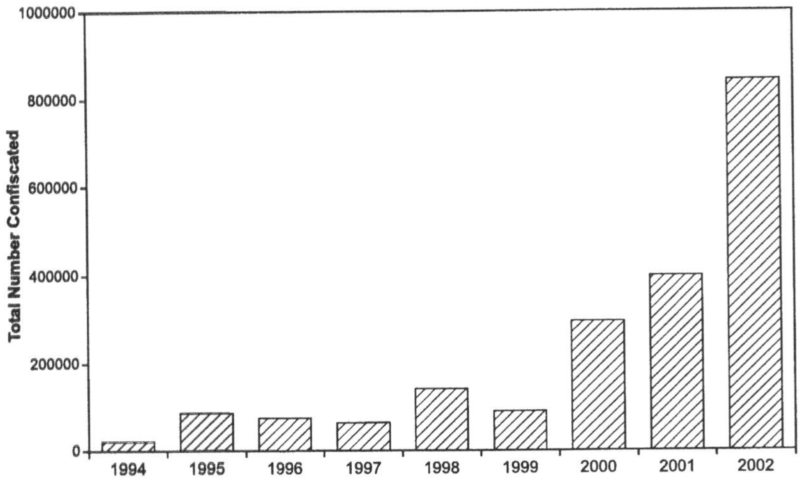

Fig. 1 Total annual confiscations of illegally caught abalone taken in South African waters from 1994 to 2002.

THE ICCAT EXPERIENCE

ICCAT was established as an independent Regional Fisheries Management Organization in 1970. In terms of its founding Convention, ICCAT is responsible for the management of all tunas, billfish and associated species (some 30 species of direct interest) throughout the Atlantic Ocean (ICCAT, 1985). Since its inception, ICCAT has had a broadly representative membership, with 23 member countries by 1988 and 32 members by 2002. These include most countries (both developed and developing) bordering the Atlantic Ocean, as well as a number of countries operating distant-water fleets in the region.

EARLY ICCAT MANAGEMENT INITIATIVES

Besides development of databases, sampling programmes and initial stock assessments, early ICCAT management initiatives naturally focused on Atlantic bluefin tuna, in response to early indications of declines in catch rates of this high-value species in the western Atlantic. Initial ICCAT minimum size and effort limitations were implemented for this resource as early as 1974. However, the primary focus in the early years was on resolving the non-member, primarily "flag-of convenience" (FOC; use of non-member flags to avoid management measures applied to member states) problem. To this end, a specific Permanent Working Group on ICCAT Statistics and Conservation Measures (PWG) was established to evaluate tuna catches by non-members and to recommend measures to monitor and control these. By comparison, little attention was paid to compliance by members, even though their catches were rapidly increasing to exceed sustainable levels in the North Atlantic bluefin and swordfish fisheries.

While ICCAT has long had the mandate to manage Atlantic tunas, it was the 1995 UN Fish Stocks Agreement (Anon., 1995) that really emphasized the obligation of those requiring Atlantic tuna fishing rights to join, or cooperate with, ICCAT. This Agreement (see Appendix 1) grants RFMOs the function of allocating fishing rights and also establishes the obligation of states to join established RFMOs. Further, it lays the legal foundation for the ensuing work of the PWG in monitoring trade and generating letters of warning to many FOC countries (notably Panama, Honduras and Belize) from 1996 onwards, followed by recommendations for import prohibitions against these countries consistent with World Trade Organization (WTO) requirements (see the ICCAT Compendium of Recommendations; ICCAT, 2002a). Most of these countries responded by participating as observers at ICCAT and removing identified FOC vessels from their registries. Remaining vessels were warned to comply with all ICCAT management measures.

However, to report Atlantic tuna catches under their flags, these countries required ICCAT quota allocations. Panama was the first to take the step of joining ICCAT in order to negotiate allocations. A number of Caribbean island states (Trinidad & Tobago, Honduras, Barbados) then joined to protect their regional tuna fisheries, followed by a few African countries (Libya, Algeria). The eventual accession of the EU to ICCAT in 1977 brought with them countries such as Italy, which also had substantial tuna fisheries, but no quotas. These EU members also demanded allocations, particularly for the Mediterranean tuna resources, and Croatia soon followed suit to protect her long-standing involvement in the Mediterranean bluefin tuna fishery.

In addition to this increase in actual membership, there has also been a dramatic increase in observers, who now almost outnumber members at annual Commission meetings. Most observers also desire allocations, but they are waiting for clarity on revision of ICCAT allocation criteria before considering membership. In effect, the Permanent Working Group's initiatives have almost been too successful, and the non-member problem has been converted into a new-member problem. All these new (and prospective new) members desire formal quota allocations, leaving ICCAT little choice but to depart from the traditional basis of past performance for allocation and to develop a revised set of allocation criteria aimed at achieving a balance between existing rights and the claims of new participants.

DEVELOPMENT OF ICCAT ALLOCATION CRITERIA

Not surprisingly, ICCAT initiatives to develop a revised basis for sharing allocations between existing and new members focused on bluefin tuna and swordfish, these being the only stocks subject to ICCAT TAC controls in 1997 (when the issue became critical), and therefore the only stocks for which new entrants required allocations at that time. It was, in fact, dispute between coastal states and distant water fleets over the development of sharing arrangements for the South Atlantic swordfish stock in 1997 that precipitated the search for a new allocation scheme within ICCAT. The recent management history of the South Atlantic swordfish stock therefore serves as a useful example to illustrate the effect of these developments.

The distribution of swordfish catches in the southern Atlantic Ocean in 1996, just before ICCAT implemented the first TAC and sharing arrangement for this stock, shows that this stock already supported a well developed fishery (ICCAT, 2000; see Figure 2). To understand the problems in trying to negotiate a sharing arrangement in the south, one needs first to look at the catch history for North Atlantic swordfish. Over the ten years from 1978 onwards, North Atlantic catches more than trebled (ICCAT, 2002b; see Figure 3). Restrictions were first mooted in 1987 when catches reached a record level, exceeding

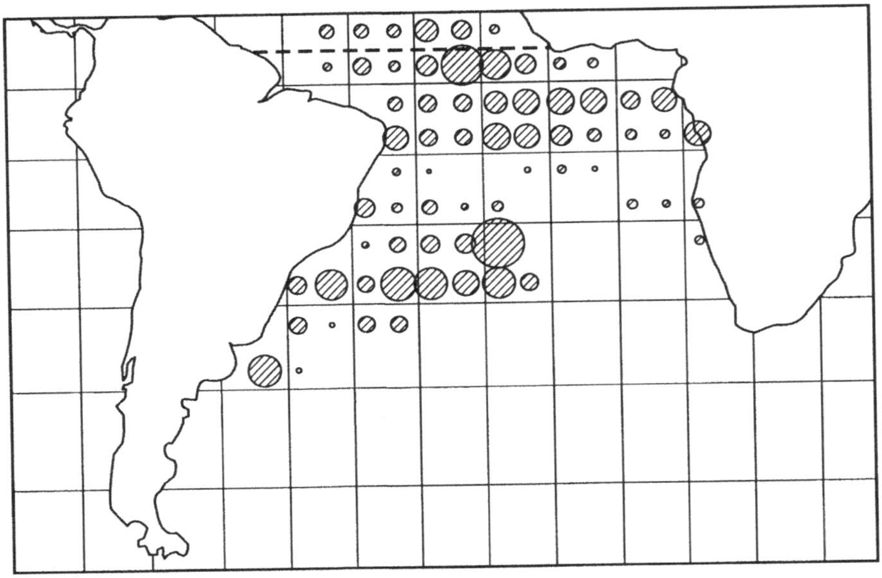

Fig. 2 Distribution of catches of swordfish in the South Atlantic in 1996, just before ICCAT established the first TAC and sharing arrangement for the resource (the dashed line shows the assumed division between North and South Atlantic stocks).

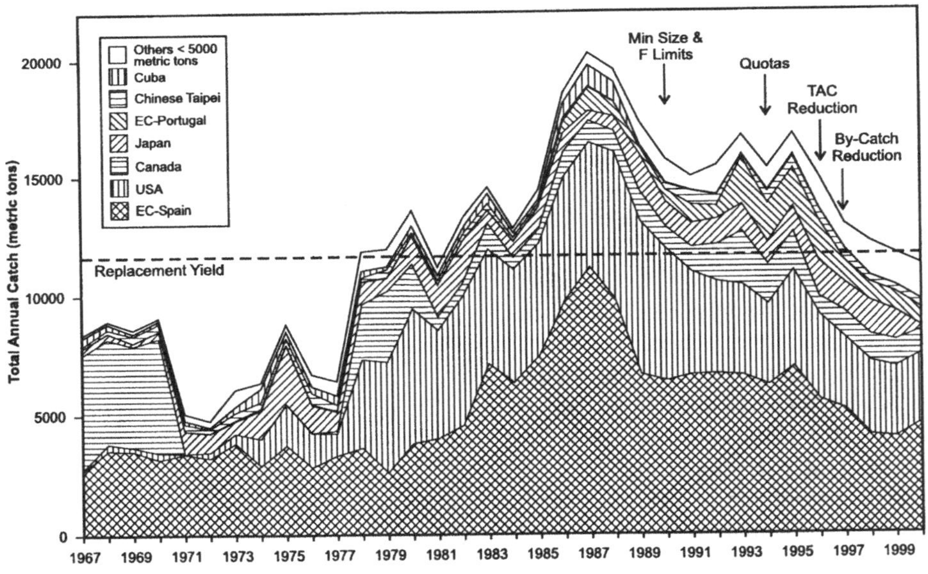

Fig. 3 Historic catches by country of North Atlantic swordfish from 1967 to 2000, showing the 2000 estimate of replacement yield (dashed line) and the years in which various management measures were introduced (after ICCAT, 2002b).

20 000 metric tons. Assessments indicating overexploitation resulted in ICCAT imposing a series of restrictions on North Atlantic swordfish from 1990 onwards, including minimum size limits, effort caps, TACs with quota allocations, and limits on by-catches. These assessments contributed to a Spanish decision to reduce their North Atlantic catches over the final years of the 1980s. However, this had important consequences for the southern stock.

In contrast, the swordfish fishery in the South Atlantic remained small prior to 1987, annual catches between 1967 and 1987 averaging just 4 300 tons, half of which was taken as non-targeted bycatch in the Japanese bigeye-tuna-targeted longline fishery (ICCAT, 2002b; see Figure 4). This situation changed dramatically after 1987. In response to declining catch rates and mean sizes, and anticipating the initial effort limitations in the North Atlantic, Spanish fleets transferred a substantial amount of fishing effort to the South Atlantic, taking 4 400 tons themselves and boosting 1988 catches to almost 13 000 tons. Japan's catch also increased, and the resulting landings in various South Atlantic ports stimulated the interest of many coastal states who had small, developing longline fisheries. Combined total catches escalated rapidly, reaching 21 750 tons in 1995, surpassing even the record catch in the North Atlantic (ICCAT, 2002b; Figure 4). Estimated replacement yield (RY) levels (estimated at 14 800 tons in 2001) were exceeded in 1989 and again in 1993.

First warnings of the need for a TAC for South Atlantic swordfish were sounded in 1996, but agreement could not be reached on allocations at that meeting, largely as a result of insistence by Brazil that the new criteria listed in the UN Fish Stocks Agreement

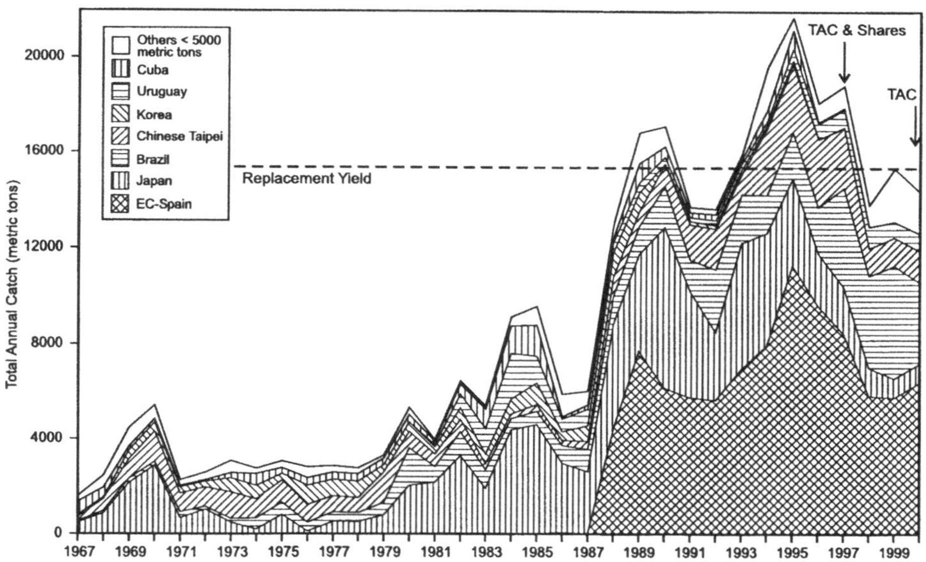

Fig. 4 Historic catches by country of South Atlantic swordfish from 1967 to 2000, showing the 2000 estimate of replacement yield (dashed line) and the years in which various management measures were introduced (after ICCAT, 2002b)

replace past performance as the basis for developing a sharing arrangement. A dedicated intersessional meeting of ICCAT Panel 4 (which is responsible for swordfish management recommendations) was therefore arranged and hosted by Brazil in 1997. At that meeting, the criteria listed in the UN Fish Stocks Agreement were used by Brazil as part of a substantial motivation to move away from past performance as the primary allocation criterion. However, no agreement was reached on criteria at all, and the South Atlantic sharing arrangement for the period 1998– 2000 was bargained in a closed meeting of Heads of Delegations.

The resultant South Atlantic swordfish sharing arrangement was considered by many South Atlantic coastal states (who had little past performance, and received no allocations) to be unacceptable, and pressure mounted for a complete review of the ICCAT quota allocation processes. A set of recommendations was finally accepted in 1998 for establishment of the ICCAT *ad hoc* Working Group on Allocation Criteria, to develop recommendations on revised ICCAT allocation procedures.

The Allocation Working Group had four meetings, three annual intersessionals in 1999, 2000 and 2001, and one special concluding session preceding the 2001 Commission meeting. The first two meetings were characterized by a stand-off between the "Group of 16" coastal states lobby (led by Brazil, Venezuela and South Africa) and the EU/Japan alliance. EU/Japan insisted that any new allocation process could not apply to stocks that had already been allocated (these being bluefin tuna and swordfish, the very reasons for establishment of the Allocation WG in the first place), and that past performance must still remain the primary criterion for new allocations. The Green Paper on a revised EU Common Fisheries Policy (which recognized the need to acknowledge fishing rights for coastal states) was issued at about that time (Commission of the European Communities, 2001), and the EU relented somewhat on their demands. The third meeting saw substantial compromise, and a comprehensive list of "ICCAT Criteria for the Allocation of Fishing Possibilities" was finalized at the fourth meeting, and adopted by the ensuing 2001 Commission meeting (ICCAT, 2002c; see Appendix 2).

Underlying the criteria is the idea that a nation must be a "responsible" member of ICCAT to receive an allocation. In all, 15 allocation criteria were adopted, reflecting a balance between ones that can be used to motivate allocations to existing participants, and those favouring new entrants (see Appendix 2). The approach of gradual change to minimize economic disruption was recognized and the guiding concept in application is "fair and equitable, ensuring opportunities to all". However, the Allocation Working Group could not agree on quantitative weightings or formulae linked to these criteria, and it was left to the ICCAT Species Panels to decide how to apply the criteria on a stock-by-stock basis as and when sharing agreements are (re)negotiated.

THE ICCAT ALLOCATION CRISIS

Although no formal objections were made to the 1997 southern swordfish allocations themselves, Brazil, Uruguay and South Africa caused substantial concern by objecting to extension of punitive compliance measures (quota reductions and import prohibitions) to South Atlantic swordfish (ICCAT recommendation COMPLY 97–8; ICCAT, 2002a) under this sharing arrangement. This arrangement provided the option for revision for 2000 and expired at the end of 2000. However, a South African attempt in 1999 to table a proposal for revision for 2000 was rejected by Panel 4 and the arrangement remained in force until it expired at the end of 2000.

With the attempts of the Allocation Working Group to develop revised allocation criteria still being hampered by dispute, efforts to negotiate a revised TAC and sharing arrangement were entirely unsuccessful at the 2000 and 2001 Commission meetings. For the first time in ICCAT history, the 2000 Commission meeting failed to negotiate a continued sharing arrangement to replace that which had expired (ICCAT, 2001). If anything, the situation at the 2001 Commission meeting was worse, with frequent informal meetings of Panel 4 being characterized by extremely confrontational, informal (closed session) efforts to implement the new allocation criteria as a basis for revising allocations. By the end of 2001, no agreement had been reached on a new sharing arrangement for South Atlantic swordfish (ICCAT, 2002d). Repeated efforts to implement the new allocation criteria failed, with existing holders refusing to decrease catches and new entrants demanding substantial allocations.

In the absence of an agreed sharing arrangement, Panel 4 adopted a recommendation in 2000, and again in 2001, calling on active South Atlantic swordfishing nations to declare their own voluntary quotas, supposedly based on their final negotiating positions at the 2000 Commission meeting (whose total was not too different from RY). However, the total of the actual declared quotas greatly exceeded RY, with two coastal states (Brazil and Namibia) substantially increasing their declarations above their stated positions at the meeting. Fortunately, however, these claims proved to be inflated and actual catches taken in 2001 remained below the recommended TAC, which had been proposed at the RY level.

The situation was similar with the most important of the ICCAT TAC limited species, Atlantic bluefin tuna. Bluefin tuna catches in the West Atlantic have been tightly controlled since 1982. However, those in the east show a similar pattern to South Atlantic swordfish, but with the added complication that catches remain well above the estimated RY (see Figure 5). Libya and Morocco had formally objected to the 1998 BFT quota allocation scheme for 1999/2000 (BFT 98–5; ICCAT, 2002a) and declared their own quotas, so contributing to this excess catch. This sharing arrangement also expired at the end of 2000 and, as with swordfish, no agreement could be reached at the 2001 meeting on a sharing arrangement to replace that for East Atlantic bluefin, despite the existence of the agreed new allocation criteria.

In an effort to increase the scope for a sharing arrangement to accommodate new members, the EU tabled a proposal for a TAC more than 30% above the estimated RY. This proposal was rejected by the USA and Canada and, again for the first time in ICCAT history, the proposal was brought to the plenary session for a vote. However, because many new members had left after presenting statements of protest early in the meeting, there was no quorum present, so no vote on the proposal was possible. Like South Atlantic swordfish, by the end of the 2001 Commission meeting, East Atlantic bluefin were also left without a TAC or sharing arrangement. However, in this case, catches continued to exceed the estimated RY, presenting a real threat of ongoing stock decline.

By the end of 2001, ICCAT was in crisis, with an increasing number of members objecting to management recommendations. Past sharing arrangements for bluefin tuna and swordfish (the only stocks subject to ICCAT TAC limits) had expired, and no replacements were in sight. Members were asked to declare voluntary quotas for southern swordfish, but the totals of those substantially exceeded scientifically recommended TAC levels. Those who had allocations made every effort to retain them, while prospective developing or new participants pushed for substantial re-allocation.

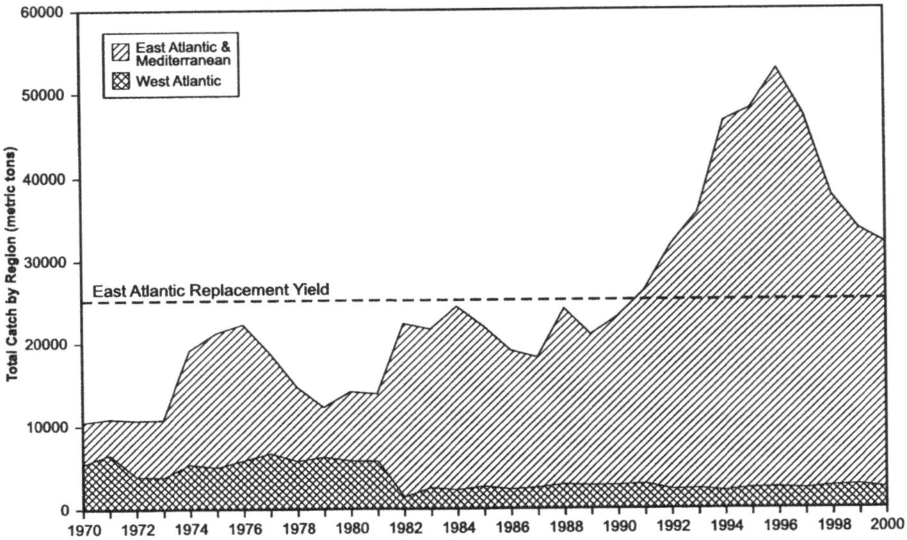

Fig. 5 Historic catches of North Atlantic bluefin tuna for the East Atlantic (including the Mediterranean) and West Atlantic components of the resource, showing the estimate of the 2000 replacement yield for the East Atlantic component (dashed line) – after ICCAT (2002b). (Note that these plots are not stacked, but are shown one behind the other. The RY line applies to the East Atlantic.)

The strategy of some major past quota-holders appeared to be to push for increased TACs above the level scientifically recommended, to accommodate new members without immediate reductions themselves, leaving it to the future to determine some slow overall phased reduction. There was virtually no sign of compromise between these opposing lobbies, and new members (primarily developing coastal states) held a blocking minority, preventing ICCAT from accepting management proposals, even if these were put to the vote.

Revised sharing arrangements for these two stocks were finally accepted and implemented at the 2002 Commission meeting (ICCAT, 2003). It is worth noting the factors that apparently contributed to the success in reaching agreement on a sharing arrangement at the 2002 meeting, after the failure of the previous two meetings. After two years of disagreement, opposing parties were essentially forced to substantiate their use of the allocation criteria they particularly favoured. Although these criteria were not formally quantified, this substantiation process made it fairly clear what weight the parties could justifiably attach to each criterion. This resulted in a slow acceptance of the merits (and perhaps inevitability?) of opposing arguments, and a resultant gradual compromise, both from existing rights-holders and new entrants. The main result of this compromise was acceptance of the need for gradual, phased change in allocations, stability of the overall scheme of change over a moderately long period (about five years) and a clear statement of the result at the end of that period.

LESSONS LEARNED

ICCAT has succeeded fairly well as an RFMO with regard to data collection, evaluation of catches of non-members and generation of stock assessments. The scientific foundation and resultant management advice are generally sound. Efforts to encourage membership, and to combat the illegal, unreported and unregulated (IUU; Kirkwood & Agnew, 2003) catch problem, have been successful beyond expectation, and a comprehensive list of allocation criteria, founded in international law, has been developed and accepted. However, this has hardly helped, and all the serious management problems are now within, rather than outside, ICCAT. The original members are having great difficulty coming to terms with the new entrants, and first attempts to apply the new allocation criteria failed spectacularly at the 2001 Commission meeting. Is it even possible to apply these non-quantitative criteria, and to obtain the cooperation of new members, without unduly disrupting existing industries, or overexploiting resources?

The main lessons evident from the ICCAT situation would appear to be:

(1) There is a clear need to provide for periodic change in high seas quota allocations, particularly to accommodate new members or fishery developments. However, this must be balanced by measures to promote stability. This dictates that change must be gradual, or phased in some agreed way.

(2) The need to accommodate new members specifically dictates a move away from past performance as the sole basis for calculating allocations. However, adoption of a broad suite of non-quantified allocation criteria is of little help. Until these have been quantified in some way, they serve only as counter-arguments in a sharing debate. Given time, however, meaningful weightings (at least to the participants) will almost inevitably evolve during such debates. It is to be expected that these weights will also change over time.

THE CCSBT EXPERIENCE

The recent history of CCSBT is of particular interest in the context of allocation for a further reason – the interweaving of scientific assessment-related issues with the allocation problem, which actually led to international litigation. Figure 6 illustrates the distribution of catches of southern bluefin tuna (SBT) over the period 1960–1993 (CCSBT, 2002). SBT is a species that spawns off Indonesia. For about their first four years of life, these fish concentrate off west and south Australia. They then distribute widely, centred on about 40° S, and ranging from the mid-Atlantic to east of New Zealand.

Historical catches of southern bluefin tuna (see Figure 7) have been dominated by Japan and Australia. Japan's longline catch has been larger by weight, but during the 1980s in particular, Australian purse-seine operations on younger fish claimed the larger catch by number (CCSBT, 2002). Concerns about this and declining longline catch rates led to an informal management arrangement being established in the early 1980s by Australia, Japan and New Zealand. By the end of the decade, the TAC to these three countries had been cut to just below 12 000 tons. In the early 1990s, these countries formalized their cooperation with the creation of the Commission for the Conservation of Southern Bluefin Tuna (CCSBT). The Commission has yet to formally agree any TAC for the species, but its member nations did concur to limit their catches to their last allocations under the preceding informal agreement. Therefore, the total catch by members did not grow during the mid-1990s after the CCSBT came into operation, but the overall

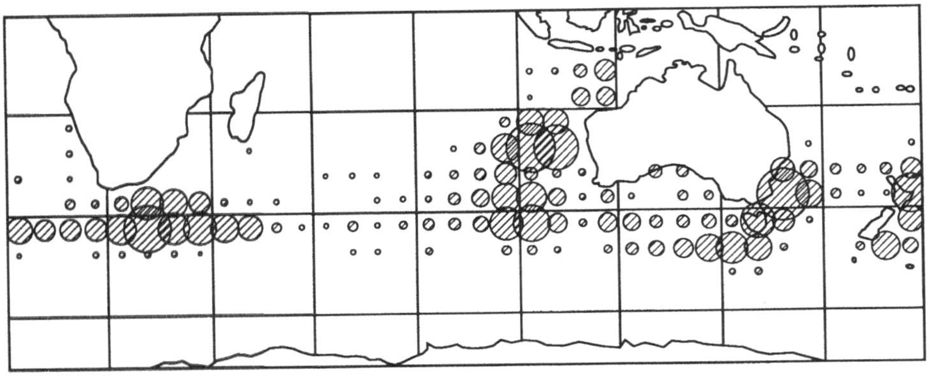

Fig. 6 Geographic distribution of catches of southern bluefin tuna over the period 1960–1993 (by 10×10°rectangle) – after CCSBT (2002)

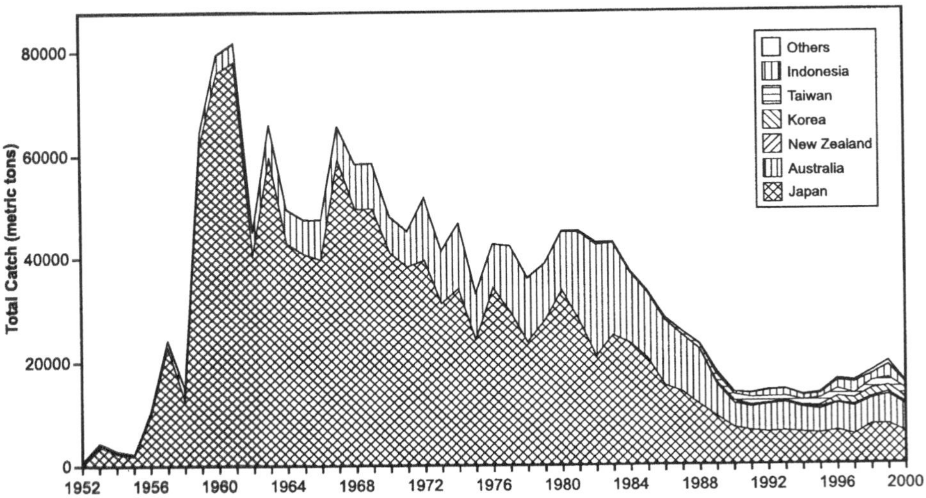

Fig. 7 Historic catches of southern bluefin tuna by country from 1952 to 2000 (after CCSBT, 2002).

catch taken increased as a result of expanding non-member operations, notably by Indonesia, Korea and Chinese Taipei.

In addition to concerns about this non-member catch expansion during the 1990s, the two major participants in the fishery found themselves subject to different economic pressures. Australia had developed a very profitable cage operation, with purse-seined bluefin being fed and fattened prior to sale to Japan. As this operation expanded to utilize virtually the entire Australian allocation for southern bluefin tuna, their industry remained reasonably satisfied with the *status quo*. Japan, in contrast, had implemented a government-subsidized 20% tuna longliner fleet reduction programme (Suisan Keizai Shinbun, 1998). Their industry was less than enthusiastic about bluefin sacrifices that appeared simply to be translated into growing catches by neighbouring countries, outside of the CCSBT, and they sought relief in the form of a TAC that allowed for a catch increase.

THE CCSBT ASSESSMENT DISPUTE

These pressures for a catch increase became entangled in a scientific dispute over the assessment of the resource. Had the catch reductions of the 1980s facilitated a recovery, which could justify such an increase? A primary, though not the only, component of this dispute concerned interpretation of catch-per-unit-effort (cpue) data. Longliner cpue had improved, but the temporal and areal extent of the fishery had reduced. This led to problems of interpretation, because cpue is related to local density, so indices of abundance need to incorporate the integration of cpue over space and time, and some of the data required for this were no longer available, following the reduction in the extent of the fishery.

At the one extreme of interpretation, the reduction in the area fished was an operational consequence of the earlier TAC reduction, longliners no longer needing to fish as widely or for as long to fill their quotas, so the cpue in months and areas (5×5° rectangles, termed "squares" in the CCSBT) no longer fished could be assumed to be the same as that in ones that continued to be fished. This so-called "constant squares" interpretation produced assessments that suggested a recovery (see Figure 8). However, at the other extreme lay the assumption that the decreased coverage by the fishery reflected a real drop in the areal distribution of the population associated with a decline in abundance. Accordingly, areas and times no longer fished should be assumed to reflect zero cpue. When integrating cpue to provide an index of abundance under this "variable squares" approach, assessments suggested no recovery (see Figure 8).

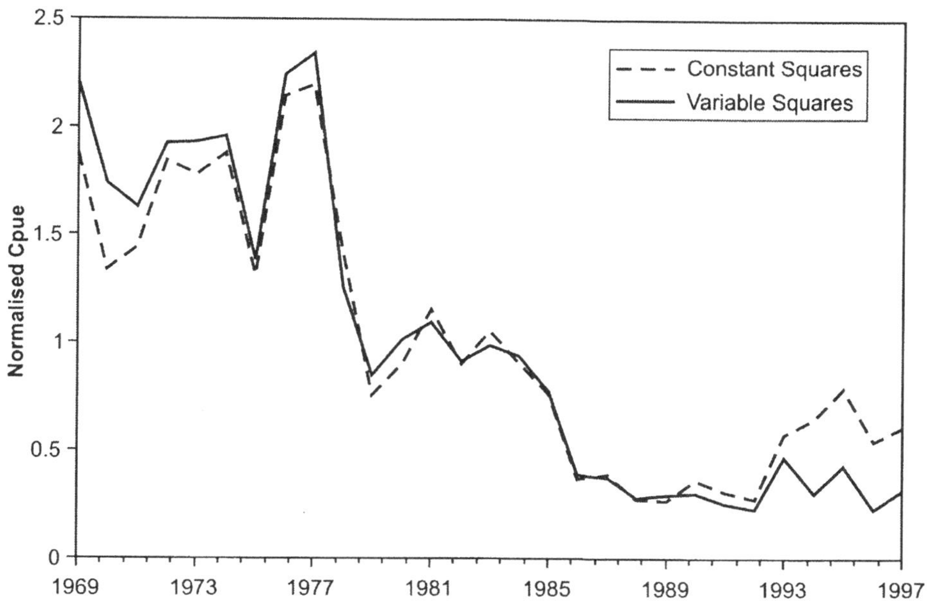

Fig. 8 The cpue trends calculated for bluefin tuna of ages 6 and 7 by application of the "constant squares" and "variables squares" assumptions (see text for details). Both series are normalized to their average value over the period shown (data from Polacheck & Preece, 1998).

In 1998 Japan unilaterally initiated an experimental fishing programme (EFP) in some of these now unfished spatio-temporal locations, to provide a basis to calibrate between the variable and constant squares interpretations. The catches made during the experiment were over and above Japan's existing allocation as recognized by the other members of the CCSBT. The methodology underlying the EFP was disputed by Australia and New Zealand, and extensive scientific attempts early in 1999 to reach consensus failed. When Japan thereafter unilaterally continued with the EFP, Australia and New Zealand served notices of claim on Japan for arbitration under the dispute provisions of the Law of the Sea Convention (LOSC). The resultant five-member Arbitral Tribunal met in 2000. Before that, however, Australia and New Zealand sought "Provisional Measures" (a form of interim injunction), in abbreviated proceedings before the International Tribunal of the Law of the Sea (ITLOS) in August 1999. ITLOS, based in Hamburg, Germany, consists of 21 permanent judges, with one *ad hoc* judge added by the plaintiffs in this instance because one of the permanent judges was from Japan.

The matter was only the second to be brought before ITLOS, and the first related to fisheries. Japan contended that the dispute arose under the CCSBT Treaty, and that that Treaty's dispute resolution provisions should govern. However, ITLOS ruled that they did have jurisdiction, and ordered Japan's 1999 EFP catches to count against their allocation for that year (International Tribunal for the Law of the Sea, 1999). However, the following year, the Arbitral Tribunal ruled that resolution of the dispute should take place under the CCSBT Treaty provisions, not the LOSC. Through this over-ruling of ITLOS on the issue of jurisdiction, the ITLOS order effectively fell away (Internet Guide to International Fisheries Law, 2000). The Arbitral Tribunal therefore never reached the stage of considering the merits (specifically the *scientific* merits) of the case.

The question of whether ITLOS itself had ruled on these scientific merits is important, because of its wider and longer term implications. ITLOS' order stated that it could not conclusively assess the scientific evidence. However, it also stated that the measures it ordered were necessary to avert further resource deterioration. Unfortunately, ITLOS relied heavily on misquotation of a key CCSBT Scientific Committee minute on stock status in motivating its decision (ITLOS stating in paragraph 71 of their order that the stock was "at its historically lowest levels", instead of "historically low levels"; Morgan, 2001).

On a later occasion, ITLOS Vice-President Judge Wolfrum reaffirmed that it was not for ITLOS to decide on the stock assessment. However, he argued that ITLOS had heard evidence that patterns in the SBT fishery were similar to those for Canadian cod, which had collapsed. He further asserted that, because there was uncertainty regarding the assessment, prudence dictated proceeding from the worst-case scenario (Wolfrum, 2000). Are these standpoints self-consistent?

The issue appears to turn on the same problem that arises with the "Precautionary Principle/Approach", which was repeatedly referenced in Australia and New Zealand's original statement of claim. The Precautionary Approach, as enunciated in Principle 15 of the UNCED Rio Declaration (Agenda 21) of 1992, states that "Where there are threats of serious or irreversible damage, lack of full scientific certainty shall not be used as a reason for postponing cost-effective measures to prevent environmental degradation." However, there will always be some degree of scientific uncertainty, so the relative plausibility of alternative scenarios has to be factored into associated decisions when considering risk–reward trade-offs. Interpreting instead on a worst-case scenario basis can lead to unnecessarily conservative decisions, because this implies that catches need to be decreased to ensure sustainability even under extreme interpretations of the available data, despite their low plausibility. Did ITLOS realize this? Was their order not tantamount to ruling on the scientific merits of the case?

THE CCSBT RESPONSE

With the ball back in its court, the CCSBT decided to depart from past practice and appoint Chairs for its Scientific Committee and Stock Assessment Group from outside member countries' scientific delegations. Furthermore, a panel of (ultimately) four external scientists was appointed to participate in these groups. The panel was instructed to compile reports, making every effort to obtain consensus among member nation scientists. Should this not prove possible, however, the panel's report would still go to the Commission, with dissenting opinions attached (CCSBT, 2000). Originally these instructions applied only to development of a SBT scientific research programme, but they have since been taken to apply generally to the panel's contributions to CCSBT Scientific Committee deliberations. The practical outcome of this has been very strong pressure towards achieving consensus, in place of the previous practice of disparate "some and others" views from the Scientific Committee forwarded to the Commission.

The present scientific consensus in the CCSBT is that recent catch levels, including catches by non-members, appear sustainable, but that there is no room for an increase. Work is in progress to develop a formal Management Procedure (*sensu*, for example, Butterworth & Punt, 1999) as the basis for future TAC recommendations. This Procedure will likely depend primarily on trends in cpue.

As for the Commission itself, Korea and Chinese Taipei have recently been persuaded to join, and reductions in their recent levels of catch have been negotiated. Disagreements remain, though not nearly as substantial as before. They primarily involve the extent to which negotiated new member catch reductions should be used to provide catch increases to existing members, or to promote stock rebuilding.

LESSONS LEARNED

The first important lesson from the CCSBT experience would seem to be that the external scientific panel (together with its empowering terms of reference) proved more successful at resolving scientific disputes than recourse to litigation. Further, meaningful appeal to, and application of, the Precautionary Approach requires quantification of an acceptable level of risk. The CCSBT has been fortunate that non-member catches do not seem to have increased sufficiently to have led to further stock decline. This has facilitated movement towards agreement on allocations following Korea and Chinese Taipei joining as new members.

However, what options remain open should further countries seek membership and allocations? Indeed this problem is now virtually upon the Commission, following an indication given in 2002 by South Africa of interest to join, suggesting at the same time that an annual allocation to them in the vicinity of 300–400 tons might be appropriate. In initial discussions of this initiative, some CCSBT members cited past performance (South Africa has a negligible past catch record) as the only acceptable basis for an allocation.

TOWARDS SOLUTIONS

The key problem in high seas fisheries allocation remains the absence of operational (or formulaic) bases to accommodate allocations to new members. ICCAT's fleshing out of the general provisions of the 1995 UN Fish Stocks Agreement remains qualitative. As such, does it serve much purpose, other than to provide a well-constructed shopping list from which countries can each choose the topics best suited to their own case to emphasize in negotiations?

The great danger here is failures to reach agreements on allocations, as a result of which the countries concerned then insist on catching what they claim, often in total above the sustainable yield level. This could lead to the meltdown situation of severe resource declines and a credibility crisis for RFMOs as the appropriate agencies for regulating high seas fisheries. Might this then fuel support for an over-hasty international reflex reaction to transfer such responsibility to an institution such as the Parties to the Convention for International Trade in Endangered Species (CITES)?

There are some generic considerations that any attempt to resolve this allocation issue needs to accommodate:

(1) The countries that initially developed a fishery – those that essentially created the asset – deserve rewards in the form of rights, though they cannot expect such original rights allocations to extend unchanged in perpetuity.
(2) Dramatic change is disruptive, and for that reason will prove unacceptable in practice; therefore, allocation changes should be gradual.
(3) As the ICCAT impasse so clearly shows, non-quantitative allocation criteria alone are not sufficient.
(4) Allocation changes must not promote an increase in the number of fishing vessels. If allocations to some countries are reduced too rapidly, surplus vessels will remain and be pressured either to continue to operate economically in the fishery by under-reporting, or to transfer their attention to, and thus probably cause problems by increasing effort levels in, other fisheries.

THE MECHANISM OF ATTRITION

As a suggestion consistent with all these considerations, the mechanism of "attrition" is proposed. The concept is that every year, each rights-holder returns a proportion of their allocation to a central authority – the relevant RFMO in high seas situations. This central authority then redistributes these returns among all members according to a negotiated process. This does not necessarily mean that existing holders lose all their contributions to the central pool – in some circumstances, they might simply be re-allocated what they had provided, so that the *status quo* remains.

The size of the proportion returned is crucial. This needs to be small for a gradual process, and also should be one that does not promote an increase in overall vessel numbers. Hence, the proportion should relate to the inverse of a typical vessel lifetime, which suggests an amount not exceeding 5% of current holdings per annum. The size of a right would show a geometric decline, so that an original right has a "half-life" rather than some maximum duration.

The benefits of such a system are that it would provide an enforced opportunity for orderly change. The advantage of awarding rights in perpetuity is that such ownership of a right creates the incentive to conserve for one's own future benefit, rather than to take unsustainably now. This is in contrast to the pressure to circumvent compliance measures if there is a high risk that what is saved today, will contribute only to somebody else's benefit tomorrow. As rights in perpetuity correspond to the limit of an attrition system as the rate of attrition becomes vanishingly small, considerations of continuity imply that, provided any attrition is sufficiently gradual, the conservation-promoting benefits of permanent ownership will remain, if slightly reduced. The design of the attrition approach is also such that the likelihood of overall vessel numbers increasing is reduced.

However, there are also some negative aspects. Some industry representatives in South Africa have opposed the concept, although it has been unclear whether their opposition stems from disinclination to forego any existing right or a more psychological reaction against prospects of possible continued down-sizing. In addition there is no track record of formal applications of such an approach. Nevertheless, the process in New Zealand whereby the Government redistributed rights to the Maoris to address Treaty of Waitangi considerations has some similarities. Initially, in terms of the Maori Fisheries Act, the Government bought back 10% of ITQ rights from industry over four years for this redistribution. However the rights then transferred came from willing sellers rather than mandatory forfeit as under an attrition system, and the eventual final settlement also included other components (Annala, 1996). Furthermore, the approach was under consideration for fisheries in New South Wales in Australia in the early 1990s, with an annual attrition rate of 2.5% proposed. The concept has recently re-emerged for their rock lobster fishery – with the appropriate name of "claw back" (James Findlay, Agriculture, Fisheries and Forestry, Australia, pers. comm.). This initiative proposes that the returned proportion be re-allocated under auction or tender, and that the funds raised be used to cover a contribution to the community for the access rights. After all, under an ownership system with a necessarily limited number of access rights, why should not the community as a whole benefit from the value of the rights (or assets) awarded to successful applicants?

Clearly also, attrition is not a complete solution. In the high seas context, at least, a formula for the redistribution of the proportions returned would still need to be developed. However, it would at least be a start, providing an overall framework for the process, and it could be applied in particular in situations where there is a pressing need for change in allocations. By limiting the debate to a smallish proportion of the TAC, the conflict would be defused to some extent. Once "allocation equity" might be generally perceived to have been attained, the mechanism could be suspended until, perhaps, similar future needs arise.

THE SCIENTIFIC ROLE

What then of the associated role for scientists? First, why any role at all? Because, ultimately, allocation pertains to some overall quantum, such as a TAC, in whose determination scientific advice has a key role to play. We advocate that fisheries scientists become more proactive. There is unsurprising confusion surrounding the Precautionary Principle/Approach, which scientists need to resolve. If non-quantitative criteria are not sufficient for allocation, surely the same must apply for effective implementation of the Precautionary Approach? This goes beyond, say, simple specification of some general biological reference points around which to develop scientific management advice. The real problem with which fisheries scientists must grapple is that the numerical values of such reference points have to be estimated in each particular case. How does one deal with the fact that structurally different model/assumption combinations can yield different answers? How might plausibility weights be assigned to such different scenarios for the evaluation of risk?

When scientific disputes arise, are arbitration panels made up of independent scientists the way to go to resolve these, rather than time-consuming recourse to legal remedies? Some four decades ago, the International Whaling Commission used an approach along these lines with their appointment of the so-called "Committee of Three" (K. Radway Allen, Doug Chapman and Sidney Holt), later expanded to Four with the addition of John Gulland (IWC, 1961). The CCSBT seems to be achieving some success with this Group's modern-day counterparts of Ray Hilborn, Jim Ianelli, Ana Parma and John Pope (although one should note that the arbitration powers accorded this group by the CCSBT go beyond the purely advisory mandate of the IWC's Committee of Three/Four).

For the future, there are suggestions of a more regular peer-review process developing between three of the Tuna Commissions: ICCAT, IOTC and CCSBT. The LOSC's dispute-resolution mechanisms envisage the possible addition of two non-voting scientific or technical experts to its courts or tribunals (LOSC Article 289). For the fisheries field, these are preferably to be chosen from a list drawn up and maintained by FAO, to which every State Party is entitled to nominate two experts (LOSC Annex VIII, Article 2). This list (http:/ www.un.org/ depts/los/settlement_of_disputes/experts_special_arb.htm) hardly seems fully developed yet, with nominees from just 15 countries. Perhaps the time is ripe for the FAO to promote it more effectively. The listed scientists could then be used directly (rather than merely as advisers to formal legal bodies) to facilitate the settlement of international fishery scientific disputes.

ACKNOWLEDGEMENTS

Victor Restrepo, Jon Sutinen and Mark Soboil are thanked for their comments on an earlier version of this document.

REFERENCES

Annala, J. H. (1996) New Zealand's ITQ system: have the first eight years been a success or failure? *Reviews in Fish Biology and Fisheries*, **6**: 43–62.

Anon. (1995) Agreement for the implementation of the provisions of the United Nations Convention on the Law of the Sea of 10 December 1982 relating to the conservation and management of straddling fish stocks and highly migratory fish stocks. United Nations conference on straddling fish stocks and highly migratory fish stocks. New York. http://www.ub.org/Depts/los/los_docs.htm (UN Fish Stocks Agreement).

Anon. (2003) Abalone under threat. *Fishing Industry News*, **February 2003**. p. 4.

Butterworth, D. S. & Punt, A. E. (1999) Experiences in the evaluation and implementation of management procedures. *ICES Journal of Marine Science*, **56**: 985–998.

CCSBT. (2000) Development of a SBT Scientific Research Program including a scientific fishing component by the CCSBT external scientists. Attachment L to: *Report of the Special Meeting*, 16–18 November 2000, Canberra, Australia.

CCSBT. (2002) Report of the Sixth Meeting of the Scientific Committee, Tokyo, Japan, 28–31 August 2001. In: *Reports of the Seventh and Eighth Years of the Commission*, CCSBT **7&8**.

Commission of the European Communities. (2001) Green Paper on the future of the Common Fisheries Policy, Brussels, 20 March 2001. 40 pp.

Hardin, G. (1968) The tragedy of the commons. *Science*, **162**: 1243–1248.

Hauck, M. & Sweijd, N. (1999) A case study of abalone poaching in South Africa and its impact on fisheries management. *ICES Journal of Marine Science*, **56**: 1024–1032.

ICCAT. (1985) International Commission for the Conservation of Atlantic Tunas: Basic Texts, 2nd Rev. Madrid, Spain.

ICCAT. (2000) Report of the Standing Committee on Research and Statistics (SCRS) (Madrid, Spain – October 11 to 15, 1999). In: *Report for the Biennial Period 1998–99, Part II (1999)*, Vol. **2**.

ICCAT. (2001) Proceedings of the 12th Special Meeting of the International Commission for the Conservation of Atlantic Tunas (Marrakech, Morocco, November 13 to 20, 2000). In: *Report for the Biennial Period 2000–01, Part I (2000)*, Vol. **1**.

ICCAT. (2002a) Compendium of ICCAT Management Recommendations and Resolutions. *ICCAT Document*, **SEC/2002/010** (available at http://www.iccat.es/Documents/Recs/Recs_eng.pdf).

ICCAT. (2002b) Report of the Standing Committee on Research and Statistics (SCRS) (Madrid, Spain – October 8 to 12, 2001). In: *Report for the Biennial Period 2000–01, Part II (2001)*, Vol. **2**.

ICCAT. (2002c) ICCAT Criteria for the Allocation of Fishing Possibilities. In: *ICCAT Report for the Biennial Period 2000–01, Part II (2001)*, Vol. **1**.

ICCAT. (2002d) Proceedings of the 17th Regular Meeting of the International Commission for the Conservation of Atlantic Tunas (Murcia, Spain, November 12 to 19, 2001). In: *Report for the Biennial Period 2000–01, Part II (2001)*, Vol. **1**.

ICCAT. (2003) Proceedings of the 18th Regular Meeting of the International Commission for the Conservation of Atlantic Tunas (Bilbao, Spain, 27 October – 4 November 2002). In: *Report for the Biennial Period 2002–03, Part I (2002)*, Vol. **1**.

International Tribunal for the Law of the Sea. (1999) Order: 27 August 1999: Southern Bluefin Tuna Cases (New Zealand v. Japan; Australia v. Japan). Request for provisional measures. 19 pp. (http://www.itlos.org/case_documents/2001/document_en_116.pdf).

Internet Guide to International Fisheries Law. (2000) Award: Southern Bluefin Tuna Case: Australia and New Zealand v. Japan. Award on Jurisdiction and Admissibility rendered by the Arbitral Tribunal constituted under Annex VII of the United Nations Convention of the Law of the Sea. (http://www.oceanlaw.net/ cases/ tuna2a.htm).

IWC. (1961) Report [of the Twelfth Meeting]. *Report of the International Whaling Commission*, **12**: 3–10.

Kirkwood, G. P. & Agnew, D. J. (2003) Deterring IUU fishing. (this volume).

Kleinschmidt, H., Sauer, W. H. H. & Britz, P. (2003) Reconciling equity and stability in the South African fishing industry. *African Journal of Marine Science*, **25**.

Morgan, D. L. (2001) A practitioner's critique of the order granting provisional measures in the Southern Bluefin Tuna Cases. In: *Current Marine Environmental Issues and the International Tribunal for the Law of the Sea*. (Ed. by M. H. Nordquist & J. N. Moore), pp. 173–213. Martinus Nijhoff Publishers, The Hague.

Polacheck, T. & Preece, A. L. (1998) Documentation of the Virtual Population analysis methods and model input used for estimating current and historical stock sizes of southern bluefin tuna: updates and fisheries 1998. *CCSBT Document*, **CCSBT–SC/9802/37**. 15 pp.

Suisan Keizai Shinbun. (1998) Far seas tuna longliner fleet reduction. November 17, 1998. Pg. 1. (In Japanese)

Wolfrum, R. (2000) The role of the International Tribunal for the Law of the Sea. In: *Current Fisheries Issues and the Food and Agriculture Organisation of the United Nations*. (Ed. by M. H. Nordquist & J. N. Moore), pp. 369–385 and 451–456 (Questions, Comments and Answers). Kluwer Law International, The Netherlands.

APPENDIX 1 EXTRACT FROM THE 1995 UNITED NATIONS FISH STOCKS AGREEMENT

Article 10: Functions of subregional and regional fisheries management organizations and arrangements

In fulfilling their obligation to cooperate through subregional or regional fisheries management organizations or arrangements, States shall:

(a) agree on and comply with conservation and management measures to ensure the long-term sustainability of straddling fish stocks and highly migratory fish stocks;
(b) agree, as appropriate, on participatory rights such as allocations of allowable catch or levels of fishing effort;
(c) adopt and apply any generally recommended international minimum standards for the responsible conduct of fishing operations;
(d) obtain and evaluate scientific advice, review the status of the stocks and assess the impact of fishing on non-target and associated or dependent species;
(e) agree on standards for collection, reporting, verification and exchange of data on fisheries for the stocks;
(f) compile and disseminate accurate and complete statistical data, as described in Annex I, to ensure that the best scientific evidence is available while maintaining confidentiality where appropriate;
(g) promote and conduct scientific assessments of the stocks and relevant research and disseminate the results thereof;
(h) establish appropriate cooperative mechanisms for effective monitoring, control, surveillance and enforcement;
(i) agree on means by which the fishing interests of new members of, or participants in, the organization or arrangement will be accommodated;
(j) agree on decision-making procedures which facilitate the adoption of conservation and management measures in a timely and effective manner;
(k) promote the peaceful settlement of disputes in accordance with Part VIII;
(l) ensure the full cooperation of their relevant national agencies and industries in implementing the recommendations and decisions of the subregional or regional fisheries management organization or arrangement; and
(m) give due publicity to the conservation and management measures established by the organization or arrangement.

Article 11: New members or Participants

In determining the nature and extent of participatory rights for new members of a subregional or regional fisheries management organization, or for new participants in a subregional or regional fisheries management arrangement, States shall take into account, *inter alia*:

(a) the state of the straddling fish stocks and highly migratory fish stocks and the existing level of fishing effort in the fishery;
(b) the respective interests, fishing patterns and fishing practices of new and existing members or participants;

(c) the respective contributions of new and existing members or participants to conservation and management of the stocks, to the collection and provision of accurate data and to the conduct of scientific research on the stocks;
(d) the needs of coastal fishing communities which are dependent mainly on fishing for the stocks;
(e) the needs of coastal States whose economies are overwhelmingly dependent on the exploitation of living marine resources; and
(f) the interests of developing States from the subregion or region in whose areas of national jurisdiction the stocks also occur.

APPENDIX 2 ICCAT CRITERIA FOR THE ALLOCATION OF FISHING POSSIBILITIES (ADOPTED BY ICCAT IN NOVEMBER 2001)

I. *QUALIFYING CRITERIA*

Participants will qualify to receive possible quota allocations within the framework of ICCAT in accordance with the following criteria:

(1) Be a Contracting or Cooperating Non-Contracting Party, Entity or Fishing Entity.
(2) Have the ability to apply the conservation and management measures of ICCAT, to collect and to provide accurate data for the relevant resources and, taking into account their respective capacities, to conduct scientific research on those resources.

II. *STOCKS TO WHICH THE CRITERIA WOULD BE APPLIED*

(3) These criteria should apply to all stocks when allocated by ICCAT.

III. *ALLOCATION CRITERIA*

A. *Criteria relating to past/present fishing activity of qualifying participants*

(4) Historical catches of qualifying participants.
(5) The interests, fishing patterns and fishing practices of qualifying participants.

B. *Criteria relating the status of the stock(s) to be allocated and the fisheries*

(6) Status of the stock(s) to be allocated in relation to maximum sustainable yield, or in the absence of maximum sustainable yield an agreed biological reference point, and the existing level of fishing effort in the fishery taking into account the contributions to conservation made by qualifying participants necessary to conserve, manage, restore or rebuild fish stocks in accordance with the objective of the Convention.
(7) The distribution and biological characteristics of the stock(s), including the occurrence of the stock(s) in areas under national jurisdiction and on the high seas.

C. *Criteria relating to the status of the qualifying participants*

(8) The interests of artisanal, subsistence and small-scale coastal fishers.
(9) The needs of the coastal fishing communities which are dependent mainly on fishing for the stocks.
(10) The needs of the coastal States of the region whose economies are overwhelmingly dependent on the exploitation of living marine resources, including those regulated by ICCAT.
(11) The socio-economic contribution of the fisheries for stocks regulated by ICCAT to the developing States, especially small island developing States and developing territories[1] from the region.

---

[1] *For the purposes of this document, the term "territories" refers only to the territories of those States that are Contracting Parties to the Convention in respect of those territories alone.*

(12)The respective dependence on the stock(s) of the coastal States, and of the other States that fish species regulated by ICCAT.
(13)The economic and/or social importance of the fishery for qualifying participants whose fishing vessels have habitually participated in the fishery in the Convention Area.
(14)The contribution of the fisheries for the stocks regulated by ICCAT to the national food security/needs, domestic consumption, income resulting from exports, and employment of qualifying participants.
(15)The right of qualified participants to engage in fishing on the high seas for the stocks to be allocated.

*D. **Criteria relating to compliance/data submission/scientific research by qualifying participants***

(16)The record of compliance or cooperation by qualifying participants with ICCAT's conservation and management measures, including for large-scale tuna fishing vessels, except for those cases where the compliance sanctions established by relevant ICCAT recommendations have already been applied.
(17)The exercise of responsibilities concerning the vessels under the jurisdiction of qualifying participants.
(18)The contribution of qualifying participants to conservation and management of the stocks, to the collection and provision of accurate data required by ICCAT and, taking into account their respective capacities, to the conduct of scientific research on the stocks.

IV. *CONDITIONS FOR APPLYING ALLOCATION CRITERIA*

(19)The allocation criteria should be applied in a fair and equitable manner with the goal of ensuring opportunities for all qualifying participants.
(20)The allocation criteria should be applied by the relevant Panels on a stock-by-stock basis.
(21)The allocation criteria should be applied to all stocks in a gradual manner, over a period of time to be determined by the relevant Panels, in order to address the economic needs of all parties concerned, including the need to minimize economic dislocation.
(22)The application of the allocation criteria should take into account the contributions to conservation made by qualifying participants necessary to conserve, manage, restore or rebuild fish stocks in accordance with the objective of the Convention.
(23)The allocation criteria should be applied consistent with international instruments and in a manner that encourages efforts to prevent and eliminate overfishing and excess fishing capacity and ensures that levels of fishing effort are commensurate with the ICCAT objective of achieving and maintaining MSY.
(24)The allocation criteria should be applied so as not to legitimize illegal, unregulated and unreported catches and shall promote the prevention, deterrence and elimination of illegal, unregulated and unreported fishing, particularly fishing by flag of convenience vessels.

(25)The allocation criteria should be applied in a manner that encourages cooperating Non-Contracting parties, Entities and Fishing Entities to become Contracting Parties, where they are eligible to do so.

(26)The allocation criteria should be applied to encourage cooperation between the developing States of the region and other fishing States for the sustainable use of the stocks managed by ICCAT and in accordance with the relevant international instruments.

(27)No qualifying participant shall trade or sell its quota allocation or a part thereof.

Management of shared Baltic fishery resources

Roberts Aps
Estonian Marine Institute, University of Tartu, Tallinn, Estonia

ABSTRACT: The International Baltic Sea Fishery Commission (IBSFC) is the regional fisheries organization responsible for rational exploitation of Baltic shared fishery resources. IBSFC activities under the framework "Agenda 21 for the Baltic Sea Region" have been crucial for the further development and implementation of management objectives and strategies for Baltic salmon, cod, sprat and herring, and other measures aimed at sustainable use of shared fishery resources. The IBSFC is entering a period of change while it attempts to adapt the stock assessment units to the structure of fish populations, and the stock assessment units to management units. Implementation of a new management system implies development of a new means of quota allocation, which is painful to many and needs considerable debate and effort. Agreement on changed allocation criteria could probably be achieved through negotiations on the basis of reciprocal compensation and side payments. The IBSFC's Contracting Parties are efficient fishing nations, and therefore it is important that members pay attention to efficiency management when negotiating shared fishery resources.

INTRODUCTION

The Baltic Sea is entirely under the jurisdiction of coastal states, and its economically most important fish stocks are shared between the countries bordering it. Shared stocks are resources that cross the Exclusive Economic Zone (EEZ) boundary into the EEZ(s) of one, or more, neighbouring coastal states (Munro, 2000).

Management of Baltic fishery resources is the responsibility of a regional fisheries organization, the International Baltic Sea Fishery Commission (IBSFC), which was established pursuant to the Convention on Fishing and Conservation of the Living Resources in the Baltic Sea and the Belts (the Gdansk Convention), a Convention signed in 1973 by Governments of Baltic Sea States. Contracting Parties agreed to promote close cooperation "... with a view to preserving and increasing the living resources of the Baltic Sea and the Belts and obtaining the optimum yield, and, in particular to expanding and coordinating studies towards these ends ..." (Anon., 1999). A schedule of activities of the Commission is drafted annually on the basis of Commission recommendations and decisions taken during its Annual Sessions. The IBSFC, established as the first international organization for the Baltic Sea, acts as the Sector Task Force.

This manuscript reviews the IBSFC's formal management framework and the measures aimed at the sustainable use and conservation of Baltic shared fishery resources, identifies some of the problems and suggests possible solutions.

FORMAL MANAGEMENT FRAMEWORK

AGENDA 21 FOR THE BALTIC SEA REGION

The IBSFC is lead agency in developing an Agenda 21 for the Fishery Sector of the Baltic Sea. In February 1997, the IBSFC, at its first Extraordinary Session since its establishment, adopted a Resolution VI on an "Agenda 21 for the Baltic Sea Region" and established a Working Group to draft such an Agenda to cover the fish resources and the associated impact of fisheries on the Baltic Sea environment.

The IBSFC contribution to Baltic 21, as it became known, was discussed at its 23rd Annual Session in 1997. Following its designation as Baltic 21 lead agent for fisheries, the IBSFC was requested to include coastal aquaculture and river and lake fisheries, all of which were normally outside its competence. The IBSFC was also asked to consider cross-sectoral issues, including the influence of environmental conditions on fisheries and vice versa.

The Sector Report on Fisheries contribution to "Agenda 21 for the Baltic Sea Region", and the Action Programme for Sustainable Development, including targets, time frames, and information on participants and financing (Chapter 3 of the Report), was discussed in depth and adopted during the IBSFC's Extraordinary Session during 1998. Due consideration was given to Priority Action 1, the development of long-term strategies for its main fish stocks: cod *Gadus morhua*, salmon *Salmo salar*, herring *Clupea harengus* and sprat *Sprattus sprattus*. It was agreed that a mortality control harvest strategy would offer good prospects of gradually achieving a balance between the harvesting capacity of fleets and the target reference points for fish stocks.

The IBSFC stressed that it would apply a precautionary approach to fisheries management, and to this end asked the International Council for the Exploration of the Sea (ICES) to suggest appropriate biological reference points to ensure that fishing took place within safe biological limits. The importance of introducing other technical measures was also recognized. However, it was conceded that frequent changes would result in extra costs for the fishing industry, and the IBSFC therefore aimed to avoid these to whatever extent was possible.

The Baltic Sea is the first region in the world to be covered by common regional goals for sustainable development (Anon., 2000). The IBSFC's Extraordinary Session in 1998 adopted a set of biological, economic and social core indicators ("Baltic 21" Indicators of the Sector Fisheries) to highlight trends in biological systems and the economics of fishing communities around the Baltic Sea.

MEASURES AIMED AT SUSTAINABLE USE OF SHARED FISHERY RESOURCES

Salmon

The IBSFC's 21st Annual Session in 1995 adopted a Resolution I on management objectives for Baltic salmon and a Resolution II concerning a moratorium on salmon fishing in all rivers and river mouths with wild salmon stocks. These Resolutions were passed because of the need to stop further degradation of wild salmon stocks and to rebuild the population of wild salmon in the Baltic Sea.

These two resolutions paved the way for the development and adoption of the Baltic Salmon Action Plan during the IBSFC's Extraordinary Session in February 1997. According to this Action Plan, the long-term objectives (to 2010) are:

(1) To prevent the extinction of wild populations, any further decrease in the number of naturally produced smolts should not be allowed.
(2) Production of wild salmon should be stimulated gradually, to attain for each salmon river by 2010 a natural production of wild Baltic salmon of at least 50% of the best estimated potential within safe genetic limits, in order to achieve a better balance between wild and reared salmon.
(3) Wild salmon populations should be re-established in potential salmon rivers.
(4) The level of fishing for salmon should be maintained as high as possible, and only restrictions necessary to achieve the first three objectives should be implemented.
(5) Releases of reared smolts and early life history stages should be closely monitored.

Based on long-term objectives, medium- and short-term strategies for Baltic salmon have been developed and adopted as part of the Baltic Salmon Action Plan. At the same time as preparing the Baltic Salmon Action Plan, close attention was paid to continuing and further promoting fishing activities.

Baltic salmon remained a focus of the IBSFC in 1998 when, at its 24th Annual Session, on the basis of the provisions of the Baltic Salmon Action Plan, it adopted a Resolution on the Principles for Salmon Enhancement Activities and listed rivers where self-sustaining wild populations should exist by 2010.

Then, in 1999 at its 25th Annual Session, the IBSFC adopted two salmon-related resolutions:

(1) Resolution XI, concerning suitable methods for salmon rearing and release, stating that regardless of rearing facilities, production of salmon should follow the principles of maintaining genetic diversity and using local salmon populations for rearing and release.
(2) Resolution XII, concerning wild salmon Index rivers and monitoring methods for the purpose of the IBSFC Salmon Action Plan for the period 1997–2010. Index rivers were to be considered the basis for monitoring the status of wild salmon populations.

At its 26th Annual Session in 2000, the IBSFC adopted another two salmon-related resolutions:

(1) Resolution XIV, on management measures to optimize the harvesting of reared salmon and to minimize the genetic impact on wild salmon.
(2) Resolution XV, on a reporting Format for the IBSFC Salmon Action Plan.

Cod

Resolution III, on the enforcement of cod Total Allowable Catch (TAC) allocations in the Baltic Sea, was adopted during the 21st Annual Session of the IBSFC in 1995. The Commission considered the uncertainties surrounding the state of cod stocks in the Baltic Sea and the Belts caused by an interruption in the submission of reliable catch data in recent years, and declared a need to respect the agreed TAC allocations in the respective economic zones by enhanced quota management and control. It was agreed that, if TACs were being exceeded, the Contracting Party would have to report the reasons to the Commission and also list the measures to be taken to prevent it in future.

Depletion of cod stocks in the Baltic continued between 1995 and 1997, and this was why the IBSFC adopted during its 23rd Annual Session in 1997 two Resolutions on Baltic cod:

(1) Resolution VII, concerning the establishment of a management strategy for Baltic cod. Contracting Parties, conscious of the need to further improve management of Baltic cod, agreed to develop a comprehensive medium- and long-term management strategy for the species. For this purpose, a special Working Group of the IBSFC was established and met in 1998. ICES was requested to provide appropriate biological limit reference points and other biological and precautionary reference points.
(2) Resolution VIII, similar to Resolution III concerning the enforcement of TACs in the Baltic Sea. Contracting Parties were requested, when TACs were being exceeded, to report the reasons to the Commission and also list the measures to be taken to prevent it occurring in future.

The parlous state of Baltic cod stocks contributed to the urgent adoption of IBSFC Resolution X on a Long-Term Management Strategy for Cod Stocks in the Baltic Sea by the 25th Annual Session of the IBSFC in 1999. The elements of the strategy are:

(1) Every effort will be made to maintain a minimum level of spawning stock biomass (SSB) greater than 160 000 t for the eastern stock and 9 000 t for the western stock.
(2) A long-term management plan will be implemented within which annual quotas will be set for the fishery on the eastern stock reflecting a fishing mortality rate (F) of 0.6, and on the western stock an F of 1.0 for appropriate age groups defined by ICES.
(3) Should the SSB fall below a reference point of 240 000 t for the eastern stock and 23 000 t for the western stock, the fishing mortality rates referred to under paragraph 2 will be amended in the light of scientific estimates of the conditions then prevailing to ensure safe and rapid recovery of spawning stock biomass to levels in excess of 240 000 and 23 000 t respectively for eastern and western stocks.
(4) For allocation purposes a combined TAC will be established; the Contracting Parties agree to further collaborate, *inter alia* through bilateral agreements, to ensure efficient management of the stocks.
(5) The exploitation pattern for cod and in particular the selectivity parameter will be improved in the light of new scientific advice from ICES, with the objective of enhancing the spawning biomass and reducing discards.
(6) Additional technical measures including, *inter alia*, further limitation on effort, restrictions on fishing days, closure of areas and/or seasons, an obligation to change fishing grounds where juveniles are particularly abundant, and special reporting requirements, will be considered.
(7) The IBSFC will, as appropriate, adjust management measures and elements of the plan on the basis of any new advice provided by ICES.

The IBSFC at its 27th Annual Session in 2001 adopted Resolution XVII on a Recovery Plan for Baltic cod.

Sprat

A Working Group established under a mandate given by the 23rd Annual Session of the IBSFC in 1997 was also requested to draft long-term objectives and strategies for the management of Baltic sprat according to terms of reference adopted by the 25th Annual Session in 1999. ICES was invited to be closely associated with this work.

At its 26th Annual Session in 2000, the IBSFC adopted Resolution XIII on a Long-Term Management Strategy for the Sprat Stock in the Baltic Sea. The elements of this strategy are:

(1) Every effort will be made to maintain a level of spawning stock biomass (SSB) greater than 200 000 t.
(2) A long-term management plan will be implemented, within which annual quotas will be set for the fishery reflecting a fishing mortality rate (F) of 0.4 for appropriate age groups defined by ICES.
(3) Should the SSB fall below 275 000 t, the fishing mortality rate referred to under paragraph 2 will be amended in the light of scientific estimates of the conditions then prevailing to ensure safe and rapid recovery of the SSB to levels in excess of 275 000 t.
(4) The IBSFC will, as appropriate, adjust management measures and elements of the plan on the basis of any new advice provided by ICES.

Herring

A Working Group established under a mandate given by the 23rd Annual Session of the IBSFC in 1997 made an attempt to draft long-term objectives and strategies for managing Baltic herring. According to the Report of the Working Group that met in Visby (Sweden) in 1999, ICES was not in a position to provide advice that could be used to inform a harvesting strategy for herring stocks in the Baltic Sea. Therefore, it was not possible to define fishing mortality limits or reference points in terms of a biomass where remedial action was needed.

Long-term objectives and strategies for the management of Baltic herring in light of the most recent scientific advice available from ICES were discussed during the meeting of the IBSFC Working Group in Turku (Finland) in 2000. Again there was insufficient recent scientific information, and formulation of a management strategy for Baltic herring was postponed again. Exactly the same occurred when the Working Group met in Warsaw (Poland) in 2001. Then, however, the Chairman of the Working Group specifically drew attention to the fact that no appropriate scientific research had been carried out on Baltic herring for the preceding 10 years; clearly, the situation called for immediate action.

The next time a long-term management strategy for Baltic herring was considered was during the meeting of the IBSFC Working Group in Cracow (Poland) in 2002. There, the Working Group agreed on the need to improve management of the herring fisheries in the Baltic Sea, acknowledging that it should be based, as far as practicable, on the advice provided by ICES. It was proposed that Contracting Parties consider the introduction of four management units for Baltic herring (a western unit in Subdivisions 22–24; a central unit in Subdivisions 25–29 + 32, but excluding the Gulf of Riga; a Gulf of Riga unit; and a northern unit in Subdivisions 30–31). (The Subdivisions are shown in Figure 1.) The Working Group further proposed that, if a new management system was implemented, allocations should in principle reflect the historical and geographical fishing patterns and be based on the data used by ICES in stock assessment.

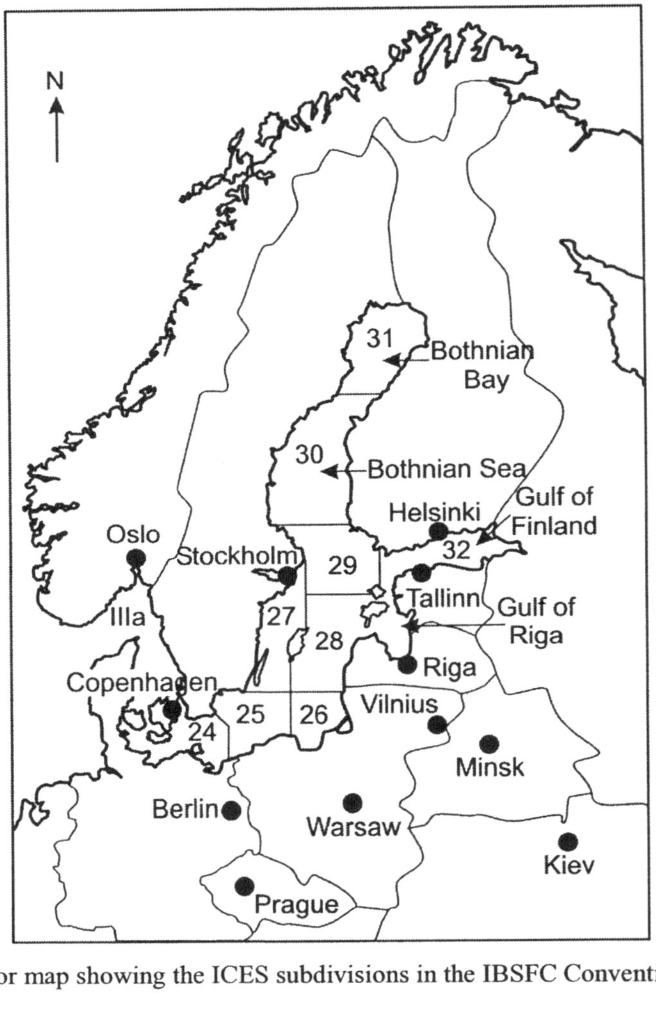

Fig. 1 Locator map showing the ICES subdivisions in the IBSFC Convention Area.

It was also agreed that future allocations be established on common principles for all management units and on average and in general reflect recent allocations agreed within the IBSFC. In terms of possible elements for a herring recovery plan, the Working Group regretted that ICES had not been able to provide scientific information on a means of protecting juvenile herring, so no proposal was made.

PROGRESS MADE

According to the Baltic 21 Triennial Report for the Sector Fisheries released by the IBSFC Secretariat at the end of 2002 and based on Baltic 21 Indicators for the Sector Fisheries, the long-term management strategies applied by the IBSFC provided a means of maintaining the total annual catch from the Baltic Sea over the period 1998–2001 at a level of 800 000–900 000 t.

The same report noted that the fisheries for Baltic sprat, Gulf of Riga herring and salmon were prosecuted sustainably, but that the status of the two cod stocks (in particular the eastern stock) was poor. The IBSFC is implementing a "Recovery Plan for Baltic Cod" (commenced in 2001) with the aim of restoring the stocks to better levels and to achieve a sustainable cod fishery. Further, in terms of the fishery for Baltic herring in the main basin (for which there was insufficient knowledge of the stock, but considered to be outside safe biological limits), the Commission is working on a new stock-by-stock management strategy aimed at ensuring sustainable fisheries.

According to the Report from 1997/98 to 2000/2001, the number of fishing vessels was slightly fewer, by about 10%. At the same time there was a trend towards increased horsepower by most countries, on average an increase of some 5%. The number of full-time fishers remained practically unchanged over the same period (1997–2001), but the database was limited because some countries still do not have a system of registering full-time fishers.

Over the period 1997–2000, the quantity of fish consumed per capita by Baltic States in general remained static.

CHALLENGES

SCIENTIFIC ADVICE AND OPENNESS ABOUT UNCERTAINITY

According to Article IX of the Gdansk Convention, it is the duty of the IBSFC to prepare and submit recommendations based as far as practicable on the results of scientific research. The Commission therefore implements its functions by seeking the services of ICES and other international technical and scientific organizations, as well as by soliciting information from the official bodies of the Contracting Parties.

Since 1998, ICES scientific advice, considered to be independent and free from political influence, has been delivered to the IBSFC according to a "Memorandum of Understanding between The International Baltic Sea Fisheries Commission and The International Council for the Exploration of the Sea". Parties agreed that ICES will provide the IBSFC with annual "standard" advice on the state and the management of its main commercial stocks (herring, sprat, cod, salmon and flatfish) according to an agreed format that contains, *inter alia*, an area overview for all stocks, information on the historical development of the fisheries and on the current state of the stocks, a short-term forecast, a medium-term consideration based on risk assessment, and a long-term consideration.

Information on yield per recruit for various levels of exploitation is normally included in the submission along with reference points such as F_{high}, F_{med}, F_{low}, F_{max} and $F_{0.1}$. Limit points are given in terms of biomass, MBAL, or fishing mortality. It is expected that a new amended Memorandum of Understanding will contain, *inter alia*, requests for information on the environment and on ecosystem interactions of relevance to management.

As mentioned previously, the IBSFC Working Group on Long-Term Management Objectives and Strategies for Baltic Cod, Herring and Sprat has had serious problems in terms of finding reliable scientific evidence for developing a long-term management plan for Baltic herring. To address this situation, the IBSFC Contracting Parties have been invited to solve it on a national level by supporting and strengthening national research capabilities to meet acknowledged scientific standards.

An important issue for the IBSFC is the firm acceptance of transparency about the uncertainty inherent in stock assessments while not using the uncertainty as an excuse for political compromise. According to Mace (1997) "Attacks on the quality or validity of scientific stock assessments are becoming commonplace but, with few exceptions, the quality of the science is usually an inappropriate scapegoat .... The fact is that stock assessments will probably always be imprecise, but the appropriate response to imprecise assessments is to manage conservatively".

INTEGRATION OF ENVIRONMENTAL PROTECTION REQUIREMENTS AND ECOSYSTEM CONSIDERATIONS INTO FISHERIES MANAGEMENT

Two regional Commissions, HELCOM and the IBSFC, arranged a joint seminar on fisheries issues and environmental protection in the Baltic Sea during 2002. That cooperation between IBSFC and HELCOM may well be the first case of common effort being made by an international fishery management organization and an environmental organization in the Baltic Sea. It was stressed "... that in HELCOM and IBSFC, there is professional competence available for both fisheries and environmental issues. For scientific advice, HELCOM and IBSFC both see ICES as the main advisory body, and the activities of ICES include work on the effects of human activities on the ecosystem in the Baltic Sea and integration of environmental and fisheries issues". The seminar concluded that integration of environmental and nature conservation issues into fishery policies and integration of fishery issues into environmental and nature conservation policies should be an ongoing process in both organizations. It was further agreed that close cooperation between HELCOM and the IBSFC would be sustained in order to cover the broad range of problems of common interest.

Subsequent to the joint IBSFC/HELCOM seminar, the Fourth Meeting of the HELCOM Monitoring and Assessment Group (MONAS) held in Warnemünde, Germany, in October 2002 considered dioxins and PCBs in the Baltic. MONAS focused on the "Community Strategy for Dioxins, Furans and Polychlorinated Biphenyls" adopted by the European Commission (EC) in October 2001 (COM(2001)593), which consists of a strategy to reduce the presence of dioxins and PCBs in the environment and another strategy to reduce the presence of dioxins and PCBs in feed and food. Attention was also paid to European Community Council Regulation No. 2375/2002 of November 2001, which establishes maximum levels for dioxin in fish.

The European Commission is leading the process and considering further development of an integrated pilot project for monitoring dioxins and PCBs in the environment and in seafood in relation to human health in the Baltic. The "Integrated Monitoring Pilot Project in the Baltic Region" will be developed within the framework of the EC's "Environment and Health" policy and implementation of the "Dioxin Strategy" and "Marine Strategy" (COM(2002)539). The IBSFC and HELCOM are already contributing to the success of the planned Pilot Project by providing an overview of existing, ongoing and planned national and international dioxin/PCB monitoring activities, and by establishing a "Baltic Regional Dioxin Network".

ASSESSMENT UNITS AND THE STRUCTURE OF FISH POPULATIONS

ICES stock assessments have revealed the complex structure of the Baltic herring stock (Anon., 2002). The herring stocks in Subdivisions 22–24 and Division IIIa (Danish Straits) – spring spawners – and in Subdivisions 25–29 and 32 (including the Gulf of Riga), i.e. a combined stock of different populations, are harvested outside safe biological limits. However, within that complex of declining stocks, herring in the Gulf of Riga are actually considered to be stable and within safe biological limits. Herring in the Bothnian Sea (Subdivision 30) are harvested outside safe biological limits and the state of the herring stock in Bothnian Bay (Subdivision 31) is still unknown.

Local Baltic herring populations have been grouped by ICES for assessment purposes in different combinations over the period 1974–2000. The main reason for creating a large assessment unit in 1990 (Subdivisions 25–29 + 32) was to try to eliminate the influence of herring migrations in the Baltic. At the same time, neglecting the differences in some basic biological parameters of herring populations resulted in increasing uncertainty within the combined assessment, and in real problems with the management advice. The optimal solution seems to be in seeking a reasonable compromise between biological knowledge and the practical constraints set by the availability of data.

MANAGEMENT UNITS AND ASSESSMENT UNITS

ICES considers cod stocks in Subdivisions 22–24 (western) and 25–32 (eastern) as separate and provides advice on them separately (see Figure 1 for ICES areas). ICES advice is to manage cod as two stocks so that the exploitation pattern can be better adapted to the dynamics of both. Cod in Subdivisions 22–24 are harvested outside safe biological limits, but the spawning biomass of this stock is currently well above B_{pa}. ICES recommended that the fishing mortality in 2002 be reduced by at least 10% to below the value of 1.0 agreed by the IBSFC in an attempt to rectify the situation. Cod in Subdivisions 25–32 are also outside safe biological limits, but the current spawning stock biomass is considered to be well below B_{pa}, and even below B_{lim}. ICES recommended no fishing for cod in the eastern Baltic in 2002.

There is similarly no satisfactory accord between assessment and management units for Baltic herring. Ongoing adjustment of herring assessment units clearly needs to be accompanied by corresponding adjustment of the management units.

In 2002, the IBSFC Strategy Working Group addressed the issue of correspondence between assessment and management units, and evaluated the advantages and disadvantages of changing the current cod and herring management units. It was noted that the currently applied TAC system for combined Baltic herring and cod stocks is workable so long as the development of the stocks is similar. At the same time it was considered that current herring stock sizes in the Baltic main basin are at historically low levels, whereas the herring stock in the Gulf of Riga is at a historical high. The two cod stocks in the Baltic have shown strongly divergent biomass trends for some time; this diverging trend requires a new management system to be evolved based on management units that correspond as closely as possible with stock assessment units.

NATIONAL ALLOCATIONS: THE KEY DEMAND IS FOR CHANGE

RELATIVE STABILITY

The IBSFC Contracting Parties undertake to give effect to any Recommendation made by the Commission during its Annual Session under Article X of the Convention, from the date determined by the Commission. At the same time, any Contracting Party may object to a recommendation within 90 days of its date of notification; in that event, that Contracting Party will not be under any obligation to give effect to the Recommendation. Objections made to a Recommendation by three or more Contracting Parties relieve the other Parties of any obligation to give effect to the Recommendation.

Formally, the ratio of annual national allocations of the agreed TAC among Fishery Zones by the IBSFC cannot be taken as reflecting any generally applicable concept, nor may it be used as a basis for TAC allocation in future. In practice, however, in order to avoid unnecessary debate, relative stability over time has been an important instrument for the IBSFC in seeking a solution on the sharing of common fishery resources. However, as it is not a formal IBSFC principle, the relative stability in national allocation proportions is based less on the TACs than on political compromise.

PROSPECTS

Shared fishery resource disputes demonstrate the limitations of the IBSFC framework in reaching consensus on setting and achieving management goals and objectives across increasingly long time-scales. There are indeed some worrying signs of growing dissatisfaction of the IBSFC Contracting Parties with the recent system of sharing common Baltic fishery resources. Two Contracting Parties objected in 2000, three in 2001, and two in 2002. Some basis for this dissatisfaction seems to lie in history, when the IBSFC created so-called "paper fish" to facilitate consensus among Contracting Parties on national allocations. The situation was also destabilized by the recent increased utilization of the TAC for herring and sprat and the dramatic decreases of the cod stock in the eastern Baltic and the herring stock in the central Baltic. The generally poor accord between assessment and management units has also created real difficulties with rational utilization and conservation of both cod and herring.

The prospects of change within the IBSFC can also be considered in the light of the results of a game theory application to fishery issues, bearing in mind that "the purpose of game theory is insight, not solution" (Casti, 1996). Where a fishery resource is shared by non-cooperative nations, the theory predicts that the stock will be fished down to a suboptimal level; an optimal solution requires cooperation (Kaitala & Pohjola, 1988). At the same time, cooperation is itself a complicated issue.

The IBSFC's recent relative stability seems now to be under threat and there is increasing likelihood that the existing agreement will be overturned. The IBSFC is therefore entering a period of difficult change, during which it has to adjust its stock assessment units to the structure of fish populations and its management units to stock assessment units. During 2002, the IBSFC Strategy Working Group recognized that increases and decreases in TAC proportional allocations had to be considered by the Contracting Parties as a consequence of the introduction of any new management system. It was further stressed that implementation of a new management system implied development of new quota allocation criteria, a painful and lengthy exercise in itself.

Simple simulations show that introduction of the new management system will change future perspectives for many IBSFC Contracting Parties: in essence, there will be winners and losers. Generally, the Contracting Parties that find themselves better off will be those that have simultaneous access to several stocks with different dynamics. In contrast, Parties with access to just one stock will be more dependent on fluctuations of that specific resource. Consequently, reaching consensus on the development and implementation of new quota allocation criteria based on best biological advice and historical catch records only is going to be extremely difficult. One way a new balance could be achieved would be through negotiation on the basis of reciprocal compensation and side payments. Kaitala & Pohjola (1988), on the basis of fishery-related game theory simulations, suggested that specific compensatory transfer or side-payment programmes would be appropriate means of increasing economic efficiency in the use of shared marine resources.

The IBSFC Contracting Parties are efficient fishing nations that now suffer as a consequence of the recent dramatic decreases in cod and herring stocks in the Baltic. As a result, and because of increasing pressure from their fishing industries, the Contracting Parties are facing great difficulty in reaching consensus on management measures based on ICES scientific advice only. Excessive fishing effort is one of the driving forces behind the reluctance of IBSFC Contracting Parties to implement the new management scheme for Baltic cod and herring on the basis of improved scientific advice immediately. Further, Lindroos & Kaitala (2000), referring to the results of a coalition game study of Atlanto-Scandian herring, showed that the possibilities for cooperation depend critically on the efficiency parameter. Their results suggest that managing efficiency is crucial in negotiations on shared fishery resources.

Finally, the European Community's likely enlargement may also be seen as a confounding factor that adds uncertainty to future prospects for the IBSFC. Within a few years, after the expected accession of a number of new members, there may be just two Contracting Parties to the IBSFC: the European Community and Russia.

CONCLUSIONS

- The activities of the IBSFC have provided valuable experience in managing the shared Baltic fishery resources over several decades. The Conference of FAO and non-FAO Regional Fishery Bodies in Rome during February 2001 considered the IBSFC as one of the three pioneer organizations in applying the ecosystem approach to fisheries (along with ICES and CCAMLR).
- The implementation of "Agenda 21 for the Baltic Sea Region" is an important driving force in further developing the management objectives and strategies for Baltic salmon, cod, sprat and herring, as well as for other measures aimed at sustainable utilization of the shared fishery resources of the Baltic. Transparency about the uncertainty inherent in stock assessments and thus in the scientific advice, as well as the integration of environmental protection requirements and ecosystem considerations into fisheries management, are important preconditions for successful management of shared fishery resources.
- Some recent IBSFC management challenges have been associated with the lack of satisfactory correspondence between stock assessment and management units, the situation contributing to difficulties in reaching consensus on national allocation of shared resources. A new balance can probably only be achieved through negotiation on the basis of special compensatory transfers or side-payment programmes.

- IBSFC Contracting Parties, as highly efficient fishing nations, must give appropriate consideration to the efficiency parameter when negotiating allocations of shared Baltic fishery resources.

REFERENCES

Anon. (1999) IBSFC Handbook of Basic Texts. International Baltic Sea Fisheries Commission, Warsaw. 55 pp.

Anon. (2000) Agenda 21 for the Baltic Sea Region. In: *Biennial Report 2000, Baltic 21 Series*, **1/2000**. Stockholm. 50 pp.

Anon. (2002) Report of the ICES Advisory Committee on Fishery Management. 2002. *ICES Cooperative Research Report*, **255** (3): 710–775.

Casti, J. L. (1996) *Five Golden Rules: Great Theories of 20th-Century Mathematics – and Why they Matter.* John Wiley, New York: 235 pp.

Kaitala, V. & Pohjola, M. (1988) Optimal recovery of a shared resource stock: a differential game model with efficient memory equilibria. *Natural Resource Modeling*, **3**: 91–119.

Lindroos, M. & Kaitala, V. (2000) Coalition game on Atlanto-Scandian herring. In: *International Relations and the Common Fisheries Policy. Proceedings of the Fourth Concerted Action Workshop on Economics and the Common Fisheries Policy, Bergen, October 2000. CEMARE Miscellaneous Publication*, **49**: 171–186.

Mace, P. (1997) Developing and sustaining world fisheries resources: the state of the science and management. In: *Developing and Sustaining World Fisheries Resources: the State of the Science and Management, Proceedings of the Second World Fisheries Congress* (Ed. by D. A. Hancock, D. C. Smith, A. Grant and J. P. Beumer), pp. 1–20. CSIRO Publishing, Collingwood, Australia.

Munro, G. (2000) On the economics of the management of "shared" fishery resources. In: *International Relations and the Common Fisheries Policy. Proceedings of the fourth Concerted Action Workshop on Economics and the Common Fisheries Policy, Bergen, October 2000. CEMARE Miscellaneous Publication*, **49**: 149–170.

The Southwest Atlantic; achievements of bilateral management and the case for a multilateral arrangement

A. John Barton
Fisheries Department, Stanley, Falkland Islands
David J. Agnew and Lynne V. Purchase
Renewable Resources Assessment Group, Imperial College, London SW7 1NA, UK

ABSTRACT: The Argentine shortfin squid (*Illex argentinus*) is one of the main commercial fishery resources of the Southwest Atlantic. Its distribution includes the fishery conservation zones of the Falkland Islands and Argentina as well as the high seas. A system for the management of shortfin squid has evolved to take account of its being both a shared and a migratory/straddling stock. Its short (annual) life cycle and highly variable migration and recruitment poses significant challenges for stock assessment. Management has developed since 1990 through the bilateral arrangement of the South Atlantic Fisheries Commission, involving Britain and Argentina. Significant progress has been made on data exchange and pre-recruit surveys, resulting in what is virtually real-time monitoring of the stock together with an ability to coordinate management action. In this paper we examine the limitations of the existing management regime, especially as regards high seas fishing. The benefits that a multilateral arrangement could bring to the management of this species, and others in the Southwest Atlantic, is discussed.

INTRODUCTION

The shortfin squid (*Illex argentinus*) is focused on here to show how management of what is both a shared and straddling stock has developed in the Southwest Atlantic Ocean. It constitutes one of the major fishing resources of the Southwest Atlantic (FAO Statistical Area 41) and is one of two species in the area that routinely make the top 20, in terms of annual catches, reported to the FAO (FAO, 1999), the other being hake, *Merluccius hubbsi*.

The shortfin squid has a life cycle of approximately one year, and biomass can vary greatly from year to year (Agnew, 2002). It has an extensive distribution and migration that includes the fishery conservation zones of the Falkland Islands and Argentina, together with the high seas (for the purpose of this paper the term "high seas" is used to denote international waters beyond 200-mile fishing zones). Significant commercial fisheries exploit shortfin squid in all three areas, most using specialized jigging vessels, although substantial catches are also taken by trawlers. Shortfin squid are a shared stock, being shared between the fisheries of the Falkland Islands and Argentina. They can also be described as a straddling stock, being fished on the high seas. The fishery has largely been prosecuted by distant water fleets, principally of Korea, Japan, Taiwan, the European Union (Spain) and latterly by an expanding Chinese fleet. Further, Argentina has developed a substantial domestic fleet of jigging vessels, about 54 operating in 2002. The Falkland Islands also have their own developing fleet of trawlers, which have limited involvement in the shortfin squid fishery. The squid fishery is of major economic importance to the Falkland Islands, and to the fleets that fish it, as well as being of increasing significance to the Argentine economy.

Shortfin squid can generally be characterized as fully exploited, though overfished in some years (the target escapement spawning stock biomass has not been met in three of the last ten years). In the space of 20 years, the Falkland Islands involvement with the shortfin squid fishery has changed from uncontrolled fishing, through the introduction of unilateral management, to the current situation of a bilateral arrangement, entailing coordinated management action by the United Kingdom, the Falkland Islands and Argentina. (It is a bilateral arrangement because the UK represents the Falkland Islands in bilateral negotiations. The UK delegation includes representatives from the Falkland Islands and any fisheries management necessary on the part of the UK is likely to be implemented by the Falkland Islands Government.)

The process of managing a short-lived volatile species of squid, fished for by a number of large distant water fishing fleets in a challenging political environment, is discussed in the following sections. The achievements of the current bilateral arrangement in managing the squid resource and other species is evaluated, and the additional benefits that a multilateral arrangement might bring are analysed.

REGIONAL GEOGRAPHY

The main fishery resources of the Southwest Atlantic, including shortfin squid, are associated with the continental shelf. The main fishery zones and FAO statistical areas are shown on Figure 1. The Southwest Atlantic is one of the few regions in the world where the continental shelf extends beyond the 200-mile fishery conservation zones. Although the shelf area beyond 200 miles is relatively small, it does support a significant high seas fishery.

The Southwest Atlantic is bounded to the south and partly to the east by FAO statistical area 48, part of the Convention area for the Commission for the Conservation of Antarctic Marine Living Resources (CCAMLR). The remainder of the eastern boundary is with FAO statistical area 47, which is covered by the newly formed Southeast Atlantic Fisheries Organization (SEAFO).

There is no currently active multilateral or regional fisheries organization for the Southwest Atlantic, although there are bilateral fishery arrangements between the UK and Argentina (South Atlantic Fisheries Commission) and between Argentina and Uruguay (Frenté Maritima) in the region of the River Plate. All countries in the region have 200-mile fishery conservation zones.

SHORTFIN SQUID BIOLOGY

A brief synopsis of those aspects of shortfin squid biology relevant to the management issues being discussed is included here for reasons of being comprehensive.

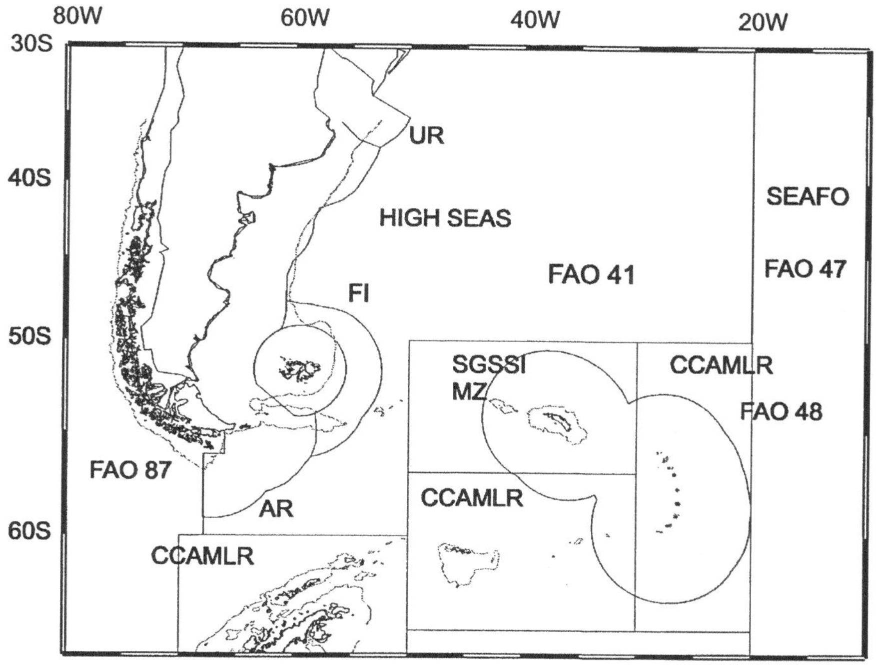

Fig. 1 Fishery conservation zones in the Southwest Atlantic (FI = Falkland Islands conservation zone; UY = Uruguayan conservation zone; AR = Argentine 200-mile conservation zone; SGSSI MZ = South Georgia and South Sandwich Islands maritime zone; SEAFO = Southeast Atlantic Fisheries Organization; CCAMLR = Convention on the Conservation of Antarctic Marine Living Resources; zones are illustrative, not definitive; 500m depth contour is shown).

Illex argentinus is distributed throughout the Patagonian shelf and in nearby oceanic waters between 22 and 54°S (Haimovici *et al.*, 1998). However, the squid has a somewhat complex population structure. Several stocks may be identified on the basis of biological data, distribution of pre-juveniles, juveniles and adults, and the area and timing of spawning (Brunetti *et al.*, 1998; Haimovici *et al.*, 1998). Analysis of maturity data indicates the presence of three or possibly four stocks, one north of 42°S, one south of 47°S and a further two stocks believed to coexist between 45 and 47°S. The last two stocks are considered to be distinct because they have different spawning times. Peak spawning of the summer spawning stock (SSS) is in January and February and of the southern Patagonian stock (SPS) in winter, July/August. They can be distinguished by a combination of size and maturity stage in February, when the main SPS fishery commences.

The fishery around the Falkland Islands and in adjacent waters of the Argentine EEZ targets squid of both SPS and SSS; of the two, the winter-spawning SPS is generally the most abundant. The main fishery operates during the feeding stages of the squids' long-distance migration throughout the period March–May. This migration originates on the winter hatching grounds of the northern shelf and continental slope east of the River Plate; the population then migrates generally south and east from the hatching areas to the feeding grounds of the southern shelf. Commercial fisheries target a mixed population of both SSS and SPS between December and June–August of the following year, fleets generally following the migration path of the squid south and east over the shelf. Early season catches (December through February) consist of varying proportions of mature and spent SSS and immature SPS. Generally, the SSS does not migrate as far south as the SPS; the two stocks are not spatially distinct and overlap to varying extent from year to year.

The population life cycle and geographic distribution of the population targeted within the fishery is linked to the oceanography. The Patagonian shelf and waters of the Southwest Atlantic south of 45°S are dominated by a distinctive pattern of surface water circulation created by opposing southerly and northerly flows of the Brazil and Falkands Currents respectively (Peterson, 1992). A confluence of poleward-moving warm water of the Brazil Current meets colder, less saline, waters of the Falklands Current flowing north towards the equator between approximately 33 and 39°S (Gordon, 1989).

Squid migrate northwards from the southern fishing grounds at the end of the austral summer (Arkhipkin, 1993). The physical location of the spawning grounds of the SPS, however, remains uncertain; it is generally believed that they spawn along the continental shelf-break within the Falklands Current (Brunetti & Ivanovic, 1992; Rodhouse *et al.*, 1995). Eggs are believed to be carried towards this confluence zone by oceanic currents, and early life stages have been found there (Brunetti & Ivanovic, 1992; Brunetti *et al.*, 1998). SPS *I. argentinus* are presumed to hatch in northern shelf waters off Brazil and Uruguay (Arkhipkin, 1993).

Paralarvae have been found in surface waters off the western boundary of the Brazil Current within the frontal zone between the shelf/slope and the Brazil Current, and in meanders of the Brazil Current, in the austral winter and spring (Brunetti & Ivanovic, 1992). Hatching success is related to favourable temperatures and associated conditions within the hatching area at the confluence of the two current systems. Abundance of squid for the subsequent fisheries to the south is associated with a low proportion of frontal waters (and a corresponding increase in the proportion of water of favourable temperature, i.e. 16–18°C) within the hatching area (Waluda *et al.*, 2001). Chang (2001) demonstrates a link between the direction and strength of onshore winds within the confluence region and subsequent recruitment strength within the fishery. Juveniles occur within the frontal region between coastal and offshore waters of the northern Patagonian shelf (Brunetti & Ivanovic, 1992).

Southward feeding migrations take place in the shallower inshore waters of the shelf, when the squid are considered to be recruiting to the fishery. The distribution of squid aggregations targeted by the fleets has been linked to specific temperature contours, characteristically between 9 and 10°C (Rodhouse *et al.*, 1995). Areas of high local abundance of *Illex* tend to be associated with warmer water; in the relatively cooler overall regime of the southern Patagonian shelf, generally dominated by colder waters of the Falklands Current, such a temperature signature may be indicative of upwelling areas, where food is plentiful (A. I. Arkhipkin, Falkland Islands Fisheries Department FIFD, pers. comm.).

FISHERIES MANAGEMENT

Annual catches of shortfin squid reported to the FAO from the Southwest Atlantic were <10 000 t until 1978. According to Csirke (1987), only then did a large-scale offshore fishery for shortfin squid develop, with 73 000 t taken in 1978 alone. Catches rose steadily

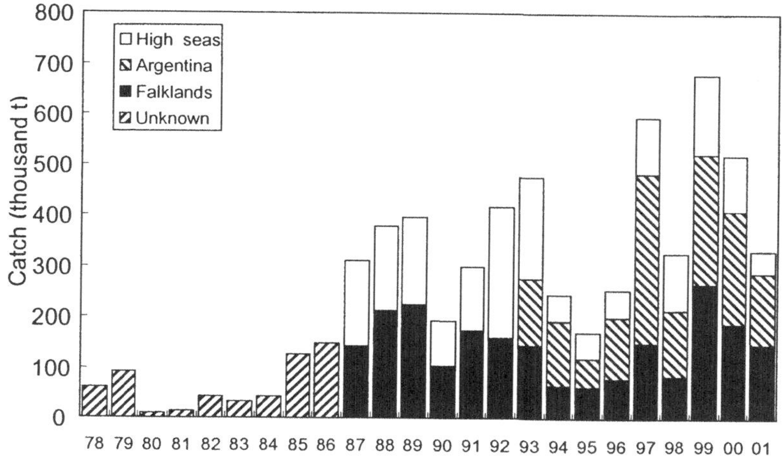

Fig. 2 Total annual catch of shortfin squid by region in the Southwest Atlantic, 1978–2001 (updated from Agnew *et al.*, 2001).

to 234 000 t by 1985, and then stayed high for many years (Figure 2), annual catches exceeding 500 000 t in some years. In the early years it was trawlers that dominated the catch, but during the 1980s there was a spectacular increase in the number of specialized squid-jigging vessels operating. Most shortfin squid are now caught by jigging vessels from Korea, Japan, Taiwan, China and Argentina, although a small trawl fishery by trawlers of the European Union and the Falkland Islands does operate.

Initially, catches were taken largely from the high seas and from the area that became part of the Falkland Islands fishery conservation zone in 1987. During the first five years of the licensed fishery in Falkland conservation zones (1987–1991), the average annual catch of shortfin squid was 170 000 t. Catches in the Argentine zone were small until 1991, when the annual total there was 46 313 t, significantly up on the previous year's 27 602 t. There had been sporadic larger catches in earlier years in the Argentine zone, but catches then rose rapidly , reaching 77468 t in 1992, 193 690 t in 1993 and a peak of 411 994 t in 1995 (Anon., 2000).

Data on catches from the high seas are sparse, and information on the number of vessels operating is only sporadically available. Some information on catches can be obtained from anecdotal reports together with daily catch reports from some segments of the high seas fleet, fishing vessels registered in the Falkland Islands for example. Other fleets may also report catches to their flag state (OFDC, 2001), but such information is not currently shared routinely.

The number of fishing vessels operating on the high seas can be significant at times. The fleet can include vessels that have licences or permits for Falkland or Argentine fishing zones, together with vessels that fish only on the high seas. For example, 111 jigging vessels from Taiwan operated in 2000, 12 in Argentine waters, 19 in Falkland zones and 80 on the high seas (OFDC, 2001). Their total catch that year was 247 857 t, of which 23 243 t came from the Falkland zone (FIG, 2002). The jiggers operating in the Argentine zone caught c. 1100 t each, i.e. a total catch of 13 200 t for the 12 vessels. By deduction, therefore, the catch by Taiwanese vessels alone from the high seas areas in 2000 would have been some 211 000 t.

Fleets have changed over the years. The mid 1980s fleet of trawlers from the former Soviet Union and other East European countries has now virtually disappeared, while Chinese jiggers have entered the fishery on a large scale. There were no Chinese jiggers licenced to operate in the Falklands' fishery until 1998, but in 2002, applications were made for 49 Chinese jiggers. Fishing vessels on the high seas are supported by a fleet of reefer vessels and trawlers that tranship catches and deliver fuel and supplies. All these support vessels probably operated in the Southwest Atlantic regardless of the outcome of the licence applications.

The annual southward feeding migration and subsequent offshore spawning migration (Arkhipkin, 2000) exposes shortfin squid to the high seas fishery at two critical periods in their life cycle. During the migration, large numbers of comparatively small squid (c. 200 mm mantle length) are caught from December to February. Later in the season, larger mature and maturing squid are caught as they move offshore to spawn (Arkhipkin, 2000). The catching of pre-spawning and spawning shortfin squid on the high seas at this stage can be particularly damaging in years where abundance is low and in years where the Falkland Islands and Argentina have closed their squid fisheries early to maintain stocks.

The annual life cycle of *Illex* leads to a certain vulnerability of the stock, in that there is no "buffer" of older age groups. The stock is entirely recruits. The stock–recruit relationship is likely to be steep (Beddington *et al*., 1990), and over most its range the relationship is likely weak. Nevertheless, the vulnerability of the fishery (and the economies of Argentina and the Falkland Islands) to low spawning stock size and recruitment failure is high. Target escapement levels have therefore been set, initially on the basis of proportional escapement and latterly on a target absolute spawning stock biomass (see below).

Fundamental to fisheries management is an understanding that recruitment variability is high, and therefore that the size of any future stock is difficult to predict. This makes management by total allowable catch particularly tenuous for short-lived species such as squid. Instead, shortfin squid management is effort-based, and the number of vessels licensed to participate in the fishery is strictly regulated. In-season monitoring, using near real-time data, is crucial to ensuring target escapement (leading to an adequate spawning stock biomass).

Around the Falkland Islands, vessels are obliged to report their catch, effort and position daily. An advantage of effort-based management is that there is no incentive to misreport the data. This is assisted by the fleet structure which, as noted above, is mostly of distant-water nature, so allowing the Fisheries Department to be prescriptive about the data that need to be reported (see also the discussion of Voluntary Restraint Agreements below). The FIFD places observers aboard vessels to collect biological data, including size and maturity. Assessments are performed by projection from pre-fishery swept-area surveys and using the multi-fleet modified deLury method proposed by Rosenberg *et al*. (1990), as modified by Basson *et al*. (1996). In this method, a log-likelihood function is used to fit a model, incorporating estimates of abundance and catchability, to a period of depletion, or declining catch rates. The method requires an estimate of the rate of natural mortality, which has been calculated at 0.06 per week for both species of squid (*Illex argentinus* and *Loligo gahi*) targeted around the Falklands using Pauly's (1980) equation (Beddington *et al*., 1990, Agnew *et al*., 1998). A similar value was obtained using Caddy's (1996) gnomonic stages approach. An alternative *ad hoc* approach uses previous estimates of catchability to calculate abundance from catch rates (Agnew *et al*., 1998). Confidence in the assessment depends on how well the depletion series conforms to the model, and generally increases as more data are added throughout the season. Assessments are tailored to the particular circumstances of each species.

THE CONSERVATION PROBLEM

The increasing catches of shortfin squid during the years 1978–1985 were derived from the high seas including waters around the Falkland Islands, because no fishery conservation zone, apart from a 3 mile territorial sea, existed then. However, an economic study on the Falkland Islands published in September 1982 (Shackleton, 1982) recommended the introduction of a 200-mile fishery conservation zone. Shortfin squid was not included among the most likely target species, but its importance became more apparent after the study was concluded.

Csirke (1987) recognized that there were some circumstances that might hinder fisheries management in the Southwest Atlantic. These included the fact that important resources like shortfin squid could be exploited both inside and beyond 200-mile fishery conservation zones, so a regional or subregional fishery management organization may be necessary to manage stocks. Csirke also highlighted the controversy between Argentina and the United Kingdom with regard to jurisdictional issues in the area around the Falkland Islands. Csirke's (1987) analysis suggested that, until 1985, shortfin squid were overall just reasonably exploited, although from 1983, with total yields very close to the maximum attainable, they should have been regarded as fully exploited. Further, Csirke indicated that, if UK evidence of a 60% increase in fishing effort in 1986 over 1983–1985 levels was real, the shortfin squid stock would have been overfished in 1986. Patterson (1985) stated that shortfin squid in the vicinity of the Falkland Islands were overexploited in 1985, assuming a unit stock. An alternative analysis (MRAG, 1985) suggested that the results for 1985 indicated almost full exploitation, but not necessarily overexploitation.

During this period, high levels of fishing activity were visible in the Falkland Islands. Numerous fishing and reefer vessels were calling to tranship in sheltered harbours, and there were local concerns about the impact of such heavy fishing activity on fish stocks and on the wider ecosystem. Falklands politicians were lobbying the UK Government to introduce a 200-mile fishery conservation zone. At the time, with total yields very close to the maximum attainable, the UK Government was striving to achieve a multilateral agreement to conserve fish stocks around the Falkland Islands (FIG, 1989). This proved unsuccessful, and effort continued to increase. As a consequence, on 29 October 1986, the UK and Falkland Islands Governments announced the introduction of the Falklands Interim Conservation and Management Zone (FICZ). This was a zone of radius 150 nautical miles.

THE ERA OF REGULATED FISHERIES

It was hoped that the introduction of the FICZ would solve the conservation problems insofar as shortfin squid were concerned. The introduction of the conservation zone and a fisheries management regime allowed fishing effort to be reduced and the collection of catch, effort and biological data to be greatly improved. The conservation target was then to ensure that the spawning stock did not fall below 40% of a level that would have occurred in the absence of fishing, i.e. a proportional escapement of 40% (FIG, 1989).

Even before the close of the 1988 shortfin squid season, less than two years after its introduction, it was apparent that the introduction of the FICZ had not solved all the problems. The 1987/88 Fisheries Report (FIG, 1989) indicates that conservation continued to be the major consideration and that the FICZ could not be viewed in isolation from surrounding waters. It went on to state that shortfin squid appeared to be particularly vulnerable to overfishing early in the year when they were fished more than 200 miles from shore. The proposed solution was to provide additional protection for the squid through a process called Voluntary Restraint Agreements.

The basis of the Voluntary Restraint Agreements (VRAs) was to use the attraction of access to the Falkland Islands shortfin squid fishery to persuade fleets to limit their effort on the high seas during key periods (Beddington *et al.*, 1990). In particular, the agreements stipulated that the vessels involved in the VRAs should: "voluntarily" not start fishing in the Southwest Atlantic south of 45°S before a particular date to avoid catching large numbers of small squid; that they should cease fishing south of 45°S on the high seas by a particular date so allowing remaining squid to spawn, and that vessel numbers should be limited. Another key provision was that all vessels involved should provide catch data from the high seas.

This style of VRA was used for the first time during the 1989 shortfin squid season (Beddington *et al.*, 1990). Initially, it had limited success in controlling fishing effort, but the acquisition of data for high seas fishing was valuable. However, the VRA process had its limitations. It had no leverage over fleets or groups of vessels that had no involvement in the Falklands fishery. Additionally, some of those who were subject to the VRAs began to break the agreements, so attempts were made to enforce them. In February 1990, a Taiwanese squid jigger, the "Fu Chun", had its Falkland Islands fishing licence revoked for allegedly contravening the VRA; its licence fee was forfeited. However, the vessel's owners applied to have the decision judicially reviewed, and the Director of Fisheries' decision in relation to this particular vessel was quashed (Milner, 2002), although the VRA process emerged intact. Monitoring the implementation of the VRAs became difficult because a number of jigging vessels took to obscuring their identification marks or taking on false identities. The jigging vessels involved in the shortfin squid fishery tend to be remarkably similar in design, so identification without an accurate name or radio callsign is difficult.

The assessment of the 1989 shortfin squid fishery concluded that, despite the introduction of VRAs, the proportional escapement for the stock south of 45°S was around 9%, well below the 40% target (Beddington *et al.*, 1989). The 1989 assessment stated that prevailing levels of effort were not sustainable in the long term. It also drew attention to the fact that extending Falkland conservation zones to 200 miles could be beneficial, albeit only increasing escapement by a small percentage. The VRAs were used until 1992, but became progressively ineffective owing to breaches of the agreements. The VRA process became completely unworkable in 1993 when for the first time large numbers of foreign squid-jigging vessels were given access to the Argentine fisheries zone. The VRA process would have required coordinated Argentine and Falkland Islands participation for it to continue.

SOUTH ATLANTIC FISHERIES COMMISSION

In the aftermath of the Falklands Conflict of 1982, the UK Government sought to resume diplomatic relations with Argentina. This was achieved in 1990 as set out in the Joint Statement[1] adopted at the second meeting in Madrid (Madrid II).

[1] *The Joint Statements and Joint Press Statements referred to here are issued simultaneously by the UK and Argentine Governments. Copies are held by the Foreign and Commonwealth Office, London*

In October 1989, a joint statement issued by the UK and Argentine Governments agreed an approach, subsequently referred to as the "formula on sovereignty" though commonly called the "sovereignty umbrella". This indicated that anything done by either the UK or Argentine Governments should not be regarded as being prejudicial to their pre-existing position on the sovereignty dispute. This paved the way for normal relations to resume, including discussion on such matters of mutual interest as the conservation of fisheries. The joint statement indicated that both Governments would set up a working group to discuss exchange of information on fisheries conservation and future cooperation on fisheries. The Madrid II joint statement reinforced this approach to Southwest Atlantic fisheries and indicated that a working group would be set up to address fishery issues.

The next step came in November 1990, with the issuing of a joint statement on the Conservation of Fisheries (Joint Statement, 28 November 1990). The two Governments agreed to open the way for cooperation in the conservation of fish stocks. This was to be achieved, first, through the establishment of the South Atlantic Fisheries Commission (SAFC), composed of delegations from both states, and second, by banning fishing in an area that would extend the Falklands conservation zone to 200 miles. The zone was duly extended through the Falkland Islands Government proclaiming (Proclamation 2 of 1990) a second fishery conservation zone; the Falklands Outer Conservation Zone (FOCZ). The FOCZ extended 200 miles from the coast in all directions except towards Argentina, so negotiation of a median line did not become an issue.

The SAFC was given the remit of:

- assessing information on fishing operations; catch statistics and analyses of the stock status of the most significant offshore species;
- making recommendations on the conservation of such species;
- making proposals for joint scientific research;
- recommending possible action for the conservation in international waters of migratory and straddling stocks;
- monitoring the implementation of the fishing ban in the FOCZ.

Table 1 sets out the chronology of SAFC meetings and their key developments. The first meeting took place in Buenos Aires on 20 and 21 May 1991 and was reasonably productive, agreeing an exchange of catch and fishing effort data. Joint scientific projects were proposed and it was agreed that, in future, SAFC meetings should be preceded by a meeting of UK and Argentine scientists, later formalized as the Scientific Sub-Committee (SSC) of the SAFC. The first meeting also considered mechanisms of communication in relation to the prevention of illegal fishing activities, which evolved later in the year into an agreement on hot pursuit. To date (early 2003), there have been 21 meetings of the SAFC plus an additional *ad hoc* meeting, and 19 meetings of the Scientific Sub-Committee. There have also been several scientific workshops, mandated by the SSC to address specific issues. This represents a considerable volume of joint UK and Argentine work on fisheries conservation in the Southwest Atlantic.

The SAFC and the SSC have provided a vital forum for addressing fisheries conservation issues in the Southwest Atlantic. The main achievements include:

- The exchange of fisheries data providing both UK and Argentine Governments with data for a wider area, contributing to real-time assessment of the shortfin squid fishery.
- Programmes of joint scientific research – Argentine research vessels have conducted research cruises on shortfin squid and southern blue whiting (*Micromesistius australis*) in Argentine and Falklands conservation zones. For the purpose, a joint UK/Argentine scientific team is embarked and the Falkland Islands meet a proportion of the costs.

Table 1 Decisions of bilateral meetings between the UK and Argentina.

Year	Dates: SAFC meeting	Dates: SSC meeting	Decisions	Notes
Implementing meetings				
1989			The UK–Argentine meeting of October 1989 set outthe framework for the normalization of future relations, referred to as the "sovereignty formula".	The FICZ was established on 29 October 1986.
1990			The UK–Argentine meeting of February 1990 agreed to exchange information on the operations of fishing fleets, appropriate catch and effortstatistics and analyses of the status of stocks of the most significant offshore species between 45 and 60°S, and to jointly assess such information and explore bilaterally the possibilities for cooperation and conservation.	
			The UK–Argentine meeting of November 1990 set up the exchange information on the most significant offshore species, assess and make recommendations to Governments for conservation, undertake joint scientific research, recommend to governments actions for the conservation of migratory and straddling stocks in international waters, and monitor the prohibition of fishing in the FOCZ.	The FOCZ was set up on 26 December 1990 following the November UK–Argentine meeting. North of the Falkland Islands it was not placed directly against the Argentine EEZ, for political reasons, and because the Argentine baselines were not fixed at that time. This created a "Gap" between the FOCZ and the Argentine EEZ.
SAFC/SSC meetings				
1991	1 (May, Argentina)		Agreement to exchange catch and effort data; proposals for joint research made.	The species that are covered by the SAFC are commonly accepted to be only *Illex*, SBW, southern hake, hoki and *Loligo* squid. Many other species are of joint concern and, according to the UK, should be included (for instance, common hake, skates and rays, Patagonian toothfish). Agreement on hot-pursuit arrangements was made independently of the main meeting. Even in this first meeting the SAFC considered the effects of third-party fishing in international waters.
	2 (December, UK)		More proposals for joint research made. experts to consider ways of harmonizing data submission.	Prior to this meeting there was a meeting of technical

Table 1 Decisions of bilateral meetings between the UK and Argentina (continued).

Year	Dates: SAFC meeting	Dates: SSC meeting	Decisions	Notes
1992	3 (May, Argentina)		Agreement to hold joint research surveys, the first to be	There was a scientific "meeting of experts" in Mar del Plata (Argentina) in 1993 in May 1992 to consider data requirements and joint research proposals. It produced a report. This was in effect a proto-SSC.
	4 (November, UK)			Throughout the period 1992–1994 the problem of the Gap was repeatedly raised by the UK.
	5 (December, Argentina)		SSC created (functions: monitor stocks, recommend conservation measures, recommend annually guidelines for total effort and average expected catch, establish a common basis for determination of total effort and to ensure that this effort does not result in recommended guidelines being exceeded); exchange of detailed near-real-time data on daily catch/effort and biological data of *Illex* during the fishing season was agreed; for the 1993 season, *ad hoc* consultations would be initiated to agree criteria on which the *Illex* fishery would be closed; Argentina limited its foreign charters to 45 vessels.	This extraordinary meeting was called in anticipation that Argentina would be opening its fishery to foreign vessels from 1993 onwards. The duties of the SSC indicate that, at this point, the SAFC was considering joint limitations oneffort (for all species, but primarily concerned with *Illex*).
1993	6 (June, UK)	1		The first exploratory joint research cruise (acoustic) took place in 1993 (*Illex*, SBW, hake).
	7 (October, Argentina)	2	Early warning system (EWS) agreed*; data exchange intensified; UK agreed to limit *Illex* effort to 1993 levels, and Argentina to 220 000 t or 80 foreign flag charters; agreed to propose long-term fisheries agreement; agreement to hold a joint pre-recruit *Illex* survey in 1994.	The 1993 meetings made significant progress on joint management of *Illex* against the background of increasing catches by foreign vessels licensed by Argentina.
1994	8 (June, UK)	3	Agreed to hold the first SBW acoustic survey in September 1994, and an Illex pre-recruit survey in February 1995.	1994 was the first year that the *Illex* fishery was closed early following the EWS. The first *Illex* pre-recruit survey was undertaken in the first half of 1994. The first SBW acoustic survey took place in September 1994. From now on, the *Illex* and SBW surveys became a feature of routine activities.

Table 1 Decisions of bilateral meetings between the UK and Argentina (continued).

Year	Dates: SAFC meeting	Dates: SSC meeting	Decisions	Notes
1994	8 (June, UK)	3		On 10 December a Joint Statement by the UK and Argentina indicated that the UK would open the FOCZ for fishing in 1994. The FOCZ had been closed from its inception in 1990 until this time.
	9 (December, Argentina)	4	Following the closure of the Gap SAFC relations were not good. The 9th meeting simply states that the report of the SSC was received, that delegations would inform their respective governments of the discussions, and that the meeting took place "in a cordial atmosphere".	The "Gap", a wedge of high seas waters between the NW corner of the FOCZ and the Argentine EEZ, was closed by the UK in late 1994. This followed UK concerns (repeatedly voiced since the early 1990s) about its function as a haven for illegal vessels poaching in Argentine and Falkland waters. In 1994 the FOCZ was extended westwards to meet the Argentine EEZ, which stopped the practice.
1995	10 (June, UK)	5	SSC agrees that a series of SBW acoustic estimates is required. SAFC encouraged improvements to the EWS, more joint surveys, and expressed concern about poaching. Agreement to discuss further the issue of high seas fishing. Agreement that scientists could publish joint papers (although this has never been actually possible).	Between 1994 and 1996 various attempts were made to draft a long -term agreement that would limit and divide effort between Argentina and the Falklands. With the adoption of the 40 000 t level and the EWS in 1997 this was no longer necessary and attempts to arrive at a long-term agreement were abandoned.
	11 (November, Argentina)	6	No major new initiatives.	By 1995 the fundamental processes were established: an SSC, joint surveys, the EWS, data exchange. Little progress was made on additional issues, such as the high seas or long-term agreements.
1996		7 (July, UK)	No major new initiatives.	
	12 (November, Argentina)	8	No major new initiatives.	
1997		9 (July, UK)	40 000 t escapement level for *Illex* agreed: "The Commission also agreed to recommend to Governments that they should make every effort to ensure that the level of *Illex* Spawning Stock Biomass at the end of the 1997 season should be at least 40 000 t."	This was a major breakthrough. Previously, there had been no generally agreed escapement level, although the UK had stated as early as 1992 that its calculation suggested 50 000 t to be a safe escapement limit.
	13 (November, Argentina)	10	No major new initiatives.	

Table 1 Decisions of bilateral meetings between the UK and Argentina (continued).

Year	Dates: SAFC meeting	Dates: SSC meeting	Decisions	Notes
1998		11 (June, UK)	No major new initiatives.	
	14 (November, Argentina)	12	No major new initiatives.	During 1998 the EU started negotiating a multilateral agreement (the SW Atlantic Multilateral Fisheries Agreement) which would tackle the high seas problem but which failed to make progress through 1999 and 2000.
1999	15 (June, UK)	13	SSC recommended reduction of SBW catches, stating that current catch levels (of the order of 100 000 t) were unsustainable.	There have been a number of workshops on SBW. An assessment of SBW in March 1995 produced ambiguous results. A joint assessment using new data in June 1999 clearly indicated a declining biomass.
	Ad hoc (September, Madrid)		Coordinated approach to preventing poaching; early establishment of a multilateral arrangement urged.	The multilateral agreement is mentioned in all SAFC statements from here on. The 1999 *ad hoc* meeting followed agreements reached between UK and Argentine foreign ministers in July 1999 (London), which directed the SAFC to consider practical ways of preventing poaching and the multilateral arrangement relating to high seas fisheries.
	16 (November, Argentina)	14	SAFC recommended reduction in SBW catches.	
2000	17 (June, UK)	15	SAFC first considered hoki.	
	18 (November, Argentina)	16	No major new initiatives.	
2001	19 (June, UK)	17	SSC agreed that shared data can be published.	
	20 (November, Argentina)	18	No major new initiatives.	
2002	21 (March, UK)	19	Catch limits of 55000–59 000 t for SBW acknowledged. proposals for high seas fishing agreements.	This meeting was called specifically to discuss new

* *Paragraph 4 of the Joint Statement of the 7th meeting states "...if a potential conservation problem were to be identified by scientists, whether at INIDEP or Imperial College, they would immediately inform the other side with a view to making a joint assessment of the state of the stocks without delay. At the same time, the scientists would inform their respective representatives on the Commission of the existence of a potential conservation problem and of the work being undertaken. The two sides would consult urgently, either via diplomatic channels or by calling an emergency meeting of the South Atlantic Fisheries Commission, in order to seek to ensure the effective conservation of the stock".*

- Joint scientific workshops to assess the status of some of the main shared species
- Implementation of an "early warning system" for the shortfin squid fishery. If UK and Argentine scientists agree that the spawning biomass is likely to fall below the conservation target, fisheries in both Argentina and the Falkland Islands can be closed early to conserve stocks. This measure has been implemented on a number of occasions, most recently in 2002.
- Recommendations to Governments on conservation measures for the most important offshore species.

For the first eight years, the SAFC settled into a routine of making slow political and rather more rapid scientific progress, but in 1999, the pace appeared to change. The management system that has evolved for shortfin squid has been relatively successful in assessing the stock and agreeing timely closure to ensure agreed escapement of squid. The surveys show good agreement with assessments made in-season (Figure 3), and are combined with projections of likely future fishing effort through the season to estimate a day on which the stock will have reached 40 000 t. As stated earlier, there is a very weak stock–recruit relationship (Figure 4), so recruitment in one year cannot be predicted from the size of the spawning stock at the end of the previous year. Although the environmental relationships described earlier do suggest that recruitment can be predicted, the models are not yet sufficiently robust to be used for stock assessment purposes. Therefore, the pre-recruit survey, in-season monitoring and the mechanism of the early warning system are really the only ways to ensure that the escapement target is achieved. However, a big unknown in the calculations is always the level of catches by vessels fishing in international (high seas) waters (Figure 5). Information is gleaned from patrols, overflights and estimates from available information about fleet sizes, etc. Any resultant estimates of high seas catches do, however, carry a high degree of uncertainty. Furthermore, catches of the southern Patagonian stock may be made on the high seas even after an agreed SAFC closure of the fishery.

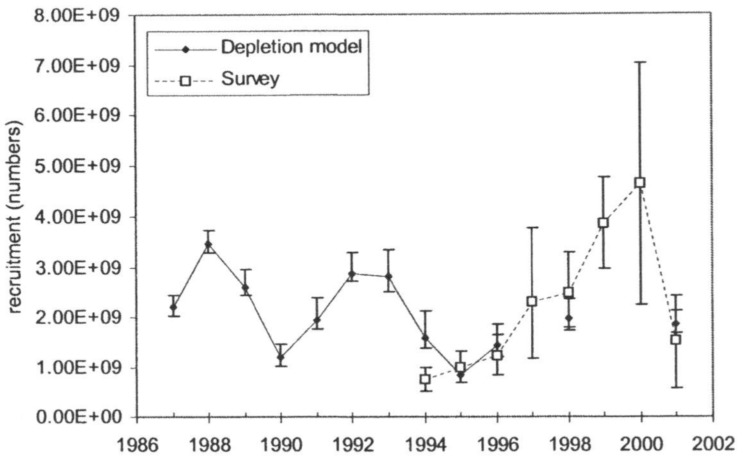

Fig. 3 Estimates of *Illex argentinus* recruitment from joint Argentine–UK pre-recruit surveys and from stock assessments using a deLury depletion model (Basson *et al.*, 1996). Depletion estimates are unavailable for 1997, 1999 and 2000.

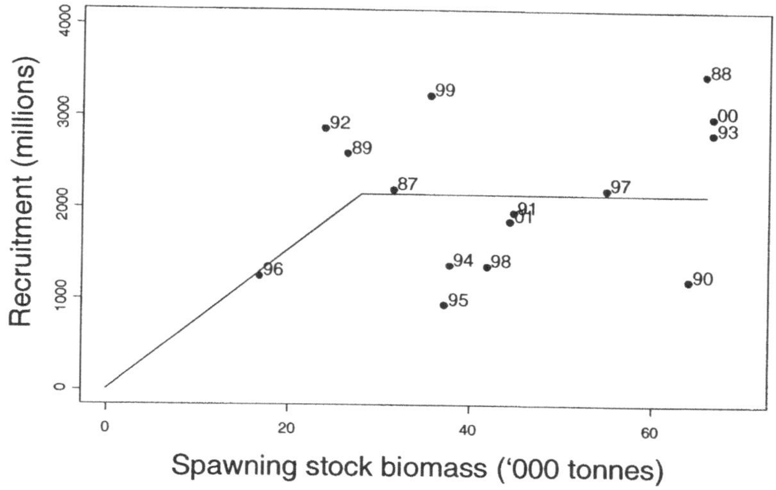

Fig. 4 Stock–recruit relationship for *Illex argentinus*.

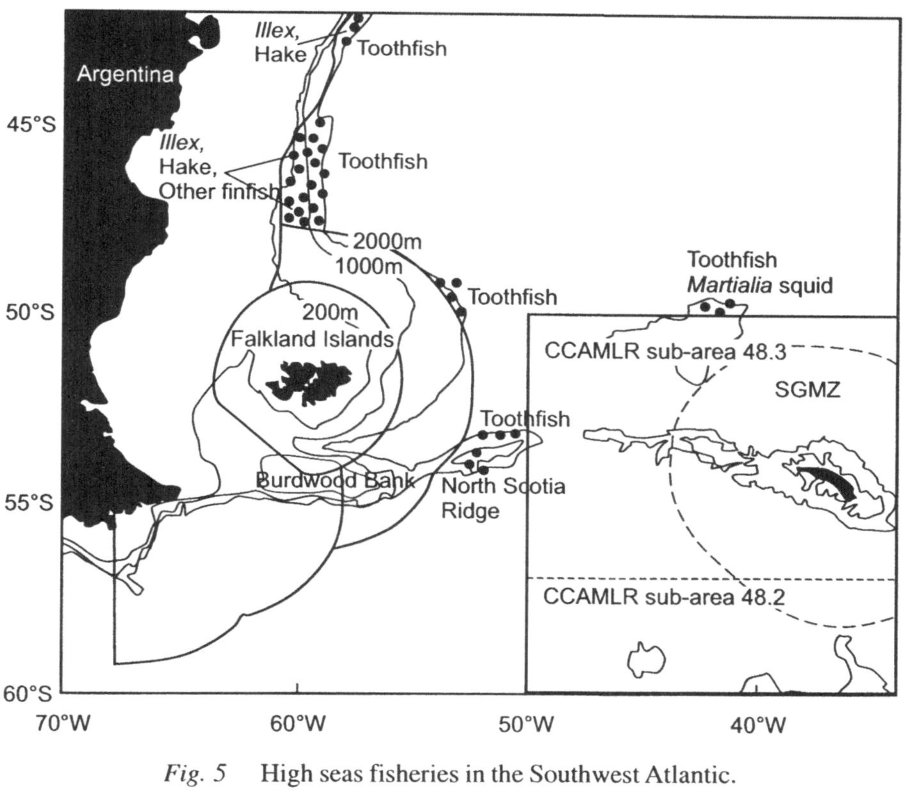

Fig. 5 High seas fisheries in the Southwest Atlantic.

This issue has been of concern to the parties of the SAFC since its inception (Table 1). Various parties have made attempts since 1990 to negotiate a multilateral arrangement, but none have been successful. The recent increases in publicity about IUU fishing (Kirkwood & Agnew, 2003) and increasing concern that the now-comprehensive methods in place in the SAFC to preserve the stock can be significantly undermined by fishing in international waters, has leant a new urgency to the discussion. The UK and Argentine Governments issued a joint statement on 14 July 1999 addressing some of the long-term aspirations of both sides. Argentina wanted to have easier access to the Falkland Islands, which the joint statement gave them. The Falkland Islands wanted to see further progress on fisheries issues. In relation to fisheries the joint statement indicated that "in the light of the shared commitment to the maintenance and conservation of fish stocks in the South Atlantic, existing levels of co-operation between the United Kingdom and Argentina will be enhanced". The joint statement was followed the same day by an exchange of letters between the UK and Argentine Foreign Ministers, restating the need for consideration of multilateral arrangements relating to high seas fisheries to be high on the SAFC's agenda.

Following the July 1999 joint statement, an *ad hoc* meeting of the SAFC took place in Madrid on 2 and 3 September 1999. The meeting discussed the issues mandated by the July 1999 joint statement and, in the joint press statement, it indicated that the SAFC agreed the need for timely adoption of longer term measures to ensure the sustainability of fish stocks in the high seas of the Southwest Atlantic. The statement went on to indicate that this could be achieved through the early establishment of a multilateral fisheries arrangement. Subsequent SAFC meetings have kept this issue on the agenda, although substantive progress has yet to be made.

CONSEQUENCES FOR FINFISH

Most of the discussion thus far has centred on *Illex argentinus*. This is the most important shared/straddling stock from the point of view of both Argentina and the UK, and was addressed first by the SAFC. However, southern blue whiting (SBW) is also important to both sides. Annual SBW catches in the early 1990s were >120 000 t. Concern about the state of this stock led to the implementation of dedicated acoustic surveys from 1994 and detailed joint stock assessment workshops in 1995 and 1999 (Table 1). These eventually led to an understanding that current levels of recruitment would only support annual catches of 55 000–59 000 t. The Falkland Islands have reduced the catch of SBW in Falkland Island waters to half this level, but this has not yet been matched by equivalent reductions in catches in Argentine waters, so catches remain at about 80 000 t.

SBW is not thought to be taken in significant quantities on the high seas (it has a more southerly distribution around South America, and occasionally has been caught in Antarctic waters). Therefore, a multilateral regime would not significantly enhance its management. There have been indications from recent research, however, that the southern Patagonian stock of SBW may mix with a smaller stock that spawns in Chilean waters (Hill *et al*., 2001). Therefore, there may well be a need for trilateral discussions on this species in the future.

There are, however, other species of commercially important fish in high seas waters as well as in Argentine and Falkland Island waters. Not all of these are covered by the most significant offshore species definition of the scope of the SAFC. Although never explicitly stated, the species scope of the SAFC is understood to cover shortfin squid, southern blue whiting, southern hake (*Merluccius australis*), hoki (*Macruronus magellanicus*) and longfin squid (*Loligo gahi*). The last of these is thought to comprise two discrete stocks, one restricted to Argentine coastal waters and one restricted to the Falkland Islands, so it is not commonly understood to be shared. The other species listed are, however, shared, and moreover are caught to various extents in high seas waters.

Although the UK delegation to the SAFC has often attempted to get other species included in SAFC talks, the attempts have thus far been unsuccessful. A multilateral regime might allow such species to be considered. For example, at the time of negotiating the SAFC, it was thought that the bulk of hake catches around the Falkland Islands would be *Merluccius australis*; hence its inclusion. In reality, more than 90% of the hake catch in Falkland Islands waters are common hake *Merluccius hubbsi* (Agnew *et al.*, 2001), and the stock is almost certainly distributed between Falkland and Argentine waters. There is significant fishing on common hake in high seas waters north of the FOCZ, so its inclusion in a multilateral regime would make sense.

Many other species are similarly distributed and caught on the high seas and in Argentine and Falklands waters, perhaps the most notable currently being toothfish (*Dissostichus eleginoides)*. Significant poaching of toothfish in the FOCZ was noted by the Falkland Islands in the early 1990s, and the species is known to be fished now right up to the edge of the FOCZ. A multilateral regime would allow this species to be controlled in the high seas waters that are currently a potential and actual source of unregulated fishing for toothfish (Kirkwood & Agnew, 2003).

PROSPECTS FOR A MULTILATERAL REGIME

The introduction of fishery conservation zones, the VRA process and the work done in the SAFC have all helped to provide the means for improving the management and conservation of shortfin squid. However, the stock remains under intense pressure, particularly in years of low abundance. There remains significant fishing effort on the species on the high seas, where virtually no conservation measures apply at present. The SAFC is committed to the early establishment of a multilateral fisheries agreement. All the factors of regional geography, straddling stocks, the statements from the SAFC and the diversity of fishing fleets involved, suggest that a multilateral arrangement for the high seas is the way forward. The setting up of a regional fisheries management organization (RFMO), using the relevant provisions set out in the UN Fish Stocks Agreement 1995 (United Nations, 1995) is an obvious possibility. That agreement developed the provisions of the Law of the Sea relating to the Conservation and Management of Straddling Fish Stocks and Highly Migratory Fish Stocks and provides a framework for implementing those provisions and for setting up RFMOs.

Whereas a RFMO seems to be the obvious solution, their use elsewhere has met with variable success. Churchill (1998) highlighted three weaknesses facing regional fisheries commissions in their management of high seas resources. These included the possibility for members to opt out of measures, the fact that some states might refuse to join the commission, and poor enforcement of commission measures. However, despite their apparent weaknesses, there really is no alternative to the RFMO model when it comes to control of high seas fisheries resources. There is no provision for extension of national jurisdiction into the high seas areas except as part of hot-pursuit operations or as RFMOs or arrangements for straddling or highly migratory stocks. Given that the imperative for the Southwest Atlantic is conservation of all resources, not just *Illex argentinus*, a broad-based RFMO such as SEAFO (Convention on the Conservation and Management of Fishery Resources in the Southeast Atlantic Ocean) would seem to be appropriate. SEAFO is the first RFMO concluded since the adoption of the UN Fish Stocks Agreement 1995, and Sydnes (2001) stresses that there is nothing in the SEAFO Convention that would prevent it from evolving into an efficient RFMO. Furthermore, although until recently other RFMOs appear to have had problems enforcing their measures, the situation may be changing. Recent developments, particularly the campaigns against IUU fishing by CCAMLR, ICCAT, CCSBT, IOTC, etc, have highlighted the need for States to act strongly and in united fashion within RFMOs to ensure that enforcement is effective. There seems, therefore, to be active development of real actions for enforcement, in particular linked to trade sanctions.

On the matter of membership and participation, we believe that, if Argentina and the UK were to act in a united fashion, there would be sufficient disincentive for third parties to opt out from either membership or management regulations. All third party distant-water fleets purchase annual licences from either Argentina or the Falkland Islands for shared, straddling and wholly contained stocks. The threat of withdrawal of licences should be sufficient to encourage participation by all parties who currently fish on the high seas, as the VRA situation showed.

Participation is usually spoken of in terms of complying with conservation measures. An equally important contribution to Southwest Atlantic fisheries conservation would be the sharing of scientific and other catch and effort data. This would considerably reduce the uncertainty currently inherent in assessments of *Illex argentinus* and *Dissostichus eleginoides*, and could potentially be achieved much sooner than a fully legal and ratified RFMO agreement, through the creation of *ad hoc* scientific working groups.

The "incentive" created by Argentina and the UK would be augmented, of course, by the obligations placed on Parties to the 1995 UN Straddling Stocks Agreement. However, of the major present or potential participants in Southwest Atlantic high seas fisheries, the 1995 Agreement has only been ratified by the UK [in respect of the Falkland Islands[2]], Uruguay and the Russian Federation. Although the UK, Argentina, Japan, Korea and Spain have signed, they have not yet ratified the agreement. Neither China nor the EU have signed or ratified, although the latter has made statements of intent.

The creation of a multilateral fisheries regime in the Southwest Atlantic would be a complicated process. In the Southeast Atlantic the negotiation of the SEAFO Convention took some five years. The tensions between coastal states and fishing nations, which had been prevalent during the negotiation of the UN Fish Stocks Agreement of 1995, also surfaced during the negotiation of the SEAFO Convention (Sydnes, 2001). Such tensions would inevitably arise in the Southwest Atlantic. Additionally there would be the complications of the sovereignty dispute over the Falkland Islands and the fact that China and Taiwan, whose political relationship is at best cool, are major participants in the fishery. Nevertheless, the sooner such a process is started, the sooner one could expect an agreement to be reached, especially if it had strong political backing from Argentina and the UK.

---

[2] *Although the UK has signed on its own behalf, ratification is pending coordination with the EU, which has not yet ratified the Agreement. See http://www.un.org/Depts/los/reference_files/status2003.pdf*

CONCLUSIONS

The past 20 years have seen the expansion of fisheries and fisheries management in the Southwest Atlantic. Conservation and management has included the introduction of unilateral fishing zones, the use of voluntary restraint agreements and the development of a bilateral fisheries commission. The latter in particular has been successful in improving the conservation of shortfin squid and other species within 200-mile fishing zones. Significant catches of shortfin squid are still taken on the high seas, and in all probability, conservation measures can only be extended to this area through a multilateral arrangement, to which the bilateral SAFC is committed. On the evidence of other regions it is likely to be some years before this is achieved, but it is likely to offer the best prospect for ensuring the conservation of shortfin squid and other Southwest Atlantic resources.

REFERENCES

Agnew, D. J. (2002) Critical aspects of the Falkland Islands pelagic ecosystem: distribution, spawning and migration of pelagic animals in relation to oil exploration. *Aquatic Conservation: Marine and Freshwater Ecosystems*, **12**: 39–50.

Agnew, D. J., Beddington, J. R., Baranowski, R., des Clers, S. & Nolan, C. (1998) Approaches to assessing stocks of *Loligo gahi* around the Falkland Islands. *Fisheries Research*. **35**: 155–169.

Agnew, D. J., Middleton, D. A. J., Marlow, T., Brickle, P. & Arkhipkin, A. I. (2001) The biology and fishery for *Merluccius australis* in Falkland Islands waters. Paper presented to the Workshop on Recruitment of Southern Hake, Chile, 28 May – 1 June 2001.

Anon. (2000) Revista Redes: annual yearbook, 1999. Vol. **13**, No. 112 (May 2000): 76–77.

Arkhipkin, A. I. (1993) Age, growth, stock structure and migratory rate of pre-spawning short-finned squid *Illex argentinus* based on statolith ageing investigations. *Fisheries Research*, **16**: 313–318.

Arkhipkin, A. I. (2000) Intrapopulation structure of winter-spawned Argentine shortfin squid, *Illex argentinus* (Cephalopoda, Ommastrephidae), during its feeding period over the Patagonian shelf. *Fishery Bulletin, U.S.*, **98**: 1–13.

Basson, M., Beddington, J. R., Crombie, J. A., Holden, S. J., Purchase, L. V. & Tingley, G. A. (1996) Assessment and management techniques for migratory annual squid stocks: the *Illex argentinus* fishery in the Southwest Atlantic as an example. *Fisheries Research*, **28**: 3–27.

Beddington, J. R., Basson, M., Tingley, G. A., Anderson, J. & Rosenberg, A. A. (1989) The *Illex* squid fishery in the Southwest Atlantic. Renewable Resources Assessment Group, Imperial College: 23 pp.

Beddington, J. R., Rosenberg, A. A., Crombie, J. A. & Kirkwood, G. P. (1990) Stock assessment and the provision of management advice for the short fin squid fishery in Falkland Islands waters. *Fisheries Research*, **8**: 351–365.

Brunetti, N. E. & Ivanovic, M. L. (1992) Distribution and abundance of early life stages of squid (*llex argentinus*) in the South-west Atlantic. *ICES Journal of Marine Science*, **49**: 175–183.

Brunetti, N. E., Ivanovic, M. L., Rossi, G., Elena, B. & Pineda, S. (1998) Fishery biology and life history of *Illex argentinus*. In: *Large Pelagic Squids*. (Ed. by T. Okutani), pp 217–231. Japanese Marine Fishery Resources Research Centre, Tokyo.

Caddy, J. F. (1996) Modelling natural mortality with age in short-lived invertebrate populations: definition of a strategy of gnomonic time division. *Aquatic Living Resources*, **9**: 197–207.

Chang, C-H. (2001) Environmental influences on the recruitment of *Illex argentinus* in the Southwest Atlantic. PhD thesis, University of London.

Churchill, R. (1998) Legal uncertainties in international high seas fisheries management. *Fisheries Research*, **37**: 225–237.

Csirke, J. (1987) The Patagonian fishery resources and the offshore fisheries in the South-West Atlantic. *FAO Fisheries Technical Paper*, **286**. 75 pp.

FAO. (1999) FAO Yearbook. Fishery Statistics Capture Production, 1997. Volume **84**.

FIG. (1989). Falkland Islands Interim Conservation and Management Zone. Fisheries Report 1987/1988. 45 pp.

FIG. (2002) Fishery Statistics Vol. **6** (1992–2001). Falkland Islands Government, Stanley, Falkland Islands. 70 pp.

Gordon, A. L. (1989) Brazil–Malvinas confluence – 1984. *Deep-Sea Research*, **36**(3A): 359–384.

Haimovici, M., Brunetti, N. E., Rodhouse, P. G., Csirke, J. & Leta, R. H. (1998) *Illex argentinus*. In: *Squid Recruitment Dynamics. The genus Illex as a Model, the Commercial Illex species and Influences on Variability*. (Ed. by P. G. Rodhouse, E. G. Dawe & R. K. O'Dor), pp. 27–58. *FAO Fisheries Technical Paper*, **376**.

Hill, S., Agnew, D., Middleton, D. & Arkhipkin, A. (2001) Review of southern blue whiting, *Micromesistius australis australis* in Falkland Islands waters. Paper presented to the Workshop on Recruitment of Southern Hake, Chile, July 2001.

Kirkwood, G. P. & Agnew, D. J. (2003) Deterring IUU fishing (this volume).

MRAG. (1985) Catch–effort analysis of squid jigging around the Falklands. Marine Resources Assessment Group, Imperial College.

Milner, A. (2002). R.v. Director of Fisheries, ExParte Fu Chun Fishery Company Limited. In: *The Revised Laws of the Falkland Islands*. (Ed. by P. A. Wooding and A. J. Fairclough), pp. 28–67 . Published by Law Reports International, Oxford. Vol. **7**.

OFDC. (2001) Annual Catch Statistics of Taiwan Deep Sea Squid Fishery, 2000 Fishing Season. Overseas Fisheries Development Council : 44 pp.

Patterson, K. R. (1985) The fishery for *Illex argentinus* in the Falkland Islands Protection Zone. Falkland Islands Development Corporation.

Pauly, D. (1980) On the interrelationships between natural mortality, growth parameters, and mean environmental temperature in 175 fish stocks. *Journal du Conseil*, **39**: 175–192.

Peterson, R. G. (1992) The boundary currents in the western Argentine Basin. Deep-Sea Research, **39**(3/4A): 623–644.

Rosenberg, A. A., Kirkwood, G. P., Crombie, J. A. & Beddington, J. R. (1990) The assessment of stocks of annual squid species. *Fisheries Research*, **8**: 335–350.

Rodhouse, P. G., Barton, J., Hatfield, E. M. C. & Symon, C. (1995) *Illex argentinus*: life cycle, population structure and fishery. *ICES Marine Science Symposia*, **199**: 425–432.

Shackleton, Lord, (1982) Falkland Islands Economic Study 1982. Command Paper **8653**. Her Majesty's Staionery Office, London: 137 pp.

Sydnes, A. K. (2001) New regional fisheries management regimes: establishing the South East Atlantic fisheries organisation. Marine Policy, **25**: 353–364.

United Nations. (1995) United Nations Conference on Straddling Fish Stocks and Highly Migratory Fish Stocks. *Agreement for the Implementation of the United Nations Convention on the Law of the Sea of 10 December 1982 Relating to the Conservation and Management of Straddling Fish Stocks and Highly Migratory Fish Stocks. UN Document*, **A/Conf./164/37**.

Waluda, C. M., Rodhouse, P. G., Podesta, G. P., Trathan, P. N. & Pierce, G. J. (2001) Surface oceanography and inferred hatching grounds of *Illex argentinus* (Cephalopoda: Ommastrephidae) and influences on recruitment variability. *Marine Biology*, **139**: 671–679.

The whole could be greater than the sum of the parts: the potential benefits of cooperative management of the Caribbean spiny lobster

Kevern L. Cochrane
Fishery Resources Division, FAO, via delle Terme di Caracalla, 00100 Rome, Italy
B. Chakalall
Subregional Office for the Caribbean, FAO, P.O. Box 631-C, Bridgetown, Barbados
Gordon Munro
Fisheries Centre, University of British Columbia, Vancouver, Canada

***ABSTRACT*:** The Caribbean spiny lobster *Panulirus argus* supports valuable fisheries along the American Atlantic coast from Brazil to Florida. The total value of landings is estimated at approximately US$500 million, and the lobster fisheries provide employment and livelihoods for tens of thousands of people throughout the coastal countries. The species occurs in waters up to 90 m deep, from North Carolina (USA) to Rio de Janeiro (Brazil). It has a pelagic larval stage lasting from 6 to 12 months, allowing for wide dispersion. The prevailing northward and westward currents probably result in a dependence of many countries on their more southerly and easterly neighbours for recruitment to their local populations, but what management there is of the fisheries for spiny lobster is carried out entirely at a national level. The status of local populations varies from fully to overexploited throughout much of its range. There has been some regional cooperation in research, mainly through the FAO and the Western Central Atlantic Fishery Commission. At a meeting in Mexico in 2000, the nations with important fisheries for the species expressed a desire to strengthen their cooperation in research and management, and to implement appropriate improvements at a national level. This was reaffirmed at a subsequent meeting in Cuba in 2002.

This paper describes some of the key features of the resource and the fisheries dependent on it. It discusses some of the existing problems in managing the fisheries and some obstacles to greater regional cooperation. Arguing on the basis of international law and using two case studies as examples, it demonstrates the need for regional management of the fisheries, if the resource and the benefits obtained are to be sustained, and evaluates some options for implementing this regional collaboration.

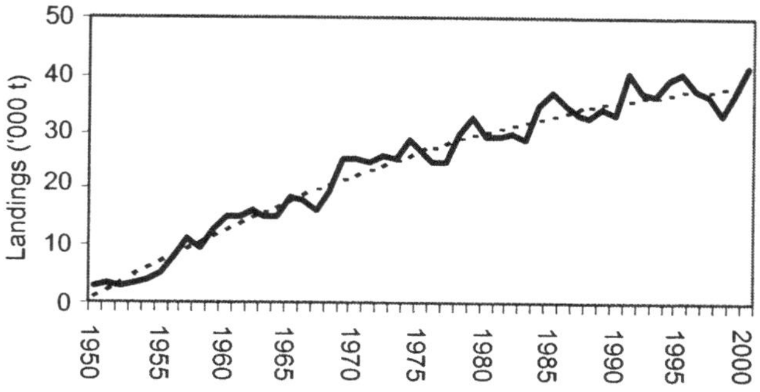

Fig. 1 Total annual landings of Caribbean spiny lobster *P. argus* recorded on FAO Fishstat: 1950–2000, and the best-fit "fisheries development" curve (see text for details).

INTRODUCTION

The Caribbean spiny lobster *Panulirus argus* supports valuable fisheries in the western central Atlantic that provide employment and high economic return. Awareness of the value of the resource has led to a steady increase in landings over the past 50 years, peaking at >41 000 t in 2000 after fluctuating around approximately 37 000 t during the 1990s (Figure 1). The landings appear to be reaching an asymptote, but those of 2000 were still the highest on record.

The species is distributed from Brazil northwards to Bermuda and North Carolina, USA, and throughout much of the Caribbean Sea and Gulf of Mexico (Cervigón *et al.*, 1993). The major lobster-producing countries of the region, as judged from landings in recent years submitted to the FAO, include Cuba, Bahamas, Brazil, Nicaragua and the USA (Table 1). In all, 25 countries submitted information on landings to the FAO, but the top 12 producers during this period accounted for more than 95% of the total recorded. At an average value of US$14.5 per kg, the mean annual landed value of the species for the period 1996–2000 would have exceeded US$500 million. Employment in the fishery runs to more than 40 000 fishers, with additional employment being provided by shore-based activities (Cochrane & Chakalall, 2001). These economic and social benefits are of considerable importance in a region that includes many developing countries.

Recognizing the importance of the fisheries, a series of four workshops has been arranged by the FAO through the Western Central Atlantic Fishery Commission (WECAFC). The workshops were held in Belize City, Belize, in 1997, in Merida, Mexico, in 1998 and 2000, and in Havana, Cuba, in 2002, and were attended by 14 countries, including all major producers (Cochrane & Chakalall, 2001; FAO, 2001a, b, 2003). They were intended to assist the countries in completing assessments of the national fisheries and local populations, to examine the status of the resource and the fisheries across the region, and to encourage greater regional cooperation in research and management. Cochrane & Chakalall (2001) reviewed the progress made through the first three workshops and examined possible modes of cooperation. This paper updates that review and in particular examines the benefits to be gained by cooperative management and, conversely, the costs of failure to do so.

Table 1 Top 12 producers of Caribbean spiny lobster as measured by average annual landings from 1996 to 2000 inclusive, and the percentage of the total average landings (37 279 t) by all countries over the same period.

Producing country	Average landings (t), 1996–2000	% of total
Cuba	9503.4	25.5
Bahamas	7947.8	21.3
Brazil	6838.8	18.3
Nicaragua	4713.2	12.6
United States of America	2763.8	7.4
Honduras	1120.8	3.0
Dominican Republic	891.6	2.4
Mexico	721	1.9
Belize	608.2	1.6
Venezuela	345.4	0.9
Jamaica	336.4	0.9
Turks and Caicos Islands	268	0.7

CURRENT MANAGEMENT AND THE STATUS OF THE RESOURCE

The most detailed information on the nature and effectiveness of management and the status of the resource across the region is still that reported in FAO (2001a, b) and Cochrane & Chakalall (2001). Those reports indicated that, of the 12 top producers, only five had limited access to the fishery in all regions and through all fishing types or fleets, while in the remainder, access was open in at least one subsector of the fishery. The countries with fully limited access were Cuba, Brazil, USA, Mexico and Venezuela. However, limited access does not necessarily mean that the allowable effort is commensurate with the productivity of the stock.

All countries that participated in the WECAFC workshops had some management regulations in place, primarily size limits (commonly minimum carapace length), and closed seasons. Details of these can be found in FAO (2001a). Cochrane & Chakalall (2001) reported that most minimum size restrictions in place would not have effectively protected an adequate proportion of the spawning stock in each country, and FAO (2001b) concluded that enforcement was a problem in many countries and that greater attention needed to be given to enforcing regulations such as closed seasons and minimum legal size.

There were substantial problems encountered at all the workshops with poor and inadequate data, and in some cases the data were insufficient to undertake any form of diagnostic analysis. Where some form of assessment could be undertaken, the overall results indicated "a resource that is being fully or over-exploited throughout much of its range" (Chapter 11, FAO, 2001a) and that there was an urgent need to control effort at appropriate levels and to implement effective minimum sizes that would be appropriate to the growth and maturity characteristics of the stock. The workshop in 2002 concluded that some progress had been made since 2000, but that many problems in implementing effective management still remained. The conclusions and recommendations arrived at in Merida in 2000 were therefore still considered to be valid.

These results are supported by a simple examination of the available data on landings of the species. The total landings available for the region, as recorded on the FAO Fishstat database, reveal a rapid increase in the past 50 years that now appears to be reaching an asymptote. The trend follows the expected progression of the generalized fishery development model described by, for example, Grainger & Garcia (1996). The model describes a progression from a rapid increase in landings under increasing effort, a levelling off and then a decline as the resource is depleted below its maximum average yield level. This sequence can be captured by a simple trinomial model, even though this form provides no functional description of the process. It was used in this study as a means of smoothing and removing the developmental trend from the time-series of landings. Fitting such a model to the total landings using the Newton non-linear routine under Solver (Microsoft Excel) to minimize the sum of squares between observed and fitted indicates that the landings could be expected to level off in approximately 2007 and thereafter begin to decline (Figure 1).

Given the different histories and economic profiles of the countries of the WECAFC area, it is also informative to examine the landings on a sub-regional basis. For practical purposes, the WECAFC Working Group postulated the existence of four separate substocks of *P. argus*, based primarily on geographical proximity, the nature of the coastal shelf and the prevailing currents (FAO, 2001a). These substocks were defined as (countries shown in parenthesis are those that have participated in the WECAFC workshops):

Southern "stock"	(Brazil, Venezuela, Dominican Republic [and including the Lesser Antilles islands]);
South central "stock"	(Colombia, Nicaragua, Honduras and Jamaica);
North central "stock"	(Mexico, Belize and southern Cuba);
Northern "stock"	(northern Cuba, USA [Florida], Bahamas, Turks and Caicos, and Bermuda).

Fitting best-fit "developmental" curves for the sub-regions suggested that landings in the north central sub-region peaked in 1984, those in the southern sub-region in 1993 and that landings in the south central sub-region are still increasing but, on current trends, would be expected to peak in 2011. The fitted curve for the downstream northern sub-region implied that landings will continue to increase there without reaching a limit, clearly an ecological impossibility but reflecting the surge in landings in the 1990s (Figure 2).

These trends need to be interpreted cautiously, especially because they are trends in total landings and therefore reflect changes in effort and the fisheries, not just changes in abundance. In the northern sub-region, the recent increases are driven entirely by increasing effort in the Bahamas, where the fishery is considered to be still underdeveloped and expanding (Deleveaux & Bethel, 2001), although recent declines have given cause for some concern (FAO, 2003). In the USA and Cuba, however, landings have stabilized in the past decade under strict management. In the north central sub-region, the resource is generally considered fully exploited. The increase in landings in the south central sub-region (Figure 2c) has been driven mainly by the fishery in Nicaragua, where there has been a resurgence in landings, or perhaps recorded landings, since the end of the civil war there in 1990 (Barnutti *et al.*, 2001). Limited data are available from this sub-region, but from the analyses that have been done, the resource is considered fully exploited, with signs of overexploitation, and in need of strict control (FAO, 2001b, 2003). The lobster fisheries of the southern region are dominated by Brazil, and landings from that

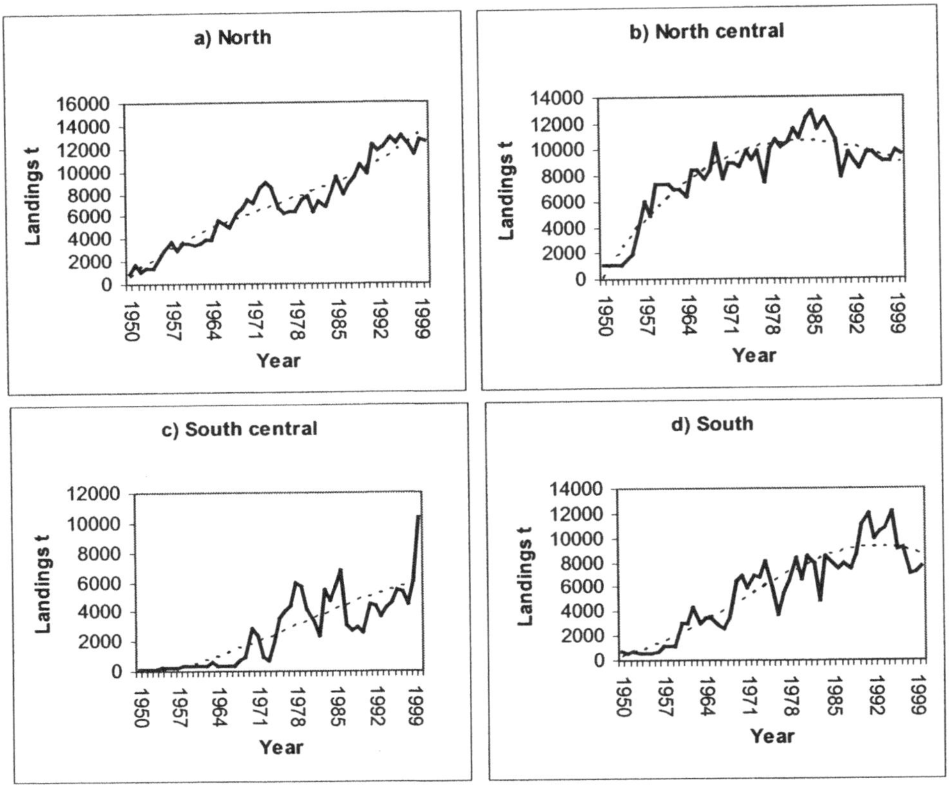

Fig. 2 Annual landings of *P. argus* from 1950 to 2000 and the best-fit fisheries development curves, by sub-region.

country have decreased dramatically since 1996. While there are management measures in place, these are not implemented effectively (FAO, 2001b) and the resource is considered to be overexploited (FAO, 2003).

This sub-regional analysis demonstrates that the trend shown by the total landings is strongly influenced by recent growth in two large producers (Bahamas and Nicaragua) and that, therefore, the overall status of the resource may be more pessimistic. In many of the main lobster-producing countries, there is a need for action at a national level to ensure that the resource is not overexploited with serious loss of socio-economic benefit (FAO, 2001b). Where countries still allow open access to the fisheries, this action should clearly include rapid implementation of limited access. In these cases and those in which access is already limited, there is also the need to reduce effective effort to levels appropriate to the productivity of the resource. The immediate social implications of such a step in many of the countries are, of course, a significant impediment to such action, and remedial action for those affected will also have to be a part of national management plans. However, as stressed by Cochrane & Chakalall (2001), failure to take the necessary management actions in the short term is highly likely to lead to even greater social and economic disruption in the long term as the resource continues to decline and to yield smaller and smaller catches.

STOCK STRUCTURE OF *P. ARGUS* IN THE WESTERN CENTRAL ATLANTIC

The pragmatic definitions of substocks used by the WECAFC Working Group are hypothetical only, but if there is to be serious cooperation in management and research, it is important that reliable information on the true stock structure of the species is provided. Limited evidence is available at present.

LIFE HISTORY AND CIRCULATION

Cochrane & Chakalall (2001) reviewed available knowledge on the life history of *Panulirus argus* and its relationship to prevailing currents. The larvae are planktonic and thought to spend between 6 and 12 months in oceanic water (Silberman *et al.*, 1994; Arce & de Léon, 2001), providing extensive opportunity for wide distribution in the Caribbean and wider Atlantic. Cruz *et al.* (2001) described variations in the patterns of puerulus settlement between Bermuda, Florida Keys, Bahia de la Ascensión (Mexico) and the Gulf of Batabanó (southwest Cuba), the four areas for which data were available. For Florida, Mexico and Cuba, there was some settlement throughout the year, whereas settlement in Bermuda was mainly in August and September. They also reported significantly different levels of settlement between the areas, but could not explain the differences other than to suggest that they could depend on variability in ocean currents and the size of the parent stock.

The predominant current flows through the Caribbean and Gulf of Mexico are towards the north and west (Figure 3) and typically vary from 20 to 250 cm s^{-1} (Silberman *et al.*, 1994; Roberts, 1997). FAO (2001b) provided a brief review on the biology and distribution of *P. argus* and referred to strong currents flowing from east to west, coupled with local gyres south of Cuba, off central America, Hispaniola and Puerto Rico, and in the vicinity of the Florida Keys. Reference was made also to a study that showed movement of drift cards released in the Gulf of Batabanó (southwest Cuba) to localities in the vicinity of the Yucatan Peninsula (Mexico), Bahamas, west of the Gulf of Mexico, Florida, and northeast and northwest Cuba. The review concluded that there was a high probability of mixing of larvae from different origins.

Roberts (1997) analysed the prevailing surface currents in the region to investigate probable patterns of distribution of larvae. In species with a larval dispersal period of only two months, substantially less than that estimated for *Panulirus argus*, he estimated an average transport distance in excess of 200 km and suggested that this would result in recruitment to any given site being drawn from an average of 3.5 other countries (range 1–7). Clearly these numbers would be substantially increased in the case of *P. argus* with its much longer larval period.

These current and dispersal patterns suggest a predominantly one-way flow, from southeast to northwest. The southern and easterm spawner populations are therefore likely to be net exporters of larvae whereas the northern and western countries are likely to be net importers, dependent on the upstream countries for at least a part of their recruitment. This would create a strong ecological link between countries sharing the resources and a need for close cooperation if use of the resource is to be optimized and equitable.

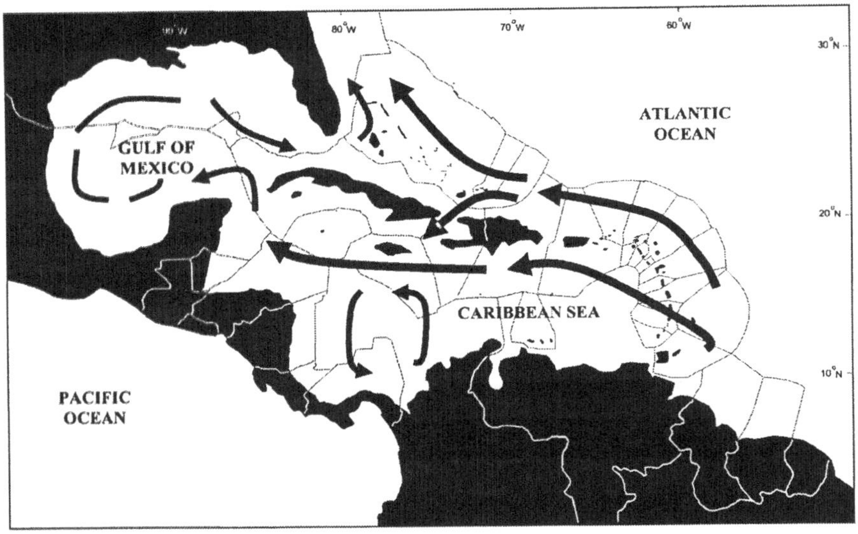

Fig. 3 Area of distribution of *P. argus,* including approximate EEZs (for illustrative purposes only), and the major prevailing currents in the wider Caribbean (after Silberman *et al.*, 1994).

GENETIC STUDIES

The implications of the combination of life history characteristics and current flows have been supported by genetic studies undertaken by Silberman and his colleagues. Using mitochondrial DNA, they investigated the genetic structure of *P. argus* and found no evidence of population subdivisions and little evidence for genetic heterogeneity within samples of lobster obtained from Panama and Venezuela in the south to Bermuda in the north. They concluded that this demonstrated a high level of gene flow between the different sites, consistent with the ecology of the species. Their results differed from earlier studies (Menzies and Kerrigan, 1979; Menzies, 1980; both reported in Silberman *et al.*, 1994) using protein isozyme electrophoresis that suggested genetically distinct populations in Jamaica and in the US Virgin Islands, whereas samples from two sites in Florida and one in Trinidad could not be distinguished. These early results, however, were not supported by three later and separate allozyme studies that failed to find heterogeneity among the sites sampled in each case (Silberman *et al.*, 1994). The bulk of the evidence therefore points strongly towards high levels of genetic flow throughout the area sampled.

A more recent study compared the mitochondrial DNA sequence variation between samples obtained from the western Atlantic and Caribbean with others obtained from Brazil (Sarver *et al.*, 1998). Those authors described genetic diversity between the two areas that was equivalent to that expected between species or subspecies. They proposed the existence of a southern and northern form of *P. argus*, and suggested that the extensive soft-bottom coastal shelf between the Brazilian coral reefs and those of the Caribbean formed a barrier between the two forms.

The information from these two studies cannot be considered conclusive and direct evidence; some quantification of a substantial export of larvae from southeast to northwest is still required before compelling arguments can be made. However, the information does raise the likelihood of two biologically distinct stocks: from Brazil and from a single wider Caribbean stock. The WECAFC working definitions should be revised accordingly and cooperative institutions would need to reflect this probability.

LANDING TRENDS

In an attempt to provide additional information on the stock structure of Caribbean spiny lobster, the time-series of landings submitted to FAO were analysed for possible correlations. This was done between the hypothetical substocks used at previous workshops and described above.

Calculating a simple correlation coefficient between the aggregate landings for each substock showed strong correlations between all sub-regions (r varied from 0.63 to 0.91). However, this correlation is likely to have been driven heavily by a similar developmental history as the fisheries across the region expanded in response to initially lightly exploited stocks and at least partially common markets. To address this problem, the developmental trends were removed from the data using simple trinomial models described earlier, and the residuals between observed and expected values estimated from the developmental models (Figure 2) were calculated for each sub-region. The residuals were then examined for correlations with other sub-regions. If there were linkages between sub-regions, e.g. recruitment in one sub-region being affected by spawner biomass upstream, a correlation could be expected.

Several significant correlations were found (Table 2), but most were negative. The only significant positive correlations were between the south and the north sub-regions with 0- and 1-year time-lags, where the lag was added in the more northerly (downstream) sub-region. The two positive correlations were also the two strongest correlations estimated. The fact that these correlations occurred between extreme upstream and the extreme downstream sub-regions is interesting, but it would be hard to use this as evidence of a direct linkage, given the lack of correlation with the two intermediate sub-regions. The results of the study by Sarver *et al.* (1998) indicating the existence of a separate stock off Brazil also suggest that these strong positive correlations do not reflect a strong link. It is therefore more likely to be a statistical artefact or, more speculatively, a result of some indirect but common driving force. In addition to the tests shown in Table 2, northerly lags of 3 and 4 years and lags in the opposite direction were also examined, with similar but even weaker results.

The results are therefore generally negative. However, this could be due to a combination of errors and incompleteness in the landings data and process uncertainty in any underlying stock/recruit relationship. They do not dispel the compelling life history arguments and the results of the genetic studies, both of which point to a high probability of the stock being shared. The absence of overt linkages between sub-stocks will, nevertheless, undoubtedly detract from the will at national level to cooperate internationally in management of the resource.

Table 2 Correlations between landings residuals after removal of developmental trends of the four sub-regions. * indicates a value of r significant with 95% confidence (~0.27).

"Stock"	North central	North central	South	South
		No lag		
Northern	1			
North central	-0.10	1		
South central	-0.38*	0.09	1	
Southern	0.60*	-0.16	-0.32*	1
		1-year lag		
Northern	1			
North central	-0.26	1		
South central	-0.40*	0.17	1	
Southern	0.51*	-0.26	-0.22	1
		2-year lag		
Northern	1			
North central	-0.29*	1		
South central	-0.31*	0.21	1	
Southern	0.21	-0.35*	-0.30*	1

THE NEED FOR REGIONAL COOPERATION

THE LEGAL BASIS

A shallow-water species such as *P. argus* is ultimately dependent on effective action within national EEZs for sustainable utilization. If each country in which the species occurs managed their populations at sustainable levels, the overall productivity of the resource should be assured. However, improved, more robust and more cost-effective utilization can be achieved by close regional cooperation (Munro *et al.*, 2003).

Shared fishery resources can be divided into two broad, non-mutually exclusive categories. The first consists of stocks that cross the coastal state EEZ boundary into the EEZs of neighbouring coastal states – referred to here as *transboundary stocks*. The second consists of stocks found both within the EEZ and the adjacent high seas – referred to here as *highly migratory* and *straddling stocks*. All existing evidence indicates that *P. argus* is strictly transboundary in nature.

The question of the management of transboundary fish stocks is addressed explicitly in the United Nations Convention on the Law of the Sea (LOS, United Nations, 1998), which achieved the status of international treaty law in November 1994. Coastal states sharing a transboundary stock, or stocks, are firmly admonished by the LOS through its Article 63(1) to cooperate in the management of the resource(s). Article 63(1) reads:

> "Where the same stock or stocks of associated species occur within the exclusive economic zones of two or more coastal States, these States shall seek, either directly or through appropriate subregional

or regional organizations, to agree upon measures necessary to co-ordinate and ensure the conservation and development of such stocks, without prejudice to other provisions of the Part."

This requirement is based on the need to manage stocks as units if management is to be effective, a fundamental concept well addressed in paragraph 7.3.1 of the FAO Code of Conduct for Responsible Fisheries (FAO, 1995).

In addition, Article 61 of the LOS, dealing with "Conservation of the living resources", in paragraph 2 requires that (United Nations, 1998):

> "The coastal state, taking into account the best scientific evidence available to it, shall ensure through proper conservation and management measures that the maintenance of the living resources in the exclusive economic zone is not endangered by over-exploitation. As appropriate, the coastal State and competent international organizations, whether subregional, regional or global, shall cooperate to this end."

These two Articles of the LOS, Article 63(1) in particular, provide the legal impetus and fundamental principles upon which regional management of *P. argus* should be based. The nature and extent of the sharing of the resource is uncertain, as a result of the unknown extent and direction of distribution of the larvae. However, there can be no doubt that it is substantial and it is likely that some countries are important donors of larvae to the region, whereas others are important recipients. All countries certainly function as both to some extent. The linkage is further strengthened by the nature of the fisheries, which use similar methods in most countries, the existence of common markets, and similarities in the problems faced by the national fisheries management agencies in executing their responsibilities in, for example, research, monitoring and enforcement. In almost all countries there is an urgent need for improved management and fishing practices, and this could be done most effectively with regional cooperation.

THE BENEFITS OF COOPERATION

To examine the benefits of cooperation in the management of transboundary stocks, we turn to analysis developed over the past 25 years by economists, which has now come to be adopted by non-economists, including marine biologists (see, for example, Caddy, 1997), and which has found its way into official publications (e.g. OECD, 1997). The analysis is discussed, in some considerable detail, in a companion paper (Munro *et al*., 2003).

The benefits of cooperation in the management of transboundary fish stocks might usefully be retitled the *costs* of non-cooperation in the management of such resources. In exploring the consequences of non-cooperation, economists bring to bear the theory of games, which is designed to analyse strategic interaction between, and among, "individuals", where the "individuals" can, *inter alia*, be states. Strategic interactions between and among states sharing a transboundary fishery resource such as *P. argus* are inescapable.

What the economic analysis predicts is that, in the absence of cooperation, optimal management at a national level will be elusive. In part, this is due to the fact that the benefits of management efforts by one state will spill over into neighbouring states. With no cooperation, the neighbouring states will simply enjoy, "free of charge" as it were, the efforts of the state undertaking the cost and expense of management. We can refer to these neighbouring states as "free riders". It is easy to show that, under non-cooperation, the best strategy for each state is to strive to be a "free rider", rather than a provider of benefits to neighbouring states. If each state strives to be a "free rider", then national resource management will obviously languish.

In some instances, non-cooperation can have the result that states will go beyond passive neglect of resource management and engage in explicit overexploitation of the resource, by becoming embroiled in "fish wars". A "fish war" has been defined by one legal expert as deliberate overexploitation of a fishery resource by one state in order to deny harvesting opportunities to another (Jensen, 1986).

The Pacific salmon treaty

An example is provided by the cooperative management of Pacific salmon by Canada and the United States, between California and Alaska. Similar to *P. argus*, the resource moves over wide distances. Moreover, also like *P. argus*, the spawning of the resource in one state has an impact on harvest opportunities in the neighbouring state. The resource is transboundary by virtue of the fact that American fishers "intercept", i.e. catch, salmon produced in Canadian rivers, e.g. the Fraser River, while Canadian fishers "intercept" salmon produced in American rivers, e.g. the Columbia River (which is primarily, but not exclusively, within the United States).

Pacific salmon are currently managed under a Treaty signed by the two countries in 1985 (Treaty, 1985). Negotiating the Treaty proved to be a difficult undertaking, extending over a 15-year period. The Treaty worked well for several years, but then came under severe stress in the early 1990s.

The motivation for initial negotiations was twofold. The first lay in the fact that both countries had the opportunity to implement enhancement projects in their major salmon rivers, such as the construction of hatcheries and the removal of obstacles confronting salmon as they swim up river to spawn. However, each country held back on implementation of major enhancement projects, for fear that the other would "free ride" on the benefits arising from its investment in enhancement facilities (Miller *et al.*, 2001) The second was the threat of recurrent "fish wars", such as happened in the early 1980s when negotiations threatened to founder. At that time, the Canadian authorities responded by actively encouraging their fishers to "intercept" aggressively Columbia River chinook and coho stocks, in an attempt to force their American counterparts back to the bargaining table (Munro *et al.*, 1998). There were numerous other, albeit less dramatic, examples of "fish war" activities.

The Treaty called for the two countries to negotiate short-term management regimes for six specific salmon fisheries, through a body to be known as the Pacific Salmon Commission. The objectives of the Commission were to conserve the resource, and to achieve an "equitable" allocation of the harvests between Canadian and American "interceptions". In the early years, a rough interceptions balance did exist. Then, climatic shifts threw the interceptions balance into disarray, leading to increasing friction between the two nations. Negotiations over short-term management arrangements broke down entirely in the early 1990s, and the threat of "fish wars" re-emerged.

The deteriorating state of the resource, particularly in the southern segment of the Treaty area, drove the two countries to attempt to revive the Treaty and eventually the two succeeded in "patching up" the Treaty by signing an Agreement in 1999. The Agreement replaces short-term management arrangements with medium- to long-term ones. Furthermore, much more emphasis is given to conservation, while less is given to "equitable" allocation of harvests (Miller *et al.*, 2001).

The original Treaty placed ceilings on the amounts of American-produced salmon that Canadians could take, and vice versa. That approach proved to be "time inconsistent" and both sides would fish to the limits of their ceilings, regardless of the states of the stocks. It has been replaced in the Agreement with an "abundance based" approach in which Canadians (Americans) will be allowed to take larger amounts of American (Canadian) produced salmon, when the stocks are increasing, but will be forced to reduce their harvests when the stocks are declining.

The theory of games also demonstrates that the scope for bargaining will be increased if provision is made for so called "side payments": essentially a transfer between players, where the transfer may be either monetary or non-monetary in form. The original Treaty was notable for its complete absence of side payments. It was argued that the resulting narrow scope for bargaining was a factor leading to the breakdown of the Treaty (Miller *et al.*, 2001). The Agreement sets a precedent for side payments in that it calls for the establishment of two Endowment Funds to be used for the purpose of restoring stocks on both sides of the border. The Funds, in their initial stages at least, are to be funded by the United States alone. Hence, there are to be implicit side payments from the United States to Canada (Miller *et al.*, 2001).

There is no guarantee that the Agreement will succeed in the long term, and it has many critics. Nonetheless, even the most vociferous critics agree that an agreement, however flawed, is better than no agreement at all and that cooperation is a necessary pre-condition for effective national resource management (Miller *et al.*, 2001).

The Pacific island states and tuna

The Pacific Salmon Agreement involved only two states, so the example of the Pacific Island States, involving 14 countries, may be more comparable to the case of Caribbean spiny lobster. The EEZs of the independent Pacific Island States encompass collectively the richest tropical tuna resource in the world, a shared resource of critical economic importance to many of the countries. As most of the resource within the Island States' EEZs is harvested by distant-water fishing nations (DWFNs) under access arrangements, the Pacific Island States realized early that, unless they cooperated, they would enjoy little economic benefit from the resource and that the DWFNs would inevitably play one Island Nation off against the other (Munro, 1991). The DWFNs with whom they had to negotiate included two economic superpowers, Japan and the USA (Munro, 1991).

Initially, achieving effective cooperation was painfully slow and appeared to be all but impossible, aggravated by distance, the number of States involved and big differences in physical size and the level of economic development. However, the 14 States do not share the resources equally and seven of them, well endowed with tuna resources, coalesced and threatened to go their own way if the others did not agree to serious cooperation. The others agreed and effectively two sub-coalitions emerged: the "haves" and the "have nots". The "haves" took the lead in formulating policy for the group, but took care to ensure that members of the other sub-coalition received a sufficient share of the benefits from cooperation to guarantee their ongoing commitment to cooperation (Munro, 1991; David Doulman, FAO, pers. comm.).

Cooperation has consisted, in the first instance, in ensuring that the Pacific Island States negotiate on a collective basis with DWFNs, so preventing DWFNs from playing off one Island State against another. Thus, the Island States have agreed among themselves on a set of minimum terms and conditions of access to be applied to DWFNs, and cooperate actively with respect to surveillance and enforcement measures. Negotiations are concerned with extracting licence fees from DWFNs, but they are also very much concerned with ensuring that the harvest regimes will conserve the resource. The dependence of the Island States on tuna may be greater than that of the Caribbean states on spiny lobster, but many of the larger lobster producers would undoubtedly feel the social and economic costs of a collapse in their spiny lobster fisheries.

OPTIONS FOR A REGIONAL ORGANIZATION FOR *P. ARGUS*

This option was discussed in some detail by Cochrane & Chakalall (2001) after the meeting of senior decision-makers in Merida, Mexico, in 2000. As discussed there, in addition to a legal obligation to cooperate on a regional or at least sub-regional basis in accordance with the UN Law of the Sea, regional cooperation would also be the most practical approach. An alternative option of a network of bilateral arrangements would clearly be impractical.

The UN Law of the Sea does not go into details about the nature of the cooperation beyond requiring that they "agree upon measures necessary to co-ordinate and ensure the conservation and development of such stocks" and that coastal States and competent international organizations cooperate to ensure "through proper conservation and management measures that the maintenance of the living resources in the exclusive economic zone is not endangered by over-exploitation." There has been good cooperation between countries at the WECAFC workshops, and there is some independent bilateral communication and cooperation in research, for example between Cuba and Mexico. Nevertheless, there is still a long way to go to meet the legal obligations. Failure to appreciate that the resource is, indeed, shared must have been an important factor in limiting the progress.

An important step in furthering regional cooperation was an agreement to establish an "*ad hoc* Working Group on Spiny Lobster" by member countries at the Ninth Session of the Western Central Atlantic Fishery Commission (WECAFC) held at Saint Lucia in September 1999, an agreement reinforced by decisions and recommendations made at the third WECAFC lobster workshop in the series, held in Merida, Mexico, in September 2000. These included the following (FAO, 2001b; Cochrane & Chakalall, 2001).

- There is a need to reduce total fishing capacity in the lobster fishery. As this is a national responsibility, it was recommended that appropriate action be taken at the national level.
- The *ad hoc* Working Group should continue its work through regular workshops to address assessment and management issues. It was recognized that emphasis be given to updating and maintaining stock assessments for management purposes, and extending the bio-economic analyses.
- It was observed that there is an absence of political will to manage the resources sustainably in some countries. This could possibly be improved through, for example, improved communication between scientists and senior decision-makers, and drawing the attention of senior decision-makers to the socio-economic importance

of the lobster fisheries as well as to the need for responsible management to maintain them.

- The users of the resource gaining benefits should be expected to contribute to the costs of research and management of the fishery.
- WECAFC should be strengthened and given a formal role in the coordination of research on and management of spiny lobster.
- A regional strategy should be developed for spiny lobster research and management.

CONSTRAINTS TO GREATER REGIONAL COOPERATION

The Merida meeting and statement were positive and indicated what appeared to be a sincere desire by the national delegations present to work together to improve national and regional management of the spiny lobster fisheries. However, since then there has been little progress, at least on cooperative initiatives, as was apparent from discussions at the most recent workshop in Havana, Cuba, in September 2002.

Fundamental obstacles to greater regional cooperation in management of the spiny lobster fisheries are the confounded problems of lack of political will and inadequate financial resources and human capacity at a national level. As discussed earlier and in Cochrane & Chakalall (2001) and FAO (2001a, b, 2003), there are serious problems in management at a national level in many countries and, where there are still substantive deficiencies in national management, it is unlikely that the necessary attention will be directed at regional activities that have less obvious direct benefits.

The most pressing of these problems is the lack of available financial resources. The last two of the recommendations summarized from the Merida statement in the previous section are the most relevant to this paper. Despite the positive intent of these recommendations, no major proposals on strengthening the role of WECAFC or of its *ad hoc* Working Group on spiny lobster were made at the meeting of WECAFC in 2001 (FAO, 2002) and the Commission remains an FAO regional body under Article VI of the FAO constitution, with no management powers. The primary concern preventing the member countries of WECAFC from taking greater control of the Commission was the financial implications of doing so. The same financial constraints are likely to hinder increasing the role and mandate of the *ad hoc* Working Group itself, or any alternative vehicle for regional cooperation.

Comparison with the two more successful examples of the Pacific Salmon Commission and the Pacific Island States discussed above reveal two additional important obstacles. In the case of the Pacific Salmon Commission, there is no uncertainty about the commonality of the shared resource. The two contesting parties are therefore both fully aware that they are fishing the same resource and that they have to cooperate if they wish to optimize management. Despite the strong genetic and life history evidence of the existence of a wider Caribbean stock, the exact nature and the degree of sharing in the case of Caribbean spiny lobster is still largely unknown. This uncertainty reduces the level of obligation by all parties to cooperate.

In the case of the Pacific Island States, they are dealing with a straddling and highly migratory stock and were galvanized into cooperative action by the presence of the distant-water fishing nations which, operating in some EEZs under agreements, were seen as a threat to the region. In contrast, the Caribbean spiny lobster is a shallow-water species found only within national EEZs, so an equivalent outside threat does not exist, at least at present, to encourage the Caribbean coastal States to work together. A further consequence of the restriction of *P. argus* to EEZs is that states are fiercely protective of their own EEZs, and are usually very reluctant to allow other nations to have any direct say in how they manage national resources.

CONCLUSIONS

The two examples of progress made in international cooperation in management of shared stocks demonstrate the costs and some benefits of non-cooperation, while economic theory demonstrates that optimal utilization of a common resource requires cooperation. To date, the Caribbean states fishing for spiny lobster have avoided "fish wars" over the stock, and they are not threatened by powerful distant-water fishing nations. However, there are worrying signs that the resource is under considerable pressure and that in many states it is currently being overfished. The problem is exacerbated by limited human and financial resources for fisheries management and by pressing social and economic needs in some countries, reducing the ability and will to implement the necessary management actions.

The life history and genetic characteristics of spiny lobster populations in the Caribbean suggest the existence of two biological stocks: those of Brazil and of the wider Caribbean. In the case of the latter, the strong gene flow suggests that widespread collapses in a country will not be limited only to that country and that the effects of large collapses would be felt throughout the sub-region, especially in those countries downstream of a collapse. For this reason especially, but also to ensure optimal utilization of the valuable resource across the region, greater efforts are needed for cooperation in regional research, and cooperation and harmonization in management of the wider Caribbean stock at least. For the upstream countries, greater cooperation in research and management would enable them to benefit by sharing expertise and capacity with neighbours, especially in areas such as research and control and surveillance. It would also recognize their contribution to other fisheries in the sub-region.

The reasons for the widespread lack of adequate action at national level are the same as those pertaining to most fisheries throughout the world: essentially the tendency for short-term social and economic considerations to outweigh ecological considerations with their attendant longer term socio-economic implications. The reluctance to strengthen regional and sub-regional cooperation is clearly an expansion of this – if there is a reluctance or inability to clean up one's own backyard, there is likely to be even less attention given to the neighbourhood as a whole. However, there can be little doubt that the adoption of a regional perspective is also hindered by the lack of direct evidence of the shared nature of the resource. At a scientific level one can infer a regional linkage with a high degree of confidence, but on the ground this is far from obvious.

Until such time as events in neighbouring countries or sub-regions are clearly seen to be impacting productivity elsewhere, there will be little incentive for countries to invest time and money in regional activities. While economic theory and experiences with other shared stocks suggest this is a suboptimal strategy, until the need for cooperation is recognized, emphasis must be placed at least on working towards responsible and effective management at a national level. However, there could also be considerable benefit through cooperating, if not at a regional level, then at least with those neighbours in reasonably close vicinity, sharing common shelf and adjacent adult biomasses. Focusing on the development of such collaborative groups of neighbours would provide at least some of the benefits of regional cooperation. It would also reduce the impact of some of the major constraints to full regional cooperation, for example reducing the number and the social and political diversity of parties in each group, the uncertainty about stock overlap within the group, and the costs of collaboration. It could also be expected that successful cooperation by neighbours would serve as a catalyst for greater regional cooperation in the future.

ACKNOWLEDGEMENTS

This paper is the second in a series arising from the WECAFC *ad hoc* Working Group on Caribbean Spiny Lobster, and all participants in the group are gratefully acknowledged for their contributions to the progress made. Nelson Ehrhardt of the University of Miami is thanked for providing useful references on the stock structure of *P. argus* and Andy Cockcroft (MCM, Cape Town) and Trevor Hutton (CEFAS, UK) for useful comments on an earlier version of this manuscript.

REFERENCES

Arce, A. M. & de León, M. E. (2001) Biology. In: *Report on the FAO/DANIDA/CFRAMP/WECAFC Regional Workshops on the Assessment of the Caribbean Spiny Lobster (Panulirus argus)*. (Ed. by P. Medley and S. Venema), pp. 17–25. *FAO Fisheries Report*, **619**.

Barnutti, R., Gallo, J., Gonzalez Cano, J., (Group Leader), Grant, S., Gutierrez, R., Irias, A., Pérez, M., Rodriguez, J., & Suazo M. (2001) Region 3: Nicaragua, Honduras, Colombia and Jamaica. In: *Report on the FAO/DANIDA/CFRAMP/WECAFC Regional Workshops on the Assessment of the Caribbean Spiny Lobster (Panulirus argus)*. (Ed. by P. Medley and S. Venema), pp. 74–90. *FAO Fisheries Report*, **619**.

Caddy, J. (1997) Establishing a consultative mechanism or arrangement for managing shared stocks within the jurisdiction of contiguous states. In: *Taking Stock: Defining and Managing Shared Resources. Australian Society for Fish Biology and Aquatic Resource Management Association of Australasia Joint Workshop Proceedings, Darwin, NT, June 1997*. (Ed. By D. Hancock), pp. 81–123. Australian Society for Fish Biology, Sydney.

Cervigón, F., Cipriani, R., Fischer, W., Garibaldi, L., Hendrickx, M., Lemus, A. J., Marquez, R., Poutiers, J. M., Robaina, G. & Rodriguez, B. (1993) Field guide to the commercial marine and brackish-water resources of the northern coast of South America. FAO, Rome. 513 pp.

Cochrane, K. L. & Chakalall, B. (2001) The spiny lobster fishery in the WECAFC region – an approach to responsible fisheries management. *Marine and Freshwater Research*, **52**: 1623–1631.

Cruz, R., Luckhurst, B. & Muller, R. (2001) Review of larval recruitment patterns and variability in spiny lobster (*Panulirus argus*). In: *Report on the FAO/DANIDA/CFRAMP/WECAFC Regional Workshops on the Assessment of the Caribbean Spiny Lobster (Panulirus argus)*. (Ed. by P. Medley and S. Venema), pp. 26–32. *FAO Fisheries Report*, **619**.

Deleveaux, V. K. W. & Bethel, G. (2001) National report on the spiny lobster fishery in the Bahamas. In: *Report on the FAO/DANIDA/CFRAMP/WECAFC Regional Workshops on the Assessment of the Caribbean Spiny Lobster (Panulirus argus)*. (Ed. by P. Medley and S. Venema), pp. 161–167. *FAO Fisheries Report*, **619**.

FAO (1995) Code of Conduct for Responsible Fisheries. FAO, Rome. 41 pp.

FAO (2001a) *Report on the FAO/DANIDA/CFRAMP/WECAFC Regional Workshops on the Assessment of the Caribbean Spiny Lobster (Panulirus argus)*. Medley, P. and S. Venema (Eds). *FAO Fisheries Report*, **619**, 381 pp.

FAO (2001b) *Report of the Workshop on Management of the Caribbean Spiny Lobster (Panulirus argus) Fisheries in the Area of the Western Central Atlantic Fishery Commission. Mérida, Mexico, 2000. FAO Fisheries Report*, **643**, 66 pp.

FAO (2002) *Report of the 10th Session of the Western Central Atlantic Fishery Commission, Bridgetown, Barbados. 24–27 October 2001. FAO Fisheries Report*, **660**: 98 pp.

FAO (2003) *Report of the Second Workshop on Management of the Caribbean Spiny Lobster Fisheries in theWECAFC Area. Havana, Cuba, 2002. FAO Fisheries Report* **643** (in press).

Grainger, R. J. R. & Garcia, S. (1996) Chronicles of marine fishery landings (1950–1994). Trend analysis and fisheries potential. *FAO Fisheries Technical Paper*, **359**: 51 pp.

Jensen, T. (1986) The United States – Canada Pacific salmon interception treaty: an historical and legal overview. *Environmental Law*, **16**: 365–422.

Miller, K., Munro, G., McDorman, T., McKelvey, R. & Tyedmers, P. (2001) The 1999 Pacific Salmon Agreement: a sustainable solution? *Occasional Papers: Canadian–American Public Policy*, **47**, Canadian–American Center, University of Maine, Orono, 60 pp.

Munro, G. (1991) The management of migratory fishery resources in the Pacific: tropical tuna and Pacific salmon. In: *Essays on the Economics of Migratory Fish Stocks*. (Ed. by R. Arnason and T. Bjørndal), pp. 85–106. Springer-Verlag, Berlin.

Munro, G. T., Willmann, R. & Cochrane, K. L. (2003) On the management of shared fish stocks: critical issues and international initiatives to address them. (this volume).

Munro, G., McDorman, T. & McKelvey, R. (1998) Transboundary fishery resources and the Canada-United States Pacific Salmon Treaty, *Occasional Papers: Canadian–American Public Policy*, **33**, Canadian–American Center, University of Maine, Orono, 43 pp.

OECD 1997. *Towards Sustainable Fisheries: Economic Aspects of the Management of Living Marine Resources*. OECD, Paris, 268 pp.

Roberts, C. M. (1997) Connectivity and management of Caribbean coral reefs. *Science*, **278**: 1454–1457.

Sarver, S. K., Silberman, J. D. & Walsh, P. J. (1998) Mitochondrial DNA sequence evidence supporting the recognition of two species or subspecies of the Florida spiny lobster *Panulirus argus*. *Marine Biology*, **120**: 601–608.

Silberman, J. D., Sarver, S. K. & Walsh, P. J. (1994) Mitochondrial DNA variation and population structure in the spiny lobster *Panulirus argus*. *Marine Biology*, **120**: 601–608.

Treaty (1985) Treaty between the Government of Canada and the Government of the United States of America Concerning Pacific Salmon. March 1985. http://www.psc.org/Treaty/TREATY.HTM (last accessed 5 February 2003).

United Nations (1998) International fisheries. Instruments with index. FAO and the Division for Ocean Affairs and the Law of the Sea, New York. 110 pp.

On the assessment and management of local herring stocks in the Baltic Sea

Evald Ojaveer, Tiit Raid and Ülo Suursaar
Estonian Marine Institute, University of Tartu, Tallinn, Estonia

ABSTRACT: VPA-based assessments of the status of three Baltic herring (*Clupea harengus membras*) populations in Estonian waters (spring-spawning herring in the NE Baltic proper, the Gulf of Riga and the Gulf of Finland) are presented. Prior to the analytical assessments, the biological data were screened. Data on long-term dynamics of mean weights-at-age indicate significant differences between the three populations. Based upon trends in year-class abundance, the populations in the Gulf of Finland and the NE Baltic proper are similar, but they both differ from that in the Gulf of Riga. The differences between the three stocks probably developed during the extraordinarily long (1979–1993) period of stagnation in the Baltic Sea. It resulted in a series of moderate/poor year classes of open sea herring whose abundance can be traced to the intensity of mixing processes in deep water in the open Baltic. The abundance of gulf herring year classes depends on local mixing processes on their spawning grounds. Hence, the recruitment dynamics differed. The study indicated that the dynamics of the three populations are significantly different. Following the principle of sustainable exploitation of marine ecosystems, the three populations should be assessed and managed separately.

INTRODUCTION

The ecosystem of the Baltic Sea, as a brackish-water body of estuarine character, is far from homogenous. Owing to the characteristic relief of its seabed, ecological subsystems have developed that remain separate as a consequence of the continuous maintenance of hydrological fronts between them (Ojaveer & Elken, 1997). In the process of adaptation as parts of these ecological subsystems, fish stocks have developed local subpopulations differing from each other in abundance dynamics, mean weights-at-age and other vital features. This sophisticated population structure complicates stock assessment and management, and the problem is critical because inappropriate assessment cannot be used as a basis for sustainable management of marine living resources.

In the 1970s and 1980s, a number of assessment units were used for assessment of Baltic herring (*Clupea harengus membras*), and the results of the assessments generally reflected the status of most of the commercially important stocks. Since the early 1990s, however, different stock units (or populations) of the main basin of the Baltic were combined into a single assessment unit (Central Baltic herring in Subdivisions 25–29 and 32, including the Gulf of Riga; ICES, 2001c). However, the annual assessments for the new unit proved unreliable and unable to provide the basis for sustainable management

(ICES, 2001a). Clearly there were shortcomings in the joint assessment in that it probably failed to follow the developments in different subunits of the combined stock adequately.

The herring as a species has a remarkable ability to adapt, and it has developed a number of infraspecific groups (Blaxter, 1985; Jorstad *et al.*, 1991), including in the Baltic Sea (Otterlind, 1962; Rannak, 1971; Ojaveer, 1988). Observations that herring in certain areas of the Baltic Sea differ in a number of conspicuous characteristics (e.g. length, weight, body proportions) from herring in other locations in the Baltic were made more than a century ago (Kessler, 1864; Heincke, 1898). Also, spring-spawning herring differ from autumn herring (Blaxter, 1958; Ojaveer, 1981), and local populations of spring-spawning herring differ significantly in their meristic characters and body proportions (Ojaveer *et al.*, 1981; Parmanne, 1990). A number of authors, among them Kändler (1942), Ojaveer (1962) and Kompowski (1971), indicated that groups of Baltic herring can be separated on the basis of their otolith characteristics. Otoliths can be used to designate catch-sampled herring into one of a number of populations.

Both direct observations (e.g. from tagging studies) and indirect information on herring migrations (e.g. determination of population affiliation of herring by their otoliths, or by analysing changes in herring monthly length-at-age data in historical time-series collected at fixed stations) have been published by a number of workers (Popiel, 1958; Otterlind, 1962; Parmanne & Sjöblom, 1982, 1983; Ojaveer, 1988; Parmanne, 1990). Those authors showed that certain herring groups (populations) have widely differing migration patterns. Rügen herring (Biester, 1979) and Swedish East Coast herring (Otterlind, 1962) perform long migrations, but most fish return to their natal areas to spawn. In contrast, the migration patterns of typical gulf herring populations inhabiting the Gulf of Bothnia, the Gulf of Riga and the eastern Gulf of Finland are generally rather weak and their migrations short (Otterlind, 1962; Parmanne & Sjöblom, 1982, 1983; Ojaveer, 1988; Parmanne, 1990). Migrations of gulf herring into the open Baltic are irregular (depending on the food reserves in the gulfs), and the herring generally return to winter in their home gulf.

In order to address the problem of stock assessment of herring in Estonian waters of the Baltic Sea (Figure 1), the principle of the priority of natural systems (including the maintenance of biological diversity at an infraspecific level; Stephenson, 1991) is invoked to ensure their sustainable management. In the area considered, this approach is possible for herring owing to the availability of results from long-term studies of population structure and population parameters (Ojaveer, 1962, 1981, 1988; Rannak, 1971; Raid *et al.*, in press). As the main problem in assessing Baltic herring stocks has been the consistency of the input data (ICES, 2001a, b, c), the biological data were initially screened. Decisions on joint or separate assessment of local stocks were supported by tests of homogeneity of input data prior to the assessment. This idea of using the results from quality-controlled biological input data to create and to revise herring assessment units in the Baltic has not been used widely before.

MATERIAL AND METHODS

The three herring stocks assessed were:

1. Herring of the northeastern Baltic proper, west of Hiiumaa and Saaremaa, spawning in the Estonian Archipelago and partly in the Gulf of Riga and the western Gulf of Finland. The feeding grounds of the stock are mainly in ICES Subdivision 29 (south) and probably Subdivision 28.

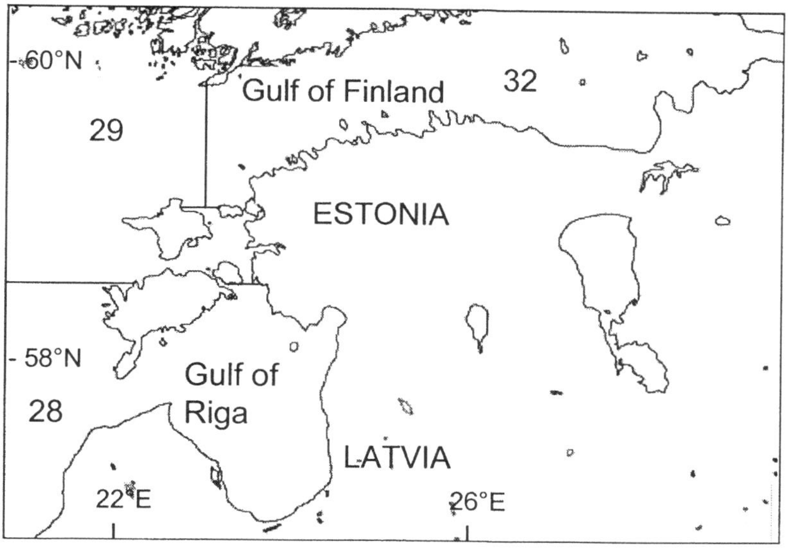

Fig. 1 ICES Subdivisions in the waters around Estonia.

2. Herring of the Gulf of Finland, spawning there. The genuine gulf herring stock inhabits the eastern Gulf of Finland (east of 26°E). Herring in the western Gulf of Finland (ICES Subdivision 32, west of 26°E) are a transition group between the gulf herring of the eastern Gulf of Finland and the open sea herring stock of the NE Baltic proper. Data on the Gulf of Finland stock used hereafter include material sampled from the eastern and western parts of Subdivision 32, i.e. including the transition group. In the absence of Finnish data, only information on herring of the east and south coasts of the Gulf of Finland has been included in the assessment. Addition of the corresponding Finnish data would likely improve the quality of the assessment. However, exchange between the portions of the stock off the northern and southern coasts is limited, so separate assessment of herring stocks off the southern and northern coasts of the Gulf of Finland has been proposed already (Parmanne & Sjöblom, 1983).
3. The Gulf of Riga herring that spends all year in the gulf, excluding a short post-spawning feeding migration of a part of the stock into the adjacent open sea in some years.

Based on otolith characteristics (Ojaveer, 1962), herring taken in mixed catches were allocated to one of the three stocks. The assessments were performed on a stock basis.

Material was collected according to standard protocols for wild fish populations (Gulland, 1977). Sampling and handling was either by the authors or by other fishery biologists competent in separating herring populations. On average, three biological samples were collected and analysed randomly each month from commercial (mainly trawl) catches of feeding or wintering herring. In addition, one biological sample was taken every five days from catches of spawning herring (length, weight, maturity stage, stomach fullness, otoliths for age determination) in the Gulf of Finland, the Gulf of Riga and the NE part of the Baltic proper (ICES Subdivision 29, south). Age determination and stock designation of individual herring (Rannak, 1971; Ojaveer, 1988) were carried out using otoliths (Ojaveer, 1962). Then, from the monthly catches of the commercial fishery in the feeding and wintering areas, the catches per five-day period on the spawning grounds and the mean weights-at-age of herring in the catches, the numbers of herring of each stock caught were calculated. Catch series for all countries in the Gulf of Riga and in the southern and eastern Gulf of Finland (excluding the Finnish fishing zone, for which no information was available throughout the study period) and for all catches in Subdivision 29 (south) were obtained in this manner. The age composition and mean weights-at-age used were those calculated by the Estonian Marine Institute.

Until 1980, autumn-spawning herring, well distinguished by their otolith structure, constituted a significant portion (up to 45%) of the herring resource of the Gulf of Riga (Ojaveer, 1988). Therefore, for all years up to 1980, catches of autumn-spawning herring were derived from the total herring catch in that gulf. In the other areas, the proportion of autumn-spawning herring was much lower (usually <5–10%), so the total herring catches were treated as if they were spring-spawners only. Annual catches of each population by age group and the mean weights-at-age of herring were used as input for Virtual Population Analysis (VPA). For the Gulf of Riga and the NE Baltic proper, totals of the monthly catches were applied throughout the period studied. However, there was a lack of data on Russian catches of herring in the Gulf of Finland in the early 1990s, so Russian catches for that period were taken as 40% of Estonian catches (the average proportion of the Estonian catch taken by Russia in the Gulf of Finland during the late 1990s).

Values of natural mortality (M) assumed by the International Council for the Exploration of the Sea Baltic Fisheries Assessment Working Group (ICES WGBFAS) for Gulf of Riga herring in the period 1970–2001 (ICES, 2002) were applied for those years, and an M of 0.15 for all age groups of the same stock was assumed for the years 1957–1969. M values for herring in the NE Baltic proper and Gulf of Finland were taken from the input data for the joint assessment of herring in Subdivisions 28, 29 and 32 for the most recent period. In all cases, the pattern of M used reflects the abundance dynamics of cod (*Gadus morhua*; the main predator of herring in the NE Baltic) as well as changes in natural mortality during the life cycle of herring (Ojaveer, 1988). Values of terminal fishing mortality (F) for Gulf of Riga herring were derived from an ICES Working Group assessment (ICES, 2002), and the F values from the ICES Working Group joint assessment of herring in Subdivisions 28, 29 and 32 (ICES, 2001a) were applied to NE Baltic proper and Gulf of Finland stocks.

Mean annual weights-at-age were computed as the arithmetic mean of the averages for each month of the year. The lengths of the time-series for each herring population considered in this analysis differ. To compare the three populations, mean weights-at-age from the period 1965–2001 and recruitment at age 1 from the period 1970–2001 were used. As stock abundance and dynamics in the three populations so obviously differed, normalized abundance estimates (the number of 1-year-old herring derived from VPA, divided by the mean value for the period 1970–2001) were applied instead of observed numbers.

Analysis of variance (ANOVA) was applied to determine the significance of differences between the time-series of weights and abundances of year classes (recruitment) of the three populations. ANOVA involves computation of sums of squares describing deviations within and between data groups. Further, depending on the degrees of freedom, the corresponding mean of sums of squares within groups (MS_W) and between groups (MS_B) were calculated. In order to decide whether the variance within the populations or between the populations was more important, F-tests were performed. F-tests were also used to compare the variances in the recruitment time-series between the three stocks (populations). The correlation coefficient between time-series was computed for pairs of populations. In general, the data conformed to the assumption of normality, considering that the objects of the statistical analysis are 31 and 37 annual mean weights by each age group and population. Because of the small size of the samples, additional data normalization was not done.

RESULTS

Data on the mean weights-at-age (Figure 2) indicated significant differences between the long-term dynamics of mean weights of the three populations. The difference was small between herring in the Gulf of Riga and the Gulf of Finland, but herring mean weights-at-age in both gulfs differ largely from those of open sea herring. ANOVA showed that the mean sums of squares between the groups (MS_B) were larger than the mean sums of squares within the groups (MS_W). The F-ratio corresponding to the analysis (total degrees of freedom, $df_T = 110$, within group degrees of freedom, $df_W = 108$, between group degrees of freedom, $df_B = 2$) was 3.08 for a 5% significance level and 4.81 for a 1% significance level (Figure 3). Consequently, the series of mean weights-at-age of the populations cannot be treated as belonging to a common set. The difference is greatest in the middle age groups, which are fully represented in the catches. The youngest age group (whose length is well below the minimum commercial length for herring imposed by Estonian National Fishing Rules), which is represented in catches only by occasional individuals, shows the least difference between populations.

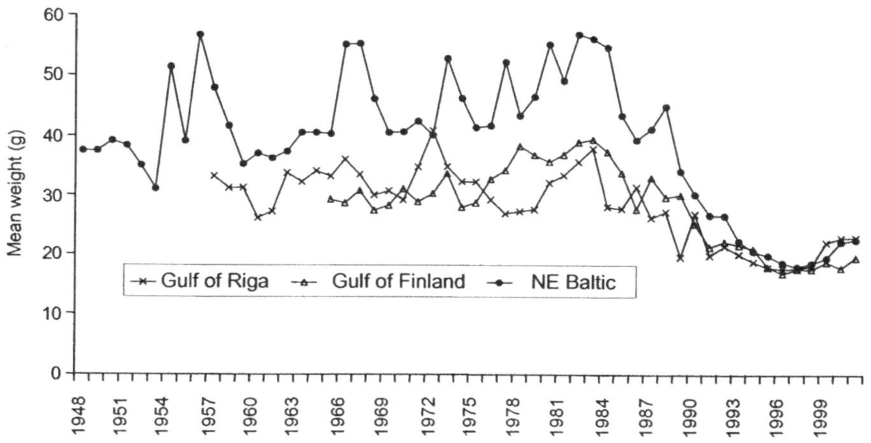

Fig. 2 Dynamics of mean weights of 4-year-old herring in the open sea of the NE Baltic proper, the Gulf of Finland and the Gulf of Riga.

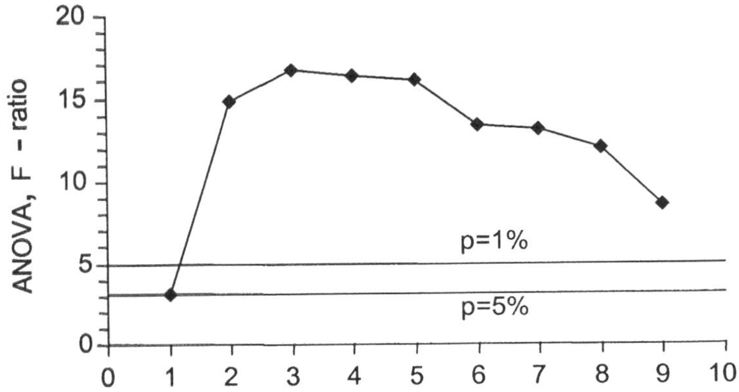

Fig. 3 Level of differences between mean weights of herring in the open sea of the NE Baltic proper, in the Gulf of Finland and the Gulf of Riga by age group.

Analysis of the data on abundance of year classes indicated a tendency for abundant year classes to form in each population during the same years. However, the relative year-class strengths of the populations differ. The variances of the annual standardized recruitment data were 0.51, 0.34 and 0.24 for the Gulf of Riga, Gulf of Finland and the Baltic proper population time-series, respectively. Comparison of the variances by F-test revealed that the difference in variance between the Gulf of Riga and the open sea population was significant at a 5% level. The differences between the Gulf of Riga and the Gulf of Finland herring were less significant, and those between the Gulf of Finland and open sea herring least significant.

Correlation between the time-series of year-class relative abundance was significant only between the populations of the NE Baltic proper and the Gulf of Finland ($r = 0.73$, $p<0.01$). There was no significant correlation between the NE Baltic proper and the Gulf of Riga populations ($r = 0.30$) or between the Gulf of Finland and the Gulf of Riga populations ($r = 0.30$). Based upon trends in year-class abundance, the populations of the Gulf of Finland and the NE Baltic proper are similar. However, they both differ significantly from the population in the Gulf of Riga (Figure 4). The Gulf of Riga population time-series shows a positive linear trend (Figure 4a), the standard error of the computed trendline slope being significantly different from zero. The computed slopes of the other populations (Figures 4b, c) are not significantly different from zero.

The results indicate significant differences in the mean weights-at-age and relative abundance of year classes between herring in the NE Baltic proper and the Gulf of Riga. Also, the difference in mean weights-at-age was significant between herring in the NE Baltic and the Gulf of Finland. Trends in the relative abundance of generations between the two last populations were similar. The dynamics of spawning stock biomass of the three populations are clearly different (Figure 5).

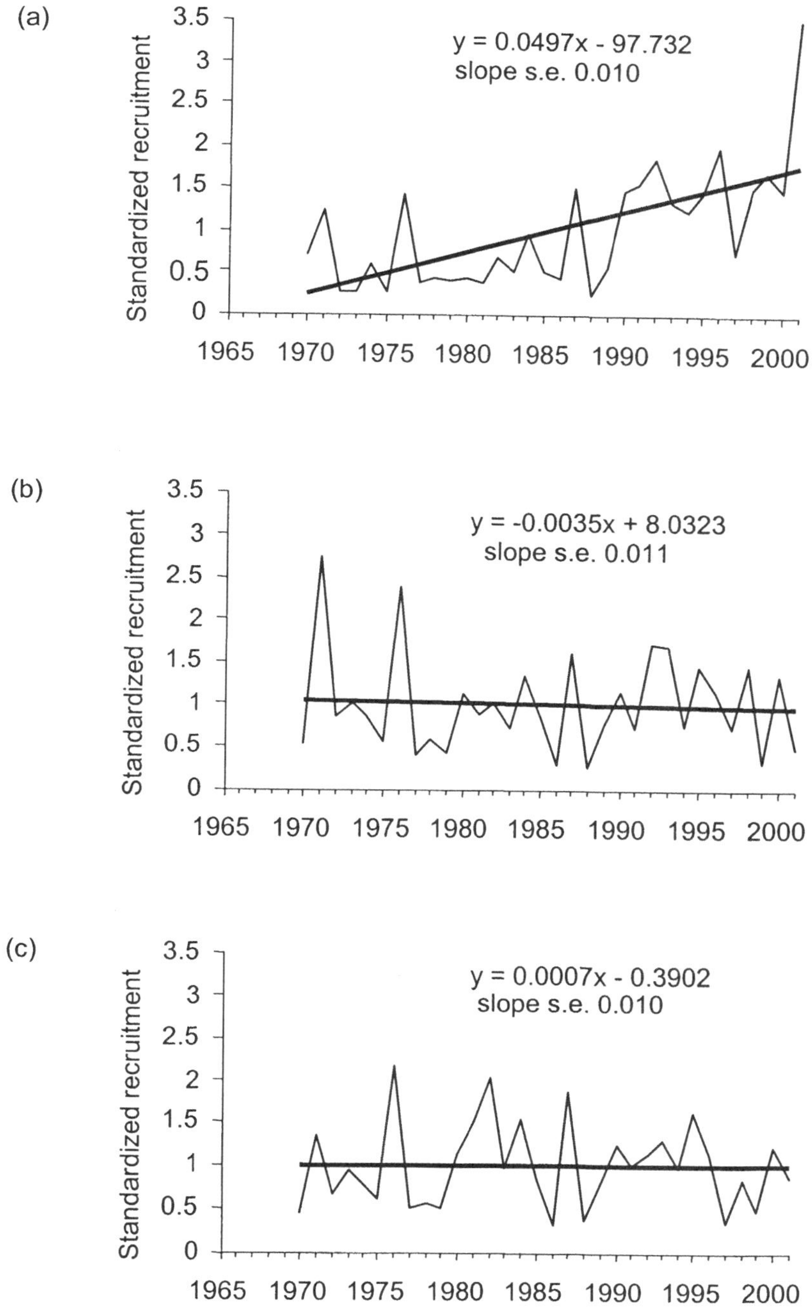

Fig. 4 The dynamics of standardized recruitment of herring in (a) the Gulf of Riga, (b) the Gulf of Finland and (c) the NE part of the Baltic proper.

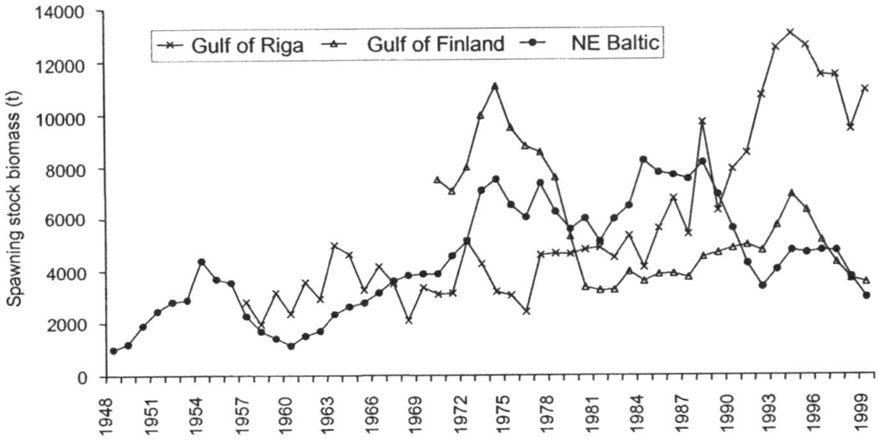

Fig. 5 Spawning stock biomass of spring-spawning herring in the Gulf of Riga, the Gulf of Finland and in the NE Baltic proper.

DISCUSSION

There are clear differences between the habitats of the local herring populations in the Gulf of Riga, the Gulf of Finland and the NE Baltic proper, resulting in specific adaptations to local conditions. In the Gulf of Riga, the average salinity at the surface is about 5 and at the bottom 6 (there is no permanent halocline), the mean temperature in summer is 18–19°C and in winter 0°C. In the NE Baltic proper the respective values are 7 and 11–12 (with a halocline), 15°C and 1–2°C (Ojaveer & Elken, 1997). The conditions in the Gulf of Finland vary between a comparatively mild, saline western portion (which is referred to as the transition area between the open Baltic proper and the eastern Gulf of Finland) and a continental eastern area of low salinity. In the western Gulf of Finland, the salinity and temperature conditions resemble those in the northern Baltic proper. The genuine gulf herring population inhabits the Gulf of Finland east of 26°E. Adaptation to local conditions is facilitated by the rather limited extent of herring migrations, especially between the gulfs and the Baltic proper (Parmanne & Sjöblom, 1982, 1983; Ojaveer, 1988; Parmanne, 1990; Parmanne *et al.*, 1997). Given the above, the smaller differences between open sea herring and Gulf of Finland herring than between both those stocks and Gulf of Riga herring are understandable. The differences clearly manifest themselves in year-class abundance, the Gulf of Finland and NE Baltic proper populations following similar trends.

Differences in the historical development of the stocks depend, in addition, on the longevity, exploitation pattern and conditions for growth. The spawning stock biomass (SSB) of the NE Baltic proper population gradually increased from the 1940s to the mid-1950s, then decreased up to the early 1960s (Figure 5). Thereafter, SSB burgeoned and remained high during the extensive influxes of saline water from the North Sea and some years after, until the late 1980s. It then decreased sharply during the less productive (stagnant) phase up to 1993. The SSB of the Gulf of Riga population increased during periods of dominant westerly winds, mild winters and warm springs. In contrast, the dynamics of spring-spawning herring SSB in the Gulf of Finland differs clearly from both of the other two populations, reflecting both the influence of environmental conditions and the exploitation pattern in its area.

The probable driver for these differences lies in the conditions during which the abundant year classes are generated in each population. In populations with spawning grounds close to the coast in the Baltic proper, abundant year classes appear during periods of higher salinity and intense water exchange between the Baltic and North Seas. Such conditions favour vertical mixing of water layers and upwelling of the nutrients required to support high biological production and, in turn, an abundance of copepod nauplii in the relative herring spawning and larval retention areas, where herring larvae switch to exogenous feeding. In the two gulf herring populations, strong year classes are formed mainly during warm springs with dominant westerly winds that promote mixing of water layers and rich biological production during the period of larval development, so enhancing their survival (Rannak, 1971; Ojaveer, 1988; Raid, 1997; Grygiel, 1999). Therefore, differences in relative herring stock abundance between various parts of the study area cannot be taken as a constant phenomenon characteristic of the Baltic. The present low level of open sea herring SSB is an extreme situation reflecting the recent (1979–1993) abnormally long period of stagnation in the Baltic Sea. Intensification of water exchange between the Baltic and the North Sea would likely enhance open sea herring SSB and bring it more in line with the SSB of herring in the gulfs.

The dynamics of mean weights-at-age indicated large periodic fluctuations in herring growth in the three areas compared here. In the Baltic Sea, the growth rate of herring is governed mainly by the abundance, composition and availability of food, as well as by temperature and the duration of the feeding period, which is much longer in the Baltic proper than in the gulfs, because at lower temperatures (in winter) herring do not feed (Kostrichkina *et al.*, 1982; Lankov, 2002, Raid *et al.*, in press).

Owing to the similarity of prevailing environmental conditions in the Gulfs of Finland and Riga, an important consideration for herring (a lower salinity prevents marine forage organisms from becoming abundant, and a relatively long fasting period in winter is caused by the lower temperatures in the gulfs than in the open Baltic), the mean weights-at-age of herring in the two gulfs are similar, but clearly different from that in the NE Baltic proper.

CONCLUSIONS

The study has indicated that local populations of herring (spring-spawning herring in the NE Baltic proper, the Gulf of Riga and the Gulf of Finland) differ significantly in many respects, supporting the need for separate assessment and management. This result needs to be borne in mind by those responsible for advising on and managing these shared stocks. However, separate assessment and management can only be accomplished if the representatives of each herring population can be distinguished in the catches and therefore if the required expertise in methods of stock differentiation is available.

REFERENCES

Biester, E. (1979) *Der Frühjahrshering Rügens – seine Rolle in der Fischerei der Ostsee und in den Übergangsgebieten zur Nordsee.* Rostock. 238 pp.

Blaxter, J. H. S. (1958) The racial problem in herring from the viewpoint of recent physiological, evolutionary and genetical theory. *Rapport et Procès-verbaux des Réunions Conseil Permanent International pour l'Exploration de la Mer*, **143**: 10–19.

Blaxter, J. H. S. (1985) The herring: a successful species? In: *Proceedings of the Symposium on the Biological Characteristics of Herring and Their Implication for Management*. (Ed. by J. R. Brett). *Canadian Journal of Fisheries and Aquatic Sciences*, **42**(Suppl. 1): 21–30.

Grygiel, W. (1999) *Rozmieszczenie i liczebność młodych śledzi i szprotów w południowym Bałtyku (lata 1976–1991)*. Morski Instytut Rybacki, Gdynia. 166 pp.

Gulland, J. A. (Ed.) (1977). *Fish Population Dynamics*. John Wiley & Sons, Chicester. 372 pp.

Heincke, F. (1898) Naturgeschichte des Herings. *Abhandlungen des Deutschen Seefischereivereins*, **2**(1): I–CXXXVI und 1–128.

ICES. (2001a) Report of the ICES Advisory Committee on Fishery Management, 2000. *ICES Cooperative Research Report*, **242**(3): 604–911.

ICES. (2001b) Report of the Baltic Fisheries Assessment Working Group. *ICES Document*, **C.M. 2001/ACFM: 18**.

ICES. (2001c) Report of the Study Group on Herring Assessment Units in the Baltic Sea. *ICES Document*, **C.M. 2001/ACFM: 10**.

ICES. (2002) Report of the Baltic Fisheries Assessment Working Group. *ICES Document*, **C.M. 2002/ACFM: 18**.

Jorstad, K. E., King, D. P. F. & Naevdal, G. (1991) Population structure of Atlantic herring *Clupea harengus* L. *Journal of Fish Biology*, **39**(Suppl. A): 43–52.

Kändler, R. (1942) Üeber die Erneuerung der Heringsbestände und des Wachstum der Frühjahrs- und Herbstheringe in der westlichen Ostsee. *Monatshefte für Fischerei*, B. **10**(2): 145–163.

Kessler, K. (1864) *Description of Fishes which occur in the Waters of the St Petersburg Guberniya*. St Petersburg. 240 pp. (in Russian).

Kompowski, A. (1971) Typy otolitów śledzi południowego Bałtyku. *Prace MIR*, **16**(A): 109–141.

Kostrichkina, E. M., Ojaveer, E. A., Jurkovskij, A. K., Rannak, L. A. & Jula, E. A. (1982) The long-term dynamics of herring growth in the Baltic Sea in connection with changes in oceanographic conditions and the food availability. *Fischerei Forschung*, **20**(1): 37–42 (in Russian).

Lankov, A. (2002) Feeding ecology of the pelagic fishes in the northeastern Baltic Sea in 1980–1990s. *Dissertations on Natural Sciences*, **6**. 192 pp. Tallinn Pedagogical University.

Ojaveer, E. (1962) Herring otolith investigations in the North-Eastern Baltic. *ICES Document*, **C.M. 1962/Herring Cttee, 121**.

Ojaveer, E. (1981) Influence of temperature, salinity, and reproductive mixing of Baltic herring groups on its embryonal development. *Rapport et Procès-verbaux des Réunions Conseil International pour l'Exploration de la Mer*, **178**: 409–415.

Ojaveer, E. A. (1988) *Baltic Herrings*. Agropromizdat, Moscow. 204 pp. (in Russian).

Ojaveer, E. & Elken, J. (1997) On regional subunits in the ecosystem of the Baltic Sea. In: *Proceedings of the 14th Baltic Marine Biologists Symposium, August 1995, Pärnu, Estonia* (Ed. by E. Ojaveer), pp. 156–169. Estonian Academy Publishers, Tallinn.

Ojaveer, E., Jevtjukhova, B., Rechlin, O. & Strzyzewska, K. (1981) Results of investigation of population structure and otoliths of Baltic spring spawning herring. *ICES Document*, **C.M. 1981/J: 19**.

Otterlind, G. (1962) Sillens/strömmingens vandringsvanor vid svenska syd och ostkusten. *Ostkusten*, **34**(1): 15–21.

Parmanne, R. (1990) Growth, morphological variation and migrations of herring (*Clupea harengus* L.) in the northern Baltic Sea. *Finnish Fisheries Research,* **10**: 1–48.

Parmanne, R., Popov, A. & Raid, T. (1997) Fishery and biology of herring (*Clupea harengus* L.) in the Gulf of Finland: a review. *Boreal Environment Research*, **2**: 217–227.

Parmanne, R. & Sjöblom, V. (1982) Recaptures of Baltic herring tagged off the coast of Finland in 1975–81. *ICES Document*, **C.M. 1982/J: 19**.

Parmanne, R. & Sjöblom, V. (1983) Differences in the state of herring stocks in the northern and southern part of the Gulf of Finland. *ICES Document*, **C.M. 1983/J: 21**.

Popiel, J. (1958) Differentiation of the biological groups of herring in the southern Baltic. *Rapports et Procès-verbaux des Réunions Conseil Permanent International pour l'Exploration de la Mer*, **143:** 114–121.

Raid, T. (1997) The effect of hydrological conditions on the state of herring stocks in the Baltic Sea. In: *Sensitivity to Change: Black Sea, Baltic Sea and North Sea* (Ed. by E. Özoy and A. Mikaelyan). *NATO ASI Series*, **21**, pp. 139–147. Kluwer Academic Publishers, Dordrecht.

Raid, T., Järvik, A., Kaljuste, O. & Lankov, A. (in press) Principal developments in the structure and dynamics of main fish stocks in the North-eastern Baltic in the 1990s within the context of environmental changes. *ICES Marine Science Symposia*.

Rannak, L. (1971) On recruitment to the stock of spring herring in the north-eastern Baltic. *Rapports et Procès-verbaux des Réunions Conseil International pour l'Exploration de la Mer*, **160**: 76–82.

Stephenson, R. L. (1991) Stock discreteness of Atlantic herring: a review of arguments for and against. In: *Proceedings of the International Herring Symposium, Anchorage, Alaska USA, October 1990*, pp. 659–666.

Fish, fisheries and dolphins as indicators of ecosystem health along the Georgian coast of the Black Sea

Akaki Komakhidze, R .Goradze, R. Diasamidze, N. Mazmanidi and G. Komakhidze
Georgian Marine Ecology and Fisheries Research Institute, Batumi, Georgia

ABSTRACT: Georgia is one of six countries abutting the Black Sea, a unique water mass of which only a small portion is supportive of the life forms commonly associated with man and fisheries. Cooperative and careful management by all its coastal nations is necessary if the Black Sea's biodiversity is to be maintained and its resources rebuilt, especially given the high levels of (often unrefined) discharge from oil and other industries, agriculture and domestic wastes. Monitoring and the establishment of early warning systems are critical, and dolphin populations and some fisheries resources may well lend themselves to use as indicators of ecosystem health.

INTRODUCTION

The semi-enclosed nature and unusual chemical composition of the Black Sea makes it one of the world's unique water bodies. It has a surface area of some 420 000 km^2 and a volume of 550 000 km^3. Of the latter, some 87–90% is contaminated by hydrogen sulphide (H_2S), meaning that it contains no life other than some micro-organisms and meiobenthic nematodes that can withstand such conditions. Only the oxygenated surface layer down to a maximum depth of 150 m supports productivity, and representatives of most taxa and virtually all its flora and fauna are found in this layer, from protozoans to mammals (Sorokin, 1982; Sergeeva, 2000). Careful management of such a water mass by the six countries that abut it is crucial or its fragile equilibrium will be irreparably harmed by anthropogenic activity. Similar caution must be applied in advising on rational and sustainable management strategies for the sea's biological resources.

The Black Sea coast of Georgia is comparatively short, just 315 km. However, with some 150 rivers of various size draining the country's interior (and in certain cases that of neighbouring countries) and emptying into the sea, its significance in terms of supporting a balanced marine ecosystem is immense. Three productive banks (Batumi, Poti-Ochamchire and Gudauti) and the Kolkheti National Park account for almost 50% of the total coastline (Figure 1).

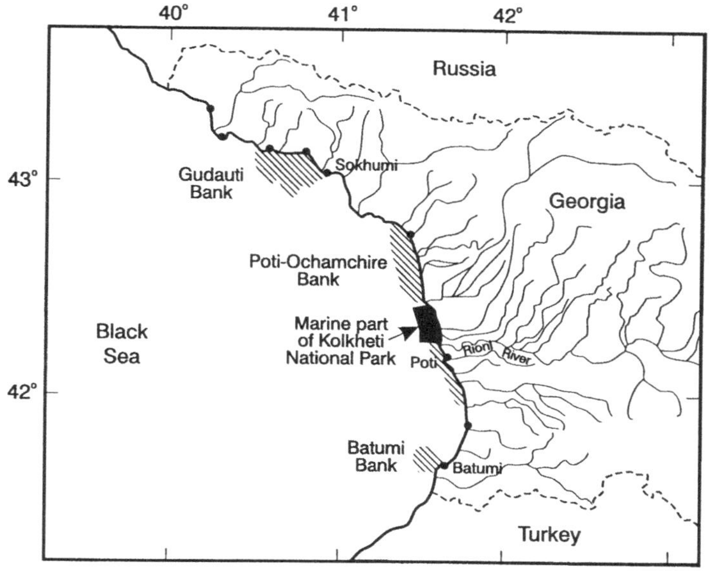

Fig. 1 The Black Sea coast of Georgia, showing the productive banks and the locality of the marine reserve. The reserve extends north from the mouth of the Rioni River for some 18 km and has an area of 15 742 ha.

The Batumi Bank is home to many representatives of typical Black Sea fauna. The Poti-Ochamchire Bank is known particularly as the area where the commercially valuable sturgeon concentrate, and is also the preferred feeding and spawning grounds of such other valuable species as turbot *Scophthalmus maeoticus maeoticus*, red mullet *Mullus barbatus ponticus* and fish of the family Mugilidae. It also overlaps the distribution areas of Black Sea anchovy *Engraulis encrasicolus ponticus*, sprat *Sprattus sprattus phalericus*, spiny dogfish *Squalus acanthias* and whiting *Merlangius merlangus euxinus*. The fact that anchovy schools congregate during winter in the Supsa canyon, where the Supsa oil terminal is situated, leads to confrontation and legal challenge by fishers, who are precluded from operating close to the terminal. The Gudauti Bank used to have large oyster beds, but anthropogenic activity has unfortunately meant that the importance of these beds has declined.

EFFECT OF POLLUTION

Man's main impact on the biodiversity and productivity of the ichthyofauna of the Black Sea is through pollution, of which two types can be highlighted. The first is domestic faecal and industrial waste pollution. None of the cities or populated areas of western Georgia have waste disposal plants, so all waste is simply discharged into the rivers that flow directly into the sea (Figure 2). The second form is oil pollution. At present, two marine oil terminals are functioning: the Batumi railway terminal, which processes more than 6 million tonnes each year, and the Supsa pipeline terminal, with an annual turnover of 5 million tonnes. Shortly too, the new Poti railway terminal, with a projected annual

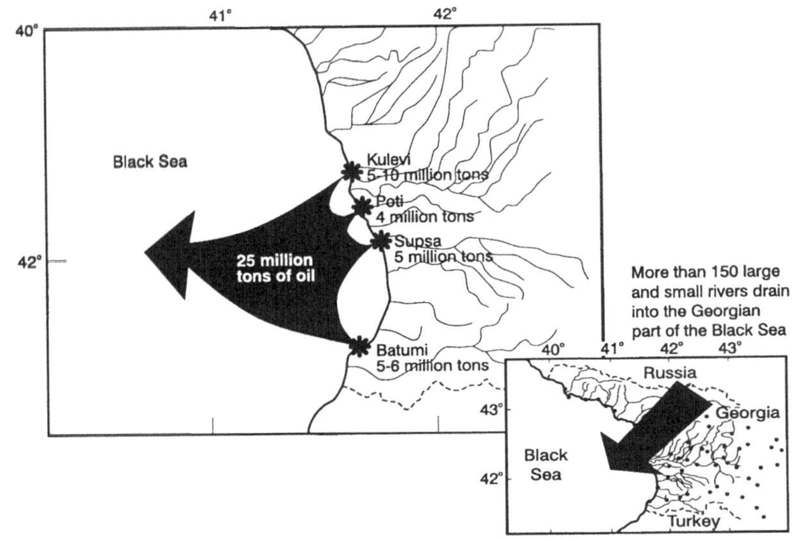

Fig. 2 The main sources of pollution along the Georgian Black Sea coast.

turnover of 4 million tonnes, will come on stream. Also being constructed at the moment is the Kulevi–Khobi railway "Terminal 2000", which will have an initial annual production of some 5 million tonnes, rising to 10 million tonnes when fully operative. Therefore, it is projected that, by the end of 2004, four modern oil terminals with a total annual capacity of 25 million tonnes will be operating along just 100 km of Georgian coastline. It does not take much imagination to realize that, even without any accident or catastrophe, a significant quantity of petroleum products will flow into the sea along the Georgian coast, likely causing wholesale damage to its living resources.

According to our data, a small concentration of dissolved petroleum products of just 0.001mg l^{-1} can seriously and negatively impact the embryonic development of fish and plankton, likely causing huge mortality. First to suffer would be fish with demersal eggs, but all species of ichthyofauna, not just benthic and coastal, will also suffer irreparable harm (Mazmanidi, 1997)

FISH AND FISHERIES

There are about 150 species of fish in the Black Sea (Svetovidov, 1964). Of these, those of major commercial importance to Georgia are anchovy, sprat, whiting, spiny dogfish, red mullet and Black Sea scad (*Trachurus mediterraneus ponticus*). Anchovy alone yield 200–300 thousand tonnes annually along the coast of Georgia, although this figure includes some Azov Sea anchovy in addition to catches from the Black Sea population. Also of commercial importance are the gastropod mollusc *Rapana thomasiana* and the black mussel *Mytilus galloprovincialis*.

The number of species exploited commercially by Georgia has changed over the years. In the 1920s and 1930s, more than 20 species were commercially important, but by the 1960s and 1970s this number had dropped to just four (Figure 3), anchovy, sprat, whiting and spiny dogfish. Nowadays, apart from red mullet and Black Sea scad, the number of

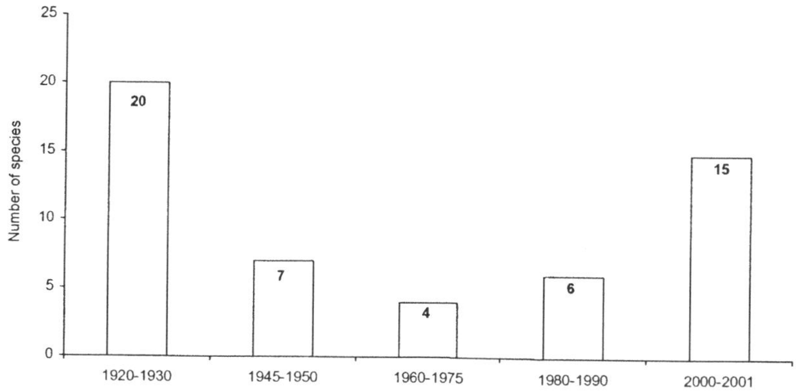

Fig. 3 Changes in the number of fish species harvested commercially along the coast of Georgia in the past 80 years

species harvested commercially has risen to about 15, including 5–6 species of goby (Gobiidae), turbot, Black Sea salmon *Salmo trutta labrax*, Kolkhean (Russian) sturgeon *Acipenser guldenstaedti*, sevryuga or star sturgeon *Acipenser stellatus* and rays (Rajidae).

Most fishing operations along the coast of Georgia are conducted from commercial fishing boats (mainly Ukrainian and Turkish) and with passive fishing gear such as traps, beach-seines and gillnets. The fixed (passive) gear fishery is the most important (mainly for turbot, sturgeon and salmon) and uses nets 5–20 m long. The gear deployed from the commercial fishing boats is mainly a modified midwater trawl suitable for operating on the sea bed.

Three phases of the Georgian commercial fishery can be identified from the catch time-series (Lobzhanidze & Chichilnitsky, 1968; Figure 4). From the 1930s to the 1960s, annual catches were generally small and fairly stable (except during World War 2, when they were very small). From the 1960s to 1990 the commercial fishery was growing; it peaked in 1985 at almost 94 000 t. Then, from 1991 onwards the fishery was in general decline, though there was a slight recovery at the end of the time-series (Figure 4). Stock declines caused by overfishing and pollution are not the only reasons for the recession in the fishing industry. A major driver was the break-up of the Soviet Union, which caused economic links to be broken and markets to be lost. Other factors contributing to the decrease were the general economic hardship in the country, which resulted in power shortages, a total lack of replacement or adequate maintenance for the outdated fishing fleet, and the decay and/or destruction of fish processing facilities.

ILLEGAL FISHING AND TRADE

Associated with the socio-economic and political instability of Georgia's past 10–12 years, which has resulted in a very high rate of unemployment and the virtual impoverishment of much of the human population, poaching of fish and shellfish has escalated. In many cases, poaching has become the only means for people to survive. In addition, some rare and valuable fish species, such as sturgeon, turbot, salmon, mullet and sciaenids, taken as bycatch in the legal industrial fishery are not reported or recorded, but instead are traded illegally. Some of these valuable species are also taken with prohibited nets, further pressurizing the already depressed resources of these vulnerable species.

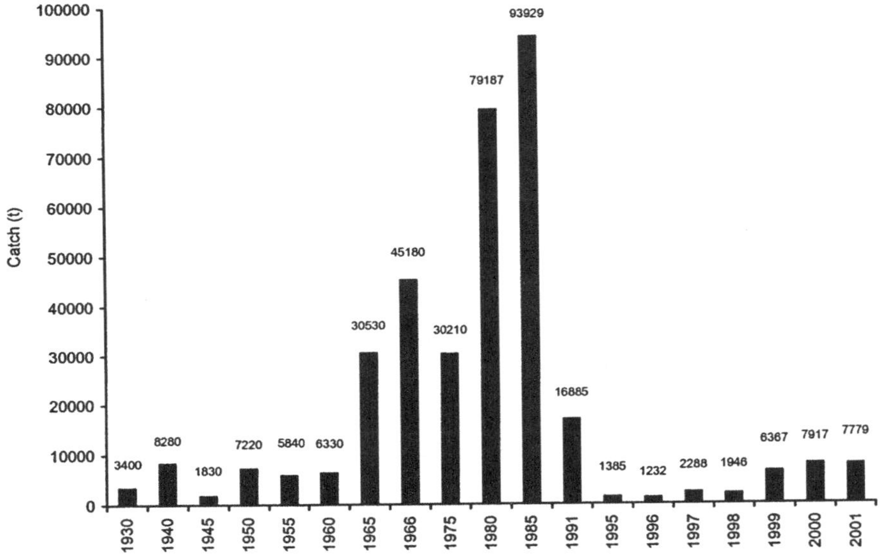

Fig. 4 Time-series of annual catches by Georgia, 1930s– 2001

Since 1990 we have been monitoring catches and trade in such valuable fish species along the Georgian Black Sea coast. From 1991 to 1996, the annual catch of such species was of the order of 200–250 tons (Goradze & Bagrationi, 1998; Figure 5). Some 36–48% of that catch was turbot, 11–18% beluga sturgeon *Hiso huso*, 8–13.5% sevryuga,

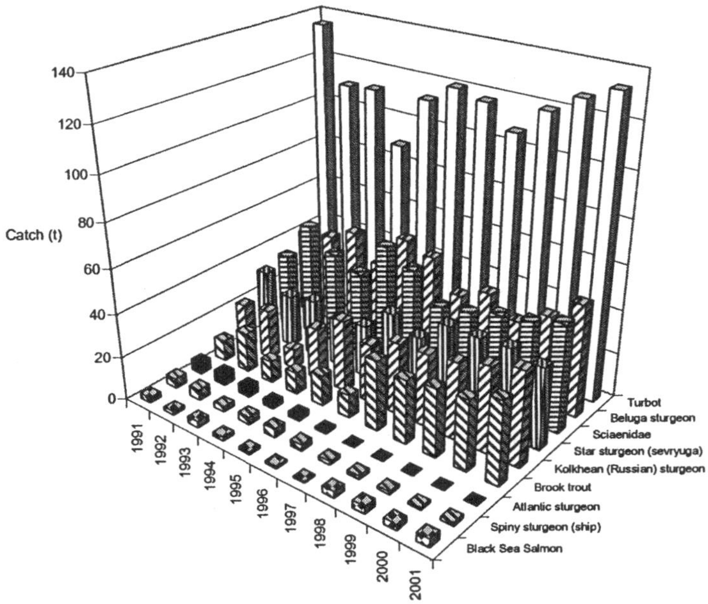

Fig. 5 Annual illegal catch of valuable and endangered fish species along the Georgian Black Sea coast

3.5–12.1% Kolkhean sturgeon, 11.8–18.5 sciaenids and 3.4–11.1 brook trout *Salmo trutta labrax morfo fario*. In addition, very small numbers of Atlantic sturgeon *Acipenser sutrio* (0.4–1.9%), Black Sea salmon (0.4–1.2) and spiny (ship) sturgeon *Acipenser nudiventris* (0.5–17%) made up the undeclared catch of such vulnerable species. Subsequently, from 1997 to 2001, the undeclared catch of these species rose to 288–355 tons. Catches of most of the above-listed species rose slightly, not, we believe, because abundance was improving, but rather because effort was increasing and monitoring success was improving. However, Atlantic sturgeon totally disappeared from the catches.

Poaching and illegal catch and trade of these valuable and endangered fish species is a year-round occurrence, but it is particularly prevalent in autumn and spring during spawning migrations and the spawning seasons. Early in spring, turbot, sciaenids and sturgeon concentrate along the coast to spawn, so they are then highly vulnerable to capture in the fixed nets and trap nets deployed by poachers. Sturgeon are also poached in rivers during their spawning migration between early March and late June.

Black Sea salmon pursue the schools of small pelagic fish to prey on them during winter, so are particularly vulnerable then to capture and subsequent illegal trade as unreported bycatch. Then, in late February as water temperatures start to rise slightly, salmon start to congregate close to shore and prepare to enter the rivers, where over a period of at least a month they become easy prey for poachers operating gillnets and trap nets (Figure 6). The density of salmon waiting to enter the rivers increases in the months following, as temperatures continue to rise, as shown by the monthly percentages of the catch (February 2%, March 13%, April 32%, May 51%, June 2%). Many salmon are also poached at their spawning areas when they are weak and easy to catch. Black Sea salmon conservation and management received a financial boost during the late 1990s when the EU TACIS programme was launched through the Georgian Marine Ecology

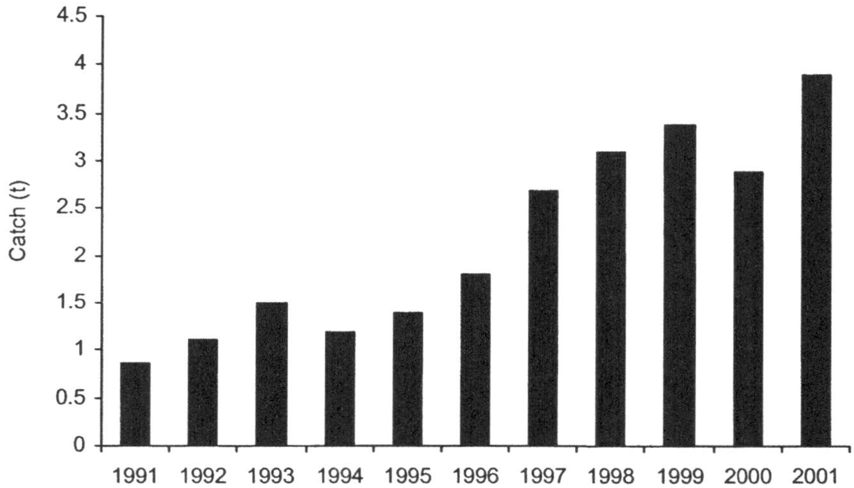

Fig. 6 Annual catches of Black Sea salmon in Georgia, 1991–2001 (Abkhazia catches excepted)

and Fisheries Research Institute. The programme focused the research efforts of scientists from Georgia, the UK, Holland, Bulgaria, Russia, Romania, Turkey and Ukraine into developing a conservation and management strategy based on shared knowledge of historic and current stock status (Goradze *et al.*, 2000).

MARINE MAMMALS AS INDICATORS OF SYSTEM HEALTH

Marine mammals can serve as a good indicator of ecosystem health, so for many years the Georgian Marine Ecology and Fisheries Research Institute has concentrated effort on studies of dolphins. Indeed, since 1932, other Black Sea institutes have been sending their material to the Georgian Institute for processing and, in 1993, the material so collected was used as the basis for a comprehensive study of Black Sea dolphins, even though much of the earlier material was associated with dolphin fishery research (Mayorova & Danilevsky, 1934).

As a fisheries resource, some 200 000 dolphins on average (including a Turkish catch of 60 000–80 000) were taken annually until the 1950s. The total population of Black Sea dolphins was then estimated to be 800 000 animals, which meant that the annual harvest did not exceed 25% of the total (Gudimovich, 1951). However, by the early 1960s, the number of dolphins in the Black Sea had decreased significantly for several reasons, notably as a result of overfishing. This triggered a decision in 1966 to ban any dolphin fishing in the Black Sea, a measure first supported by the USSR, Bulgaria and Romania, and later by Turkey (Shevalev & Tsuladze, 1975). This moratorium had a positive influence on the dolphin population, in terms of total numbers of animals and also the number and size of schools, as revealed by a joint expedition of Russian, Georgian and Ukrainian scientists in 1984 and 1985 (Komakhidze & Mazmanidi, 1998).

Direct fishing, however, is not the only activity that impacts dolphin populations negatively; (over)exploitation of their fish prey also has a negative impact either through removing the prey or in incidentally causing mortality of the dolphins in the nets. With this in mind and with financial support from the European Union, Black Sea coastal countries carried out a two-year study of dolphin mortality in 1997 and 1998. In Georgia, the study took in a 90 km stretch of coast between the mouths of the Rivers Chorokhi and Rioni. To assist in observations, the project team established a network of volunteers among people who either lived or worked at the coast. During the two years of the study, 17 dolphins were taken as bycatch in other fishing activities, 70% of them harbour porpoises (*Phocoena phocoena*) and the balance common dolphins (*Delphinus delphis*). Most of the bycatch was in the fixed nets set for turbot, but a few were drowned during purse-seining for anchovy. Harbour porpoises dominated the bycatch for two reasons, the sheer size of the population and the fact that they only inhabit the coastal fringe. The third species of dolphin common in the area, bottlenose (*Tursiops truncatus*), was not taken as bycatch during the study, but it is known from stranding data that they do occasionally fall foul of fishing gear (Figure 7a). Most of the bycatch (direct and stranding) of dolphins (46 and 50% respectively) is made in spring (Figure 7b, c), when the fixed net fishery for (spawning) turbot is at its peak.

Georgian scientists, fishers and volunteers still provide information on dolphin school size (small – 4/5 to 20–39; medium – 50–60 to 150–200; large – 400–500 to 700–1000). Interpretation of such subjective data is not unanimous, but the consensus is that the number of dolphins has not declined, but seems to be recovering.

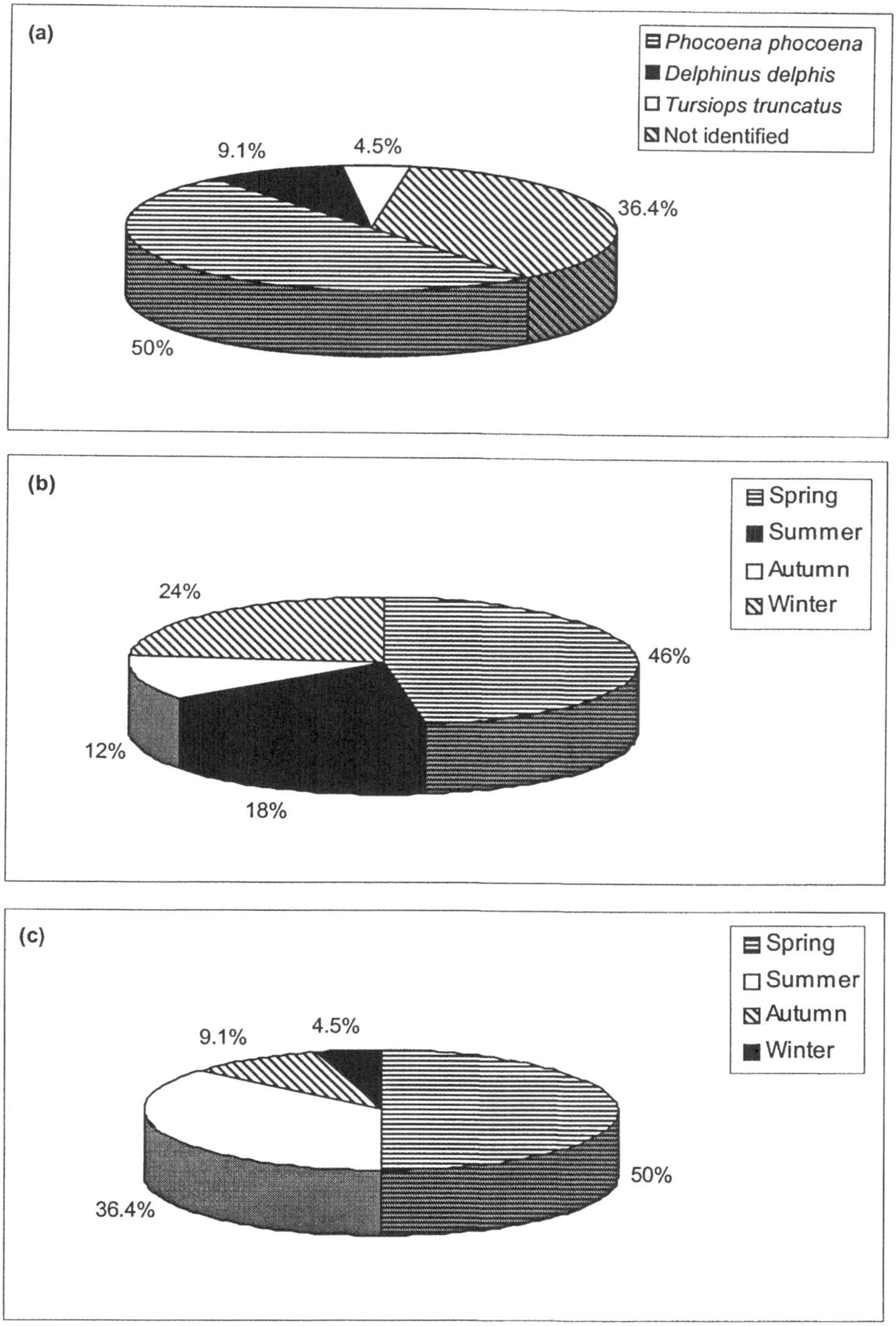

Fig. 7 (a) Species composition of stranded dolphins, and the seasonal occurrence of dolphin (b) by-catch and (c) strandings along the Georgian coast, 1997–1998

CONCLUDING COMMENTS

- To maintain biodiversity within productive areas where commercial fish stocks are subject to extensive harvest is difficult. Coastal zone monitoring of the Black Sea is therefore crucial, nowhere more so than within and adjacent to the borders of the Kolkheti National Park marine reserve.
- The main direction of the coastal currents is south to north, so the whole Georgian coastline (including marine reserves) and much of the coastline beyond Georgia's international boundary is at risk of serious pollution risk from oil, industrial, agricultural and domestic waste and run-off. Regular eco-toxicological monitoring of the water column and the seabed is therefore vital if early warning signs of potentially damaging levels of oil, heavy metals, bacteriological or pesticides are to be sought (Mazmanidi & Komakhidze, 2001).
- Rational and wise use of existing Black Sea biological resources needs regular assessment of an ecological and economic nature, not only for and by Georgia, but also for and by its neighbours.
- Among the various solutions to the current critical situation in Georgia are a need to improve the economic standing of its citizens; the careful and urgent development of the agriculture and food industries, including fisheries and aquaculture as priority sectors; an improvement in the level of environmental legislation and enforcement; and a raising of public awareness.
- Georgia cannot manage its coastline and resources in isolation. Serious and sustained efforts need to be made to work in association and sharing common aims with its neighbours around the Black Sea. Management of shared resources is always difficult, but difficulty should not preclude every effort being made to find common grounds and indicators of system health, and for trying to legislate uniformly within the whole Black Sea. For instance, common and special efforts must be made to conserve dolphins, especially from mortality in fixed nets and in monitoring their status in the Black Sea. Similar multinational action is necessary to preserve and rebuild sturgeon and salmon stocks. The price of failure to heed these warnings is ecosystem catastrophe and the possible economic failure of several Black Sea countries.

REFERENCES

Goradze, R. & Bagrationi, D. (1998) Artifical reproduction: a feasible way of conserving the disappearing Black Sea fishes. In: *Conservation of the Biological Diversity as a Prerequisite for Sustainable Development in the Black Sea Region*, pp. 397–407. *NATO ASI Series 2: Environmental Security*, **46**. Kluwer Publishers, Dordrecht.

Goradze, R., Komakhidze, A., Solomon, D., Bagrationi, D., Gadaeva, M. & Goradze, I. (2000) The biology and status of the Black Sea salmon *Salmo trutta labrax*. Batumi. 33 pp.

Gudimovich, P. K. (1951) The Black Sea fishes and fishery. Overview of the Georgian Department AzCherNIRO. 18 pp.

Komakhidze, A. & Mazmanidi, N. (1998) Biodiversity Activity Centre, Batumi, 1998, Black Sea Biological Diversity Georgia. Georgian National Report. Black Sea Environmental Series, **8**. United Nations, New York. 167 p.

Lobzhanidze, S. I. & Chichilnitsky, I. M. (1968) Georgian Fisheries Industry. Sabchota Sakartvelo, Tbilisi. 13–36.

Mayorova, A. A. & Danilevsky, N. N. (1934) Materials on biology of the Black Sea dolphins. *Reports of the Georgian Scientific Fisheries Station*, **1**: 181–210.

Mazmanidi, N. D. (1997) *The Black Sea Fish Ecology and Fish*. Ajara Publishers. 147 pp.

Mazmanidi, N. D. & Komakhidze, A. M. (2001) Necessity of the complex monitoring of the Black Sea oil pollution. Problems of preservation and use of the bioresources of the Azov–Black Sea basin. Material of the International Scientific Conference, Rostov-on-Don. 128–129.

Sergeeva, N.G. (2000) To the question of the biological diversity and deepwater benthos of the Black Sea. *Marine Ecology*, **50**: 57–62.

Shevalev, A. E. & Tsuladze, L. E. (1975) 50 issues on the dolphin. Publications "Sabchota Adjara", Batumi. 53 pp.

Sorokin, U. I. (1982) *The Black Sea*. Nauka, Moscow. 216 pp.

Svetovidov, A. N. (1964) *Black Sea Fishes*. Nauka, Moscow.

The role and determination of residence proportions for fisheries resources across political boundaries: the Georges Bank example

Stratis Gavaris
Fisheries and Oceans Canada, 531 Brandy Cove Road, St. Andrews, NB E5B 2L9 Canada
Steven A. Murawski
National Marine Fisheries Service, Northeast Fisheries Science Center 166 Water Street, Woods Hole, MA 02543-1026 USA

ABSTRACT: Fishery resources considered transboundary for management purposes are those with spatial distributions that extend beyond a political boundary and which display movement and migration across that boundary at some life stage. The distribution and the nature of the migration are important considerations in determining management strategies. Consistent management for cod, haddock and yellowtail flounder on Georges Bank, which are considered transboundary and are subject to directed fisheries in both Canadian and USA waters, promises benefits. Annual and seasonal distributions across the boundary may also be influenced by total abundance, environmental factors, distribution of fishing effort and substock structure of the various resource species. Age-specific and seasonal distribution and migration patterns were examined to obtain a responsive annual signal of the distribution of the resource relative to the boundary. A simple empirical procedure was employed to utilize information from annual bottom trawl surveys conducted at different times of the year.

INTRODUCTION

There is a long history of dispute pertaining to claims over access rights and utilization privileges of lucrative fishing grounds between sovereign nations, as well as between regions, fleets and individual fishers within nations. Limitation of fishing freedoms can be broadly classified as territorial exclusions and non-territorial restrictions (Koers, 1973). Non-territorial restrictions are most often achieved through arrangements arrived at by regional fisheries management organizations or by multilateral negotiation between affected parties. The declaration of Exclusive Economic Zones (EEZ) by coastal nations in the mid 1970s redefined the extent of territorial exclusion, effectively ascribing property rights to coastal nations out to 200 miles (McRae & Munro, 1989). The establishment of EEZs greatly simplified management of resources that were entirely contained within a single jurisdiction.

Whereas the Convention on the Law of the Sea (United Nations, 1995) prescribed that suitable measures be taken to coordinate and ensure the conservation of transboundary stocks, i.e. those that were not entirely within a single jurisdiction, it essentially left these situations unsettled. Anadromous stocks, which were covered by special provision, were an exception. Management of transboundary stocks that straddle the EEZ and/or are highly migratory was left to regional fisheries management organizations to resolve. Complications relating to management of these stocks were initially considered to be of limited importance, but this assumption has proven false (Munro, 2001). Transboundary stocks that are contained entirely within EEZ boundaries are referred to as shared stocks. Caddy (1998) argued that a large number of resources thought to be proprietary were actually shared with other coastal nations. These situations have been left to bilateral or multilateral negotiation between contiguous coastal nations. International law and convention provide guiding principles for dealing with the issues of managing shared stocks (Harwood, 1998), but offer few details on the solutions.

Harvest policy options for each case of shared resources is unique and depends on the combination of distribution and movement of fish between jurisdictions and the exploitation rate allowed by other parties (Gulland, 1980; Walters, 1998). However, case studies can provide important lessons. This paper chronicles recent developments between Canada and the USA for managing shared resources on Georges Bank and explores the basis for formulating a consistent management framework.

FOUNDATION FOR COOPERATION

The fisheries by Canada and the USA in the Gulf of Maine area of the northwest Atlantic represent one of the many cases of shared resources left for resolution in the wake of the declaration of EEZs. The desirability of cooperative management was recognized very early and deliberations progressed rapidly. A proposed East Coast Fisheries Bilateral Agreement was negotiated by 1979, but dissatisfaction with the terms of the agreement led to intense lobbying from fishing interests, and it was never ratified. The dissatisfaction arose out of the ambiguity about property rights. Both nations claimed a disputed zone on the eastern Georges Bank that occurred because of intersecting claims within the 200 mile limits of both countries. This led to conflicting views about the appropriate entitlements to the benefits from the resources, both between nations and between fishing fleets within nations. The USA and Canada agreed to refer the boundary dispute to the International Court of Justice. In October 1984, the Court delivered its judgement, and the international maritime boundary between the USA and Canada was established (Figure 1).

The activities of fishers from the USA and Canada were subsequently constrained to their respective territories. The boundary effectively established national access and property rights. However, it did not resolve fisheries management concerns for those resources where the movements of the fish exposed them to exploitation by both nations. Several fishery resources on Georges Bank are considered transboundary. For a stock to be a candidate for transboundary management it must satisfy two criteria:

- the spatial distribution of the resource must extend beyond political boundaries;
- there must be notable movement and migration of fish across the boundary at some life stage.

The first requirement is tautological. The second recognizes that the full benefits of unilateral management actions on the portion of a resource within national boundaries could be compromised by the absence of cooperation. Gulland (1980) noted that these

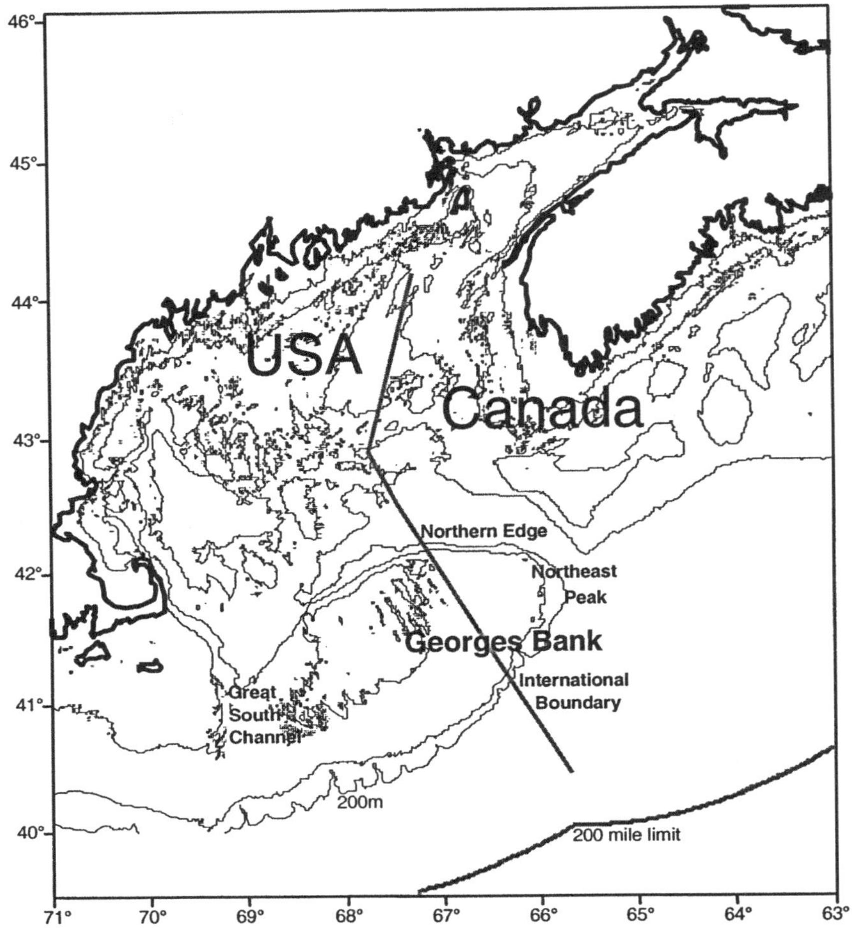

Fig. 1 Gulf of Maine area showing the international boundary between Canada and the USA across Georges Bank.

transboundary–migratory stocks commanded the greatest demand for cooperative management. The enactment of the international boundary between Canada and the USA, by establishing exclusive access rights to fishing grounds, thereby heightened the prominence of resource distribution and movement of fish as a determining factor for the resolution of sharing arrangements.

While several species are caught in the fisheries by Canada and the USA on Georges Bank, the active directed fisheries for cod, *Gadus morhua*, haddock, *Melanogrammus aeglifinus*, and yellowtail flounder, *Limanda ferruginea*, were considered a first priority. Their distributions span the maritime boundary between the USA and Canada and the fish migrate across the boundary, though the movements of yellowtail flounder are less pronounced. Progress on cooperation between Canada and the USA for managing these resources is adhering to the archetypal trajectory (Gulland, 1980), with cooperation in research first, followed by collaboration in setting management measures, and finally leading to compatible administrative arrangements. Fishery and survey information were exchanged and shared from an early stage. Reciprocal participation in assessment meetings was eventually replaced by joint assessment reviews with the formation of the Transboundary Resource Assessment Committee in 1998. The common perspective arising from the assessment evaluations about worsening stock status and heavy exploitation led to collaborative conservation efforts and gave new urgency to making progress on administrative arrangements for management.

CONSISTENT FISHERIES MANAGEMENT

The aim of fisheries management is to govern human activities in a manner that achieves stated objectives. The development of consistent management for shared resources relies on compatibility in the values that are upheld by the governance institutions (Gulland, 1980). An important ingredient in the progress for the Georges Bank stocks was the cultural similarity in the fisheries of Canada and the USA on the northeast coast of North America and the shared history in approaches to management (Halliday & Pinhorn, 1996). In this section, this similarity is examined in the context of the hierarchy of planning elements involved in fisheries management, but first, the implication of consistency is examined.

The broad aspirations of fisheries management planning are referred to as general objectives. The means to achieve the general objectives and make the plan operational are called strategies, e.g. keep fishing mortality moderate. Implementation of the strategies is through explicit regulatory actions called management measures, e.g. specification of Total Allowable Catch (TAC). In this framework for fisheries management planning, what does "consistent" management mean? It was recognized from the outset that the fisheries management measures employed by the USA and Canada with respect to these transboundary resources differed. In the recent past, the USA has relied largely on input controls (i.e. fishery effort restrictions) supplemented by area closures and gear restrictions, while Canada has principally used output controls (i.e. catch limits). It was not considered essential that both nations employ the same management measures, only that the respective measures had the same effect. Therefore, "consistent" management was taken to mean agreement in the general objectives and concordance in the operational strategies; it did not require similarity in the management measures employed.

Objectives typically include conservation of the ecosystem and individual harvested resources as well as aspects related to social and economic goals. Initial discussions focused on conservation objectives. Although the general conservation objectives in the USA and Canadian fishery management plans that govern the fisheries for cod, haddock and yellowtail flounder on Georges Bank are not identical, their intent is largely compatible. In essence, the general objective is to ensure that fishing does not reduce the productivity of the resources or induce modifications to ecosystem structure or function that are difficult or impossible to reverse. Also, rebuilding depleted resources is implicit in the Canadian plans and explicitly specified in the USA plans. Hence, achieving consistent management between the two countries involves development of common operational definitions for strategies.

Strategies for achieving the conservation objective address three types of concerns: a) impacts on target species, b) impacts on incidentally caught species, and c) impacts on habitat. While it seems inevitable that discussions by the USA and Canada will eventually encompass all three aspects, first attention has been given to target species impacts. Management considerations regarding impacts on incidentally caught species and on habitat are proceeding independently in Canada (DFO, 2002) and the USA (e.g. Cargnelli *et al.*, 1999), with reciprocal awareness of respective directions in anticipation of harmonizing the approaches in future.

The principal strategies used by both Canada and the USA in addressing target species impacts are preserving population components, maintaining fishing mortality at a moderate level, managing the size/age of capture, and minimizing disturbance during spawning. At a community level, preservation of key predator-prey relationships through conservative harvesting of prey species is an implicit tactic. There was consensus that Canada and the USA should pursue deliberation on these strategies to develop consistent operational specifications. To advance this agenda, the Transboundary Management Guidance Committee (TMGC), consisting of representatives from fishing industry, management and science was formed in 2000.

RESOURCE SHARING

Management of shared resources also implies establishment of a framework for sharing the benefits from the common resource (Caddy, 1998). The TMGC turned its attention first to the task of resolving national entitlements of the harvest. While other mechanisms for extracting economic rent from common resources are possible, the advantages of allocated quota systems (Christy, 1976) have resulted in the prevalence of this approach. The resource-sharing arrangement developed by the TMGC was also founded on an allocated quota system. International law and convention prescribe some broad criteria to be considered in determining allocations, but leave specifics to be resolved (Harwood, 1998). Some of these guidelines are fairly general and pertain more to straddling stocks than to shared resources where political boundaries between states have been established.

For transboundary stocks, principles of resource sharing may include consideration of two aspects:

- access to resources occurring or produced within national boundaries;
- historical participation in exploitation of the resources.

The latter gives recognition to traditional involvement and investment in development of a fishery. The former has emerged from the effective property rights attached to the EEZ as well as from the significance given to the extent that stocks occur in areas under national jurisdiction (United Nations, 1995), and has been associated with biological zonal attachment. During initial deliberations of the TMGC, it was affirmed that fishers are entitled to the resources in their nation's respective waters. The catch records were used for assessing historical participation in exploitation. Both principles were incorporated in the TMGC sharing proposal, with the weighting of the historical participation decreasing over an initial phase-in period, eventually to rely primarily on resource distribution. The next section elaborates how the available information was used to characterize the resource distribution relative to the boundary and in relation to development of a sharing arrangement. The subsequent section gives detailed results for Georges Bank cod, haddock and yellowtail flounder and elucidates how the information about resource distribution and migration was incorporated into an incipient sharing proposal for consistent management of these resources.

RESOURCE DISTRIBUTION CONSIDERATIONS

The nature of the movement and migration of fish is an important consideration in developing effective management. Gulland (1980) identified implications to collaborative management attributable to movements associated with phases of the life history, regular seasonal migrations and temporal shifts in distribution. These types of movement and migration are discussed below in the context of their impact on development of a resource sharing arrangement.

While movement and migration during the pelagic egg and larva life history stages may be of some consequence, particularly in relation to understanding the relationship between spawning stock and recruitment, the benthic juvenile and adult life stages, when they are vulnerable to fishing, are of most immediate concern. Differential spatial distribution patterns of younger ages may exist (Overholtz, 1985; Van Eeckhaute *et al.*, 1999), but their contribution to the biomass should be relatively minor, particularly for populations subject to moderate rates of exploitation. It was considered adequate therefore to ignore age- and size-specific distribution patterns and to develop, from each survey, annual estimates of the total survey biomass on each side of the boundary within the designated management units.

The biomass estimates from distinct surveys represent synoptic snapshots of resource distribution at specific times during a year. The proportion of the biomass on respective sides of the boundary from each survey constitutes a sufficient statistic of resource distribution from that survey for the purpose of sharing arrangements. The issue of how to combine the results of multiple surveys requires understanding of seasonal movement patterns. Ideally, information on the resource distribution throughout the year, if it was available, could be integrated over the year to get an annual average. With only discreet survey snapshots, it becomes necessary to determine how much of the biological year each survey represents. If directed migrations are not a major feature, each survey can be viewed as an equally representative and independent observation of the average annual resource distribution. A simple average of the available surveys in any year then provides an estimate of the resource distribution that makes best use of all the data. If directed migrations are an important feature, each survey can be associated with the season it best represents. A simple average of surveys made during the same season can then be combined. Seasonal results can subsequently be combined, taking into account the duration of the seasons.

Having integrated the information into an annual signal, a reliable near-term (1–3 year) forecast of resource distribution is desired for establishing national entitlements. Annual survey observations display considerable dispersion. Some of this is due to real, but unpredictable, fluctuations in resource distribution, which do not tell us much about near-term forecasts. Another component of the dispersion is due to statistical sampling variation. One means of removing both unpredictable fluctuations and sampling variation is to apply a "smoother". Smoothers are a "descriptor" of observed data, and extrapolation beyond the data requires judgement. Following some debate on whether resource distribution during a period of relative abundance or resource distribution from the recent past was more relevant, it was determined that a technique that established current trends and was responsive to changes would provide the most pertinent results for contemporary resource sharing. In order to avoid imposing model assumptions that could be challenged, it was resolved to apply the smoothers to the available time-series and that the results for the latest observation would be accepted as the near-term forecast.

Alternative computational sequences could have been used. The smoother can be applied to each of the indices to obtain smooth indices from which resource distribution for each survey can be derived and subsequently combined (i.e. smooth then combine). Alternatively, the resource distribution for each survey may be derived using the observed data, subsequently combined, with the smoother applied as the final step (i.e. combine then smooth). Applying the smoother at some intermediate step is also possible. Initial investigation indicated that smoothing the proportions was more practical owing to the challenges of fitting occasional extreme outliers when working with absolute values of biomass. The "combine then smooth" procedure was viewed as being subject to fewer complications and offering a more transparent process.

Various techniques have been developed and employed to smooth data, including regression, time-series analysis, splines, kernel estimators and locally weighted regression. A technique that made few assumptions about the form of the trend and about the distribution of errors around the predicted trend was preferred. Further, the method had to be simple, transparent and obviously responsive to the data, rather than dependent on underlying assumptions. Regression was ruled out because of the requirement to specify the form of the trend. Time-series analysis was ruled out because the data did not appear to follow regular patterns and the time-series was short. Splines, kernel estimators and locally weighted regression have many similarities and do not require specification of a parametric form for the relationship. A robust locally weighted regression algorithm (Cleveland, 1979) has gained particular favour, is easily implemented and is transparent in how the data are manipulated. This technique, referred to as LOWESS, was therefore adopted.

Application of any smoother, including non-parametric types like LOWESS, involves some subjective judgement. LOWESS requires two subjective inputs: a) the fraction of data used to obtain the "smooth" at any point, referred to here as the smoothing parameter, and b) the number of iterations for robustness. Available guidelines suggest that the smoothing parameter be between 20 and 80%, with 50% as a reasonable compromise. Experience indicates that most of the benefits associated with the robustness pass-through are achieved in two iterations. It was recommended that the default of two robustness iterations be used, and results using 30, 50 and 70% as the smoothing parameter be explored. A convenient feature of LOWESS was that the smooth of the percentage on the Canadian side and the smooth of the percentage on the USA side are exact complements. This is useful, because it is desirable that the two smooth series add to 100% in each year.

RESOURCE DISTRIBUTION: GEORGES BANK COD, HADDOCK AND YELLOWTAIL FLOUNDER

The designation of units for management entails a compromise between the biological realities of stock structure and the practical convenience of analysis and policy making Gulland (1980). For yellowtail flounder, Canada and the USA use a common management unit consisting of the entire bank east of the Great South Channel (Figure 2). The resource distribution of yellowtail flounder across the boundary was examined relative to this management unit. For Georges Bank cod and haddock, the USA employs a management unit consisting of the entire Georges Bank and extending south and west of Cape Cod. Canada, however, uses a management unit consisting of only the eastern portion of Georges Bank. The TMGC agreed that, for the purpose of developing a sharing proposal for cod and haddock, the resource distribution across the boundary would be limited to only the eastern portion of Georges Bank (Figure 2).

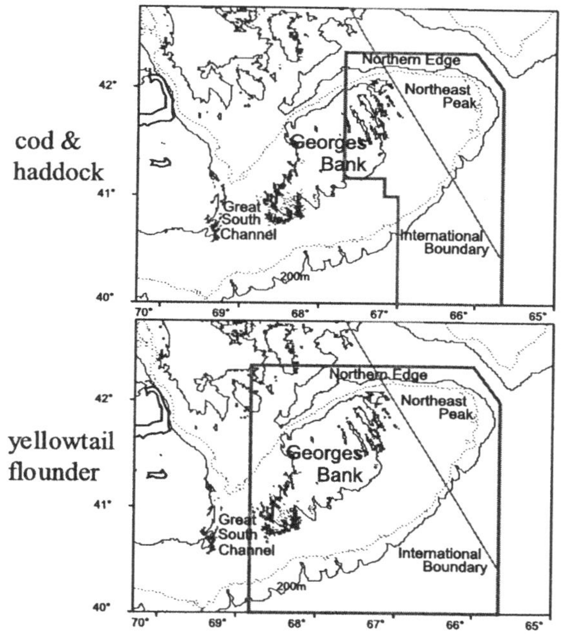

Fig. 2 The resource distribution of cod, haddock and yellowtail flounder across the boundary was considered relative to the management units illustrated.

Surveys of Georges Bank have been conducted by the USA National Marine Fisheries Service (NMFS) each autumn (October) since 1963 and each spring (April) since 1968, and by Fisheries and Oceans Canada (DFO) each spring (February) since 1986. All surveys used a stratified random design. For the NMFS surveys, two vessels have been used during the time-series, and a trawl-door change occurred in 1985. Vessel and door conversion factors, derived experimentally from comparative fishing (Forrester *et al.*, 1997), have been applied to the survey results to make the series consistent. Additionally, two different trawl nets were used on the NMFS spring surveys, a modified Yankee 41 during the period 1973–1981 and a Yankee 36 in other years, but no conversion factors are available.

Swept area biomass, considered a relative index of abundance, was computed and apportioned to the USA and Canadian sectors in each year. As the survey design was based on randomization within strata, the data were post-stratified to USA and Canadian zones within the existing survey strata. Figure 3 depicts the strata and strata sections on each side of the international boundary within the management units that were used in the analysis. Estimates of biomass for strata that were divided by a management unit boundary or the international boundary were readily calculated, unless there were no observations within the stratum section. On the few occasions where no observations were available in a stratum section, density and distribution patterns from adjacent areas and years were used to derive values. Although this procedure required judgement, the biomass contributed by the derived values was generally small and did not unduly influence results. Particularly when combined in averages over years, the effect of these adjustments diminishes. The results were then used to aggregate the biomass on the USA and Canadian sides of the boundary within respective management units.

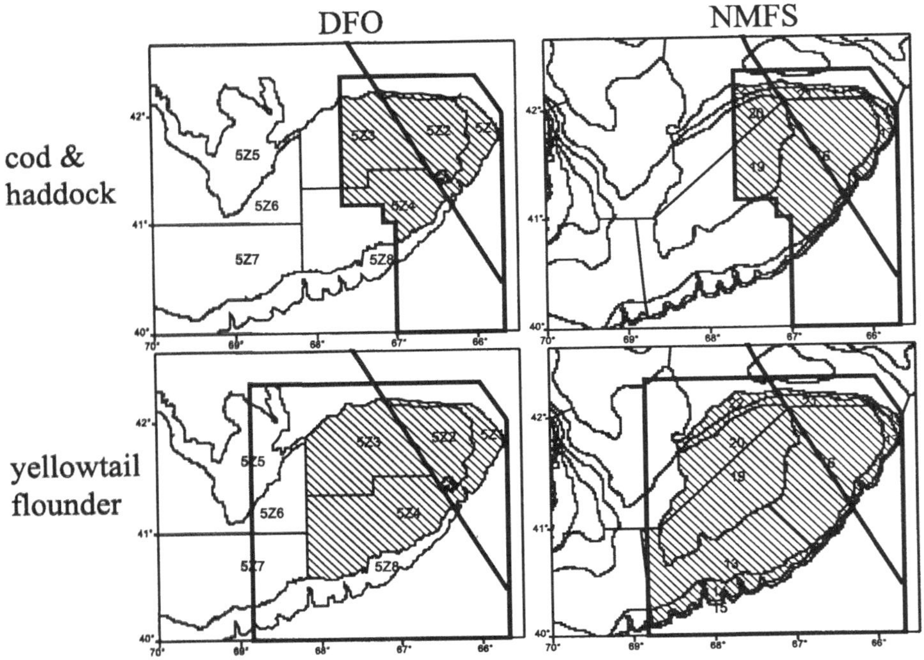

Fig. 3 The survey designs were post-stratified to accommodate the international boundary and the management unit borders. Strata boundaries (thin black lines) with strata labels are shown. The hatched area represents the strata and strata sections that were used to approximate the respective management units (thick black lines).

The abundance of cod declined in the mid 1980s and cod biomass on the USA side declined disproportionately, particularly evident in the NMFS autumn survey (Figure 4). Most of the cod biomass during the NMFS spring survey and the DFO survey was on top of the Bank in shallower water, NMFS strata 16 and 19 and DFO stratum 5Z2. During the NMFS autumn survey, the deeper slope strata have always been more important for cod, but after the late 1980s more of the biomass shifted from the top of the Bank to its deeper slopes, particularly in deeper water on the Canadian side of NMFS strata 17 and 21. The percentage of cod on the Canadian side during the NMFS spring and DFO surveys was lower than the percentage during the NMFS autumn survey, and the difference has become more pronounced since 1980. This pattern is consistent with previous observations from the fishery, and from surveys that indicate a southwesterly migration of cod during the winter/spring period associated with spawning and subsequent return migration northeastwards. The spawning period is rather protracted, but it peaks during March (Smith, 1983), near the time period when the DFO and the NMFS spring surveys are conducted. Accordingly, for cod, the DFO and the NMFS spring surveys in each year were averaged first to characterize the distribution during winter/spring. The result was then averaged with the NMFS autumn distribution percentages, thereby giving equal weight to the winter/spring and summer/autumn periods. Prior to 1987, when the DFO survey was initiated, the NMFS spring survey alone was used to characterize the winter/spring period.

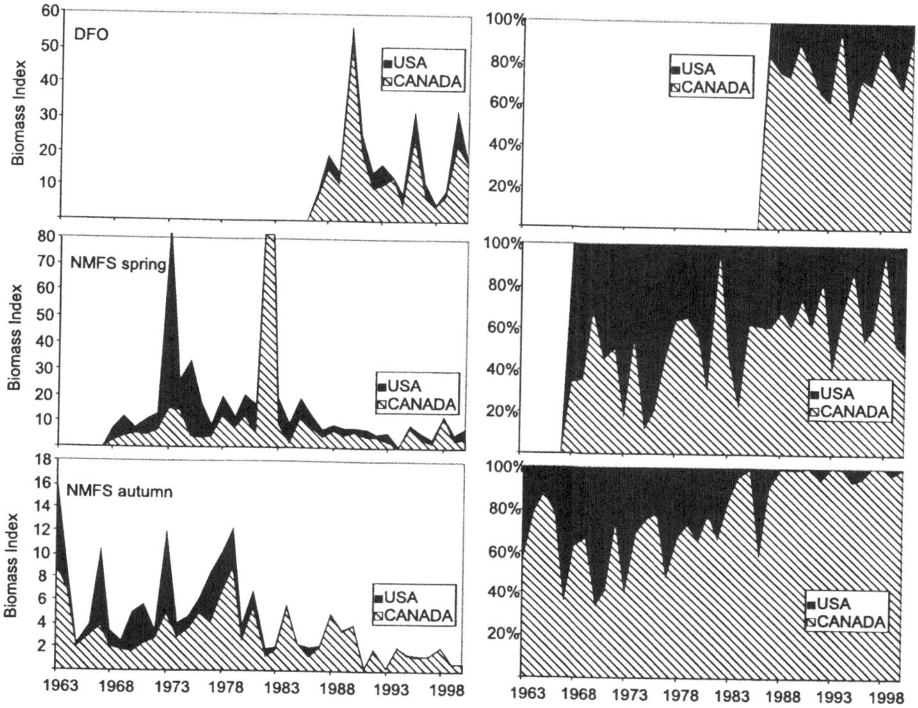

Fig. 4 Relative indices of biomass and percentage resource distribution in relation to the international boundary for cod on eastern Georges Bank shown as stacked area charts.

Haddock abundance was highest during the early 1960s, was high again in the late 1970s and increased during the 1990s (Figure 5). The biomass on the USA side was exceptionally high during the 1960s. As with cod, haddock biomass is concentrated on top of the Bank during the NMFS spring survey and DFO survey, though perhaps to a lesser extent. With the exception of the 1960s, haddock biomass is concentrated in the deeper slope strata during the NMFS autumn survey. Mature haddock also display a seasonal migration (Van Eeckhaute *et al.*, 1999) associated with spawning. During late winter and early spring, there is a southwestward movement, with peak spawning in April, followed by a return northeastward migration. This suggests that seasonal averaging may be appropriate. However, the percentage of haddock on each side of the boundary from the DFO survey was somewhat intermediate between the NMFS autumn and NMFS spring survey results. An unequivocal association between seasons and surveys was not evident. A simple averaging of the available survey distribution percentages was therefore used to represent the annual pattern.

Yellowtail flounder abundance was high in the 1960s, declined and remained low during the 1970s and 1980s and increased during the 1990s (Figure 6). The biomass on the USA side was highest during the 1960s. During all three surveys, yellowtail flounder biomass is concentrated on the southwest flank of the Bank in shallower water, NMFS strata 13 and 16 and DFO strata 5Z2 and 5Z4. NMFS stratum 19, a shallow stratum near the middle of the Bank, was important during the 1960s only. Yellowtail flounder are thought to be relatively sedentary during their juvenile and adult benthic stages

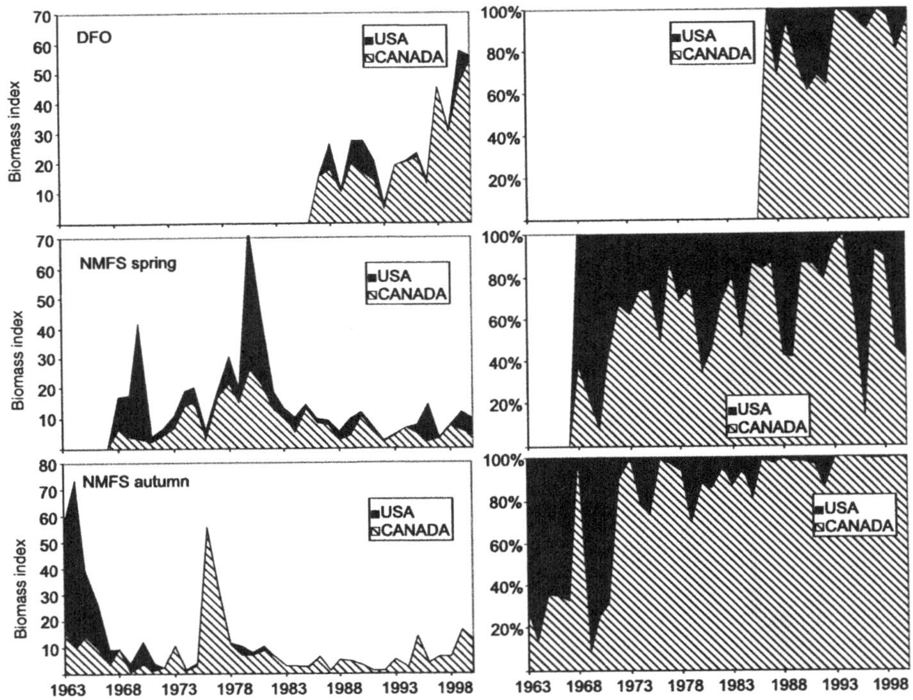

Fig. 5 Relative indices of biomass and percentage resource distribution in relation to the international boundary for haddock on eastern Georges Bank shown as stacked area charts.

(Royce *et al.*, 1959; Lux 1963). Ontogenetic migration patterns of Georges Bank yellowtail flounder are not well understood, because the age structure of the stock has been truncated over most of the past four decades as a consequence of heavy exploitation. The percentage of yellowtail flounder on each side does not show significant differences among the three surveys. A simple average of the available survey distribution percentages in each year was therefore used for yellowtail flounder.

The limitations of using three survey "snapshots" to interpret a continuous diffusion process leaves any scheme for combining the surveys subject to debate. It was therefore recognized that negotiations might be invoked where the interpretations were equivocal. Both nations agreed that a simple average of all available surveys was appropriate for yellowtail. The initial Canadian position was to treat cod and haddock similarly, by averaging the NMFS spring and DFO surveys before combining with the NMFS autumn results, whereas the initial USA position was to take a simple average of all available surveys. The compromise reached was to average the NMFS spring and DFO surveys before combining with the NMFS autumn survey for cod, where the seasonal pattern was most evident, and to take a simple average of available surveys for haddock.

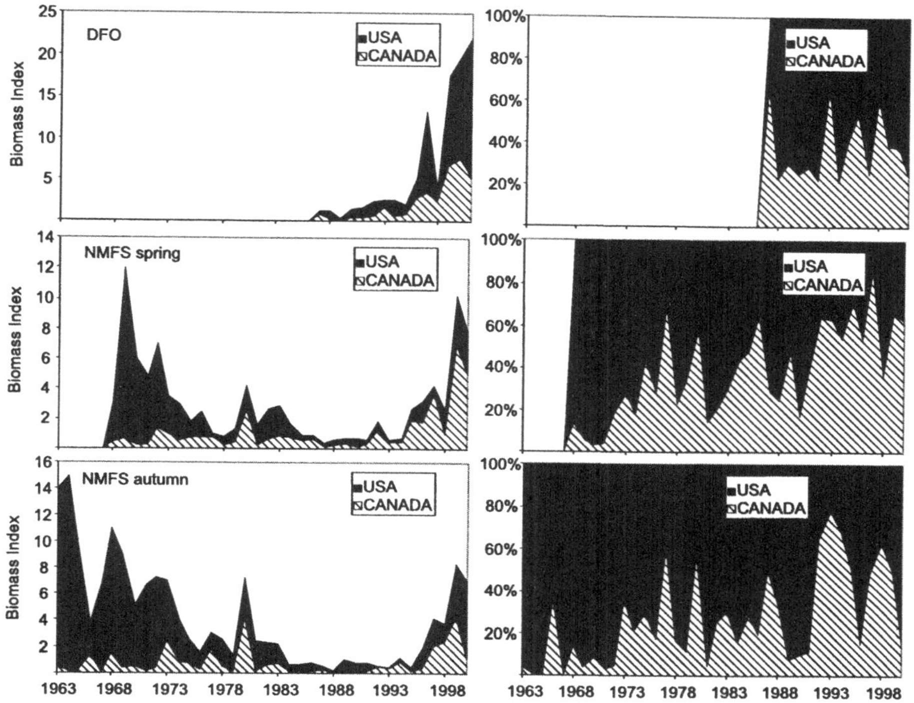

Fig. 6 Relative indices of biomass and percentage resource distribution in relation to the international boundary for yellowtail flounder on Georges Bank shown as stacked area charts.

The annual percentage on each side of the boundary was calculated from 1968, when the NMFS spring survey was initiated. As the percentage on the USA side is the complement of the percentage on the Canadian side, the two adding up to 100%, results are described for percentage only on the Canadian side. The observed percentage of cod on the Canadian side was about 50% in the early 1970s, progressively increasing to about 80% by 1990, where it has since remained (Figure 7). Results using 30, 50 or 70% smoothing parameters showed little difference. The observed percentage of haddock was variable in the early 1970s, fluctuated with a gradually increasing trend from about 60 to 80% until 1990, and has subsequently remained near that value. The smooth results show considerable divergence in the early 1970s, but afterward the 50 and 70% smoothing parameter options are virtually indistinguishable and the 30% option is not very different either. The observed percentage of yellowtail flounder on the Canadian side was very low in the early 1970s, fluctuated at about 20% during the 1970s, gradually increased through the mid 1990s, but has recently declined. The results using smoothing parameters of 50 and 70% were similar and the 30% option followed the same general trend, but with slightly more of the fluctuation.

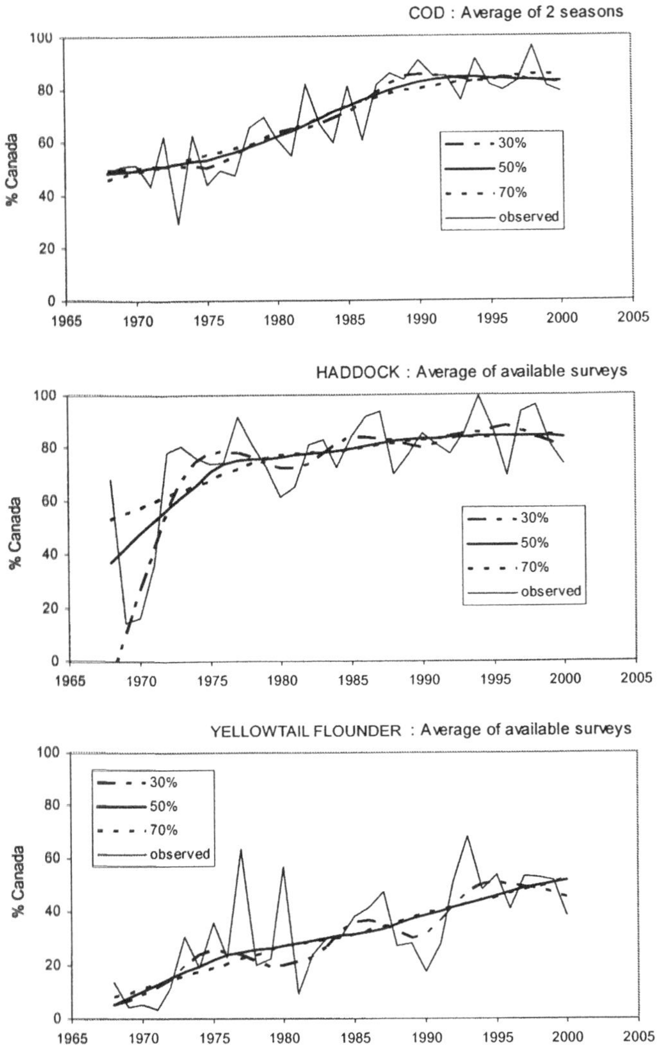

Fig. 7 Observed annual percentage and smoothed trends of proportion of eastern Georges Bank cod and haddock and of Georges Bank yellowtail flounder on the Canadian side of the international boundary.

Overall, it was concluded that the responsiveness of the 30% smoothing parameter, with an implied 10-year time window, was desirable. While the oscillations displayed by the 30% option appeared at times to reflect "unpredictable" fluctuations, it was considered that greater emphasis on the more recent observations was preferable. For subsequent application, as more years of data are added, it was considered appropriate to keep the "time window" constant. Accordingly, future calculations will be done using a 30% smoothing parameter on the most recent 33-year time-series, such that the time window remains 10 years.

SUMMARY AND CONCLUSIONS

Cooperative arrangements between Canada and the USA for the cod, haddock and yellowtail flounder resources on Georges Bank were pursued because it was considered that effective unilateral management could be compromised. These stocks are classic examples of shared resources that display transboundary–migratory properties. Of the types of movements that could impede unilateral management, the most compelling arguments in this case pertained to the regular seasonal migration patterns, particularly for cod and haddock. Ontogenetic patterns were not vital because the species are exploited exclusively as adults. Random dispersion, while obviously operative, did not appear to be a major consideration at the scale of concern for the sharing arrangement and is probably overwhelmed by the redistribution associated with the seasonal migrations. Given the nature of many fisheries, it seems likely that regular seasonal migration patterns resulting in movements across political boundaries would be the principal cause that results in classification of a resource as shared and requiring cooperative management. The most obvious exception is the case where a resource occupies different political territories at various ontogenetic phases, and fisheries harvest the resource at multiple life stages.

Progress towards agreement on cooperative administrative arrangements was greatly facilitated by the common fisheries culture and fisheries management history shared by Canada and the USA. In both Canada and the USA, greater differences between management practices on the Pacific and Atlantic coasts are evident. Even within the Atlantic coast, the ties between fishers in the northeastern USA and southwestern Canada are more apparent than similarities with adjacent areas within each nation. Administrators capitalized on this feature, encouraging discussion and consultation on a local scale. In contrast to management issues surrounding straddling/highly migratory stocks, shared stocks involve contiguous nations, which are likely to have more in common. Therefore the resolution of cooperative arrangements for shared stocks may be more readily achieved, and the prospects for success may be enhanced further by dealing on a more regional and local scale.

It was important to understand clearly what is meant by "consistent management" between the USA and Canada on Georges Bank, because this requirement had to be translated into operational terms. Consideration was greatly facilitated by asking the question in the context of a rational planning framework for fisheries management. The framework establishes a hierarchy of objectives, strategies and management measures. It was obvious that compatibility of broad objectives and establishment of common strategies were absolutely essential to consistent management, but implementation using the same management measures was not a requirement. Acceptance that consistent management allowed for divergence in management measures was an important realization that eliminated what appeared to be insurmountable obstacles. Therefore, the key for achieving consistent management lay in development of common operational definitions for the strategies.

Consistent management of shared resources is founded on common operational strategies. However, administrative arrangements for dividing the benefits of harvesting the resource are a critical element. While territorial allocation serves a conservation purpose of maintaining a roughly uniform exploitation rate across the range of the resource, considered desirable for maintaining population components and substructure by avoiding localized depletion, it is looked upon principally as a means for establishing equitable entitlements. The failure of the proposed agreement in 1979 is testament to the advantages of establishing clear exclusive access rights to fishing grounds. This conveys effective property rights to nations, but these rights are imperfect because of the motility of fish. Nevertheless, owing to the significance attributed to the extent that stocks occur in areas under national jurisdiction (United Nations, 1995), biological zone attachment persists as an important principle, and practical approaches for its determination can be developed. Accordingly, resource distribution played a prominent role in determining the national sharing arrangements for the fisheries on Georges Bank and is likely to be a dominant consideration for most shared resources where clear territorial boundaries have been established.

Despite the prominence of resource distribution as a principle for determination of sharing arrangements, the interests of the participating nations must be given consideration. Historical participation and investment in development of the fishery are perhaps the most prevalent interests. Such claims, however, are commonly viewed as diminishing in relation to the time elapsed after establishment of boundaries, and perhaps eventually expiring. Examples of dealing with such matters include the phasing out of distant water fleets after declaration of an EEZ (McRae & Munro, 1989). For groundfish on Georges Bank, the TMGC proposed that resource distribution be given 60% of the weighting in the sharing formula for the inaugural year of 2003, with the remaining 40% of the weighting being given to resource utilization, based on the catch history during the period 1967–1994. The weighting for resource distribution would progressively increase to 90% by 2010 and remain at 90% thereafter. This proposal acknowledges the investment made by nations in developing the fisheries, but recognizes that fishers are entitled to the share of the resource occurring in their respective waters.

Methods other than simple calculations of resource distribution percentages from surveys could have been pursued as a basis to inform the process. Information from the fisheries might be used to glean inferences about resource distribution. Model reconstructions of population development that used length and age structure for fishery information as well as for the survey indices could have been explored. However, only direct observations derived from bottom trawl survey results were considered, in order to avoid invocation of controversial assumptions regarding migrations. Ongoing tagging studies that would permit annual evaluation of migration patterns are not available and are not being considered owing to practical limitations. It is recognized that the three available synoptic surveys may not be ideally spaced over the year and are insufficient fully to characterize the intricate nature of fish diffusion. Research surveys are expensive to conduct, however, and it is unlikely that enhanced survey activity will be forthcoming in the near future. Investigation of the potential use of fishery catch rates may be considered. Nevertheless, interpretation of fishery catch rates is fraught with caveats associated with the impacts of ever-changing management measures on the behaviour of fishers and with the lack of randomization and complete coverage of the range of stocks. In recent years, governments and the fishing industry have partnered to conduct industry surveys. Perhaps well-designed industry surveys, targeted at characterizing the movements of principal species, may offer advantages for enhanced understanding of distribution and migration. Despite the information deficiencies, the availability of credible fishery independent surveys to serve as the basis for the evaluation of resource distribution was an important factor for making progress.

Gulland (1980) noted the importance of recognizing shifts in distribution over a longer temporal scale. Non-stationarity in biological processes and properties challenges administrative arrangements with respect to time consistency (Munro, 2001). Given observed changes in the proportions of the resources occurring each side of the boundary, and the lack of definitive explanations for these changes, a sharing proposal that is flexible to such shifts minimizes the potential for inconsistent management. The method used to summarize resource distribution had to be suitable for application to near-term projections. Therefore, it was important to capture persistent trends in shifting distribution without emphasizing ephemeral fluctuations. Further, it was essential that the results were driven by the observed data, with minimal impact from conditioning assumptions. Adoption of a transparent empirical procedure that summarized recent trends in resource distribution proved to be a valuable attribute for acceptance of the results.

Although scientific analyses played a key role in guiding sharing agreement decisions, it was important to acknowledge and recognize the limits to which judgement could be informed by the observed data. One example was the equivocal interpretation of the seasonal representation of distribution patterns in the three surveys. Another example was the subjective choice of the smoothing parameter, although this choice did not have substantial influence on the general pattern, except when the observed series showed extreme fluctuation for haddock in the early 1970s. Perhaps some of these inadequacies may be overcome, for example by use of randomization testing as the basis for selecting a smoothing parameter, but it must be admitted that there will always be gaps in the knowledge that can be gleaned from the observations. The negotiation process must be respected as a legitimate resolution mechanism during deliberations on administrative arrangements for shared stocks when objective conclusions are not forthcoming from the analyses.

Development of a proposal for a sharing arrangement was only the beginning of a process to establish consistent management of shared resources. Canada and the USA are actively engaged in deliberations aimed at agreement on a common fishing-mortality-rate-based harvesting strategy. Concurrently, discussions on establishing fishery monitoring standards for measurement of performance against plan objectives are being conducted. While these developments may adequately address concerns about the impact of fishing on target species, consistent management also encompasses aspects related to impacts on incidentally caught species and habitat. Undoubtedly, success in establishment of consistent harvesting of target resources will lead to discussions aimed at mitigating any negative effects of fishing on the structure and functioning of the Georges Bank ecosystem.

ACKNOWLEDGEMENTS

We thank Fred Serchuk, Rob Stephenson and an anonymous reviewer for their constructive suggestions. We also thank the organizers of the symposium for giving us the opportunity to present these results. Finally, we are appreciative of the effort and commitment by the members of the TMGC to the process of developing consistent management.

REFERENCES

Caddy, J. F. (1998) Establishing a consultative mechanism or arrangement for managing shared stocks within the jurisdiction of contiguous States. In: *Taking Stock: Defining and Managing Shared Resources*. (Ed. by D. A. Hancock), pp. 81–123. Australian Society for Fish Biology and Aquatic Resource Management Association of Australia Joint Workshop Proceedings, Darwin, June 1977. Australian Society for Fish Biology, Sydney.

Cargnelli, L. M., Griesbach, S. J., Berrien, P. L., Morse, W. W. & Johnson, D. L. (1999) Essential fish habitat source document: haddock, *Melanogrammus aeglefinus*, life history and habitat characteristics. *NOAA Technical Memorandum NMFS–NE*–**128**.

Christy, F. T. (1976) Limited access systems under the Fishery Conservation and Management Act of 1976. In: *Economic Impacts of Extended Fisheries Jurisdiction*. (Ed. by L. G. Anderson), pp. 141–156. Ann Arbor Science, Ann Arbor, Michigan.

Cleveland, W. S. (1979) Robust locally weighted regression and smoothing scatterplots. *Journal of the American Statistical Association*, **74**: 829–836.

DFO (2002) Fisheries management planning for the Canadian eastern Georges Bank groundfish fishery. DFO Maritime Provinces, Regional *Fisheries Status Report* **2002/01E**. (http://www.mar.dfo-mpo.gc.ca/science/rap/internet/fsr-2002.htm).

Forrester, J. R. S., Byrne, C .J., Fogarty, M. J., Sissenwine, M. P. & Bowman, E. W. (1997) Background papers on USA vessel, trawl, and door conversion studies. *SAW/SARC 24* Working Paper **Gen 6**. Northeast Fisheries Science Center, Woods Hole, MA.

Gulland, J. A. (1980) Some problems of the management of shared stocks. *FAO Fisheries Technical Paper*, **206**: 22 pp.

Halliday, R. G. & Pinhorn, A. T. (1996) North Atlantic fishery management systems: a comparison of management methods and resource trends. *Journal of Northwest Atlantic Fishery Science*, **20**: 143 pp.

Harwood M. (1998) Biting the allocation bullet - allocation in international fisheries. In: *Taking Stock: Defining and Managing Shared Resources*. (Ed. by D. A. Hancock), pp. 125–130. Australian Society for Fish Biology and Aquatic Resource Management Association of Australia Joint Workshop Proceedings, Darwin, June 1977. Australian Society for Fish Biology, Sydney.

Koers, A. W. (1973) *International Regulation of Marine Fisheries; a Study of Regional Fisheries Organizations*. Fishing News Books, London.

Lux, F. E. (1963) Identification of New England yellowtail flounder groups. *Fisheries Bulletin, US*, **63**: 1–10.

McRae, D. & Munro, G. (1989) Coastal state "rights" within the 200 mile exclusive economic zone. In: *Rights Based Fishing*. (Ed. by P. Neher, R. Arnason & N. Mollet), pp. 97–112. Kluwer, Dordrecht.

Munro, G. R. (2001) The United Nations fish stock agreement of 1995: history and problems of implementation. *Marine Resource Economics*, **15**: 265–280.

Overholtz, W. J. (1985) Seasonal and age-specific distribution of the 1975 and 1978 year-classes of haddock on Georges Bank. *Northwest Atlantic Fisheries Organization Scientific Council Studies*, **8**: 77–82.

Royce, W. F., Buller, R. J. & Premetz, E. D. (1959) Decline of the yellowtail flounder (*Limanda ferruginea*) off New England. *Fishery Bulletin US*, **53**: 169–267.

Smith, W. G. (1983) Temporal and spatial shifts in spawning of selected fish and invertebrate species in the Georges Bank region. *NEFSC Sandy Hook Laboratory Reference Document*, **83–08**. 22 pp.

United Nations (1995) United Nations conference on straddling fish stocks and highly migratory fish stocks. Agreement for the implementation of the provisions of the United Nations Convention on the Law of the Sea of 10 December 1982 relating to the conservation and management of straddling fish stocks and highly migratory fish stocks. *U.N. Doc. A/Conf.*/**164/37**.

Van Eeckhaute, L. A. M., Gavaris, S. & Trippel, E. A. (1999) Movements of haddock, *Melanogrammus aeglefinus*, on eastern Georges Bank determined from a population model incorporating temporal and spatial detail. *Fishery Bulletin, US*, **97**: 661–679.

Walters C. (1998) Stock assessment and harvest policy design for shared resources. In: *Taking Stock: Defining and Managing Shared Resources*. (Ed. by D. A. Hancock), pp. 51–61. Australian Society for Fish Biology and Aquatic Resource Management Association of Australia Joint Workshop Proceedings, Darwin, June 1977. Australian Society for Fish Biology, Sydney.

Integrating climate variation and change into models of fisheries yield, with an example based upon the southern Newfoundland (NAFO Subdivision 3Ps) cod

John G. Pope
Norwegian College of Fisheries, Universitetet i Tromsø, Tromsø, Troms, Norway.
Correspondence address: NRC (Europe) Ltd, The Old Rectory, Burgh St Peter, Norfolk NR34 0BT, United Kingdom

ABSTRACT: A delay-difference model is formulated that updates the biomass and abundance of a fish stock on an annual basis. It may be used for simulating time-series of annual yield and also for providing estimates of steady state yield. The approach is a variant of Deriso's delay-difference model. The formulation is length-based and provides matrix-based annual recurrence relationships for numbers and biomass, together with apparently "nuisance" values of "biolength" and "bioarea" of the stock. The model allows a better description of size-based selection and is constructed with explicit parameters describing growth, natural mortality rate and recruitment processes. Hence, it is well suited to estimating long-term steady states of the yield of fish stocks under combinations of harvest rate and selectivity regime. It is also able to provide estimates of how long-term yield might react to climate change. An example of the use of the method is provided based upon the cod stock of NAFO subdivision 3Ps. The fishery for this stock is in a post-moratorium stage. This follows the serious stock declines and subsequent harvesting moratoria experienced by many of the groundfish fisheries of Atlantic Canada during the late 1980s and early 1990s. Management is seeking viable long-term harvesting plans for this and other Atlantic groundfish stocks. Investigation of how changing ocean climate, as manifest by temperature, affects the yield of this stock indicated that the main effect was through recruitment of small fish to the fishery. The fitted stock-recruitment relationship indicates a complex interaction of temperature with stock size in the determination of recruitment. Compared with the effect of colder temperatures, increased temperature appears to increase recruitment at low stock size but to decrease it at high stock size. Temperature and the ratio of the average weight of fish in the catch to the average weight of fish in the stock both have a profound influence on the level of harvest that the stock can sustain; increases in either allow higher sustainable harvest rates. The method facilitates the simulation of the performance of harvest rules that are being considered for the management of this stock. These simulations indicate that placing a sensible upper limit on the Total Allowable Catch may be an effective way of achieving a sustainable harvest for a stock whose stock-recruitment function is modified by extended variations in its temperature regime.

INTRODUCTION

Long-term global climate change is considered likely (IPCC, 1995), and interdecadal changes in ocean climate are widely observed (Dickson *et al.*, 1994). These are particularly apparent in the Northwest Atlantic (Colbourne *et al.*, 1994). The effect of climate change has been a subject of particularly intense research for North Atlantic cod (ICES, 1994). Climate changes must be expected to alter the growth and recruitment of fish and hence the yield of fish stocks. Sea temperature has often been linked with these life history factors (e.g. Brander, 1995; Planque & Fredou, 1999). By contrast, there are relatively few examples of models of fisheries yield that contain both fishing effects and climatic effects. Walters & Parma (1996) show one approach to the problem, but there seems to be a need for a simple model to be developed that could incorporate climatic factors into equations for predicting yield on short- and long-term time frames.

The delay-difference model of Deriso (1980) is simple and formulated with explicit parameters for growth and recruitment. It might therefore be used for the purpose, but it requires certain approximations to growth and to fisheries selection, which may not always be appropriate. Consequently, it is used here as the starting point for the development of a model that strives to overcome these limitations. Development of this model seems an appropriate way of celebrating the centenary of the Lowestoft Fisheries Laboratory that has been the birthplace of significant works on fisheries yield (Beverton & Holt, 1957) and of the effects of marine climate on fisheries (Cushing, 1982).

To illustrate how such models might be used in providing management advice for fisheries, they are applied to topical management questions concerning the cod stock of NAFO subdivision 3Ps. This stock has a number of subcomponents found in the bays of southern Newfoundland and on the offshore banks north of the Laurentian Channel. The stock is recovering after the severe decline it and other Northwest Atlantic cod stocks suffered in the late 1980s and early to mid 1990s – declines that probably resulted from the combination of too much fishing pressure in a period (from the perspective of cod stocks) of climatic downturn. The stock was under moratorium from 1993 to 1997. The rather remote, fisheries-dependent economy of southern Newfoundland requires that the stock be rebuilt and then harvested at a sustainable level while maintaining as far as possible the fisheries communities, economy and infrastructure of the region. This requires that long-term plans be developed that emphasize stable catch rather than stable exploitation rates or yield maximization.

Long-term management plans for 3Ps cod are under development by an independent body, the Fisheries Resource Conservation Council (FRCC, 2002), established by the Canadian Minister of Fisheries to provide him with conservation advice on Atlantic groundfish. The plans will be subject to consultation with Government and Industry. The advice on level of Total Allowable Catch (TAC) and other conservation measures is made by a consensus decision of the FRCC, after briefing by Department of Fisheries and Oceans (DFO) scientists and after consultation with managers, the fishing industry and other stakeholders. Such broad consultations are healthy in weighing all evidence about fish stocks, in increasing understanding of the need for conservation, and in increasing the transparency of the process to the industry and in facilitating their buy-in to conservation plans. However, a process that includes consultation cannot provide a precisely defined management procedure that might be exhaustively tested by stochastic simulations, e.g. the International Whaling Commission's Revised Management Procedure for Whales (Kirkwood, 1992) or the management of a number of fish stocks (Cochrane *et al.*, 1998). Nevertheless, the management process may be informed by estimates of long-term yield under various possible changes in climate to identify viable exploitation levels. It may also be informed by simulations of possible management approaches made under possible variations in climate. These will help suggest the management rules that are most viable.

To facilitate these needs, this paper provides estimates of long-term yield under conditions of possible variations of mean temperature and considers the viability of possible management rules against a background of medium-term climatic variation. To do this requires that the effects of temperature on the growth and recruitment of 3Ps cod be estimated. Examination of these factors therefore forms an additional aim of this paper, but one conducted in a preliminary fashion for illustrative purposes, because properly this is an investigation requiring a paper in its own right.

MATERIAL AND METHODS

MODEL

The development of the length-based delay-difference model is shown in the Appendix. It is based upon the following principal assumptions and approximations:

1. The catch is taken mid-year and growth occurs after this time.
2. Growth in length follows a recurrence (Ford–Walford) relationship.
3. Weight at length is the product of length cubed and a condition factor assumed constant over all lengths.
4. Total numbers and biomass are the sum of the numbers and the sum of the product of all fish older than an age of first capture c.
5. Biolength and bioarea are the sums of the products of numbers and length or length squared respectively for all fish older than the age of first capture.
6. Different harvest rates may act on total numbers (Hn), biolength (Hl), bioarea (Ha) and biomass (Hb) to allow for size-selective harvesting.
7. The stock-recruitment relationship is assumed to have a simple analytical form such as the Ricker or Beverton & Holt relationships.
8. Spawning stock biomass is taken to be equal to (or strongly correlated with) biomass of age c and older.
9. For the purpose of year-to-year simulations, changes in the parameters of the growth recurrence relationship are the same at all sizes (i.e. all cohorts experience the same change in von Bertalanffy K and L_∞ from year to year). Other parameters may also change interannually.
10. For the purpose of calculating long-term steady states under a new regime, the parameters are assumed to have values characteristic of the regime.

These assumptions were chosen to provide a simple formulation. They lead to a simple matrix-based annual update procedure for the state-vector of numbers, biolength, bioarea and biomass (see Appendix equation A6). The recurrence relationship for these four factors may be solved analytically for its steady state when recruitment is determined by simple stock relationships such as the Ricker or Beverton & Holt forms (see Appendix equations A7 and A10).

Temperature data for the period 1950–1999 for near-surface waters and waters 20, 50 and 75 m deep for offshore NAFO subdivision 3Ps are given in Colbourne (2000). Updated series were kindly provided by E. Colbourne (pers. comm.) as annual and as five-year-centred averages. Colbourne advocated their use in a running average form owing to large variances in the annual results. Consequently three-year-centred averages were formed of the temperatures measured in near-surface waters and waters of 20 m depth over St Pierre Bank.

Recruitment data were calculated from catch-at-age data (Brattey *et al.*, 1999), using a simple cohort analysis with the natural mortality rate on all ages taken to be 0.2 per annum. Owing to the full moratorium on the fishery from 1994 to 1996 and changed partial recruitment patterns in recent years, only approximate terminal F values could be used for initiating the cohort analysis. Hence, recruitment estimates for the last years of the series are suspect and are only taken for the period 1959–1995.

Commercial catch weight-at-age data are available from Brattey *et al.* (1999). These are available for ages 3–14 for the period 1959–1998. Unfortunately, the catch weights at age given are of a constant (average) form prior to 1976, so annual data are only available from this source from 1977 to 1998.

Measurements taken on Canadian DFO spring groundfish surveys of NAFO subdivision 3Ps (see Brattey *et al.*, 2001) provide estimates of weight at age (ages 3–9, years 1978–2001) and length at age (ages 1–12, years 1972–2001). Brattey *et al.* (2001) indicate that the values for some of the youngest and oldest ages are missing or resulted from low sample sizes, so these were omitted from subsequent analysis.

Condition factor (used in Appendix equations A3 and A6) was calculated as the average value for ages 3–9 for the years 1978–2000. These estimates were calculated from research length and weight data (Brattey *et al.*, 2001). The possibility of these estimates being affected by temperature was considered by making regressions of the average annual values on the temperature series given above.

The von Bertalanffy growth parameters K and L_∞ used by the model were estimated from research catch length-at-age data. In fitting these parameters, the possibility that they were linear functions of exp(Temperature) was considered using the centred three-year running average temperature series described above. The full equation considered was

$$L(a+1,y+1) = \alpha \exp(\text{temp.})L(a,y) + \beta L(a,y) + \delta \exp(\text{temp.}) + \gamma \tag{1}$$

where the α, β, δ and γ in this and subsequent equations are the regression coefficients. In this formulation, only statistically significant terms (as judged by an ANOVA of sequentially fitted terms) were retained.

As harvesting of the 3Ps cod stock starts at about age 3, the model requires an estimate of L(3,y). These inputs might be affected by temperature in the preceding years. This possibility was considered by regressing the L(3,y) inputs on the average temperature that fish of age 3 would have been subject to in their first three years of life (taken as the –2 year lag) of the temperature series referenced above.

The stock-recruitment relationship (used in Appendix equations A9 and A10) was estimated in the form given below, assuming normal errors:

$$\text{Ln}(R) = \alpha\text{temp.} + \beta \text{Ln}(B') + \delta B'\exp(\text{temp.}) + \gamma \tag{2}$$

The statistical significance of fitting successive terms in this equation was tested using ANOVA. Where β was found to be close to 1.0, it was then adjusted to 1.0 (to achieve a Ricker formulation of the stock-recruitment curve) by taking an offset of Ln(B') from Ln(R) and recalculating the regression, omitting the βln(B') term. The logarithmic form of fit, shown in equation (2), was chosen to conform with the typically log-normal nature of recruitment data. Stock-recruitment relationships were fitted both to the entire time-series (1959–1995) and to the two half time-series (1959–1976 and 1977–1995) to check that any relationships found were consistent over time.

SIMULATION OF LONG-TERM YIELD

Long-term steady state yield for 3Ps cod was calculated using the model described in Appendix equation A6 with estimates of the relevant growth, recruitment, initial size and condition parameters (temperature related as appropriate), calculated as indicated above. The logarithmic stock-recruitment relationship (equation 2) fitted was modified with an additional constant of –3.0*M to account for mortality between notional "0" group recruitment and recruitment at age 3 (Rc). It was further corrected by the $\exp(\sigma^2/2)$ multiplier required to convert from a geometric to an arithmetic mean, the σ^2 term being taken as the residual mean square of the Ln(recruitment) regression.

Calculation of long-term steady state yields (Appendix equations A7 and A10) were made for various ratios of the weight of fish in the catch to the weight of fish in the stock (see Appendix equations A11–A14). These ratios specify the size selectivity. For simplicity, the harvest levels for bioarea (Ha) and biolength (Hb),

$$Ha = Ca/Ba \quad \text{and} \quad Hl = Cl/Bl \tag{3}$$

were set using the rule that

$$Ha = Hn^*\{Hb/Hn\}^{2/3} \text{and} \quad Hl = Hn^*\{Hb/Hn\}^{1/3} \tag{4}$$

Using these rules, calculations of long-term yield were made for average temperature anomalies of +0.5°C, 0.0°C and –0.5°C and for a range of harvest levels.

SIMULATION OF THE APPLICATION OF LONG-TERM HARVEST PLANS TO 3PS COD

Long-term harvest control plans are being developed for 3Ps cod, but are not yet agreed. Current management has been pragmatic ("seat of the pants") following the restart of the fishery and has the main objective of stock recovery followed by catch stability. As the management process involves consultation with industry, it is likely that any agreed long-term plan would have subjective elements. Hence, it will be difficult to define a plan as a set of control rules. Nevertheless, simulations of how possible harvest rules might work in the long term would clearly inform this process, so four rules for setting TAC levels in 5 000 t steps (the current practice of choice for 3Ps cod) are investigated. These four rules were all used to simulate management of the 3Ps cod stock in accordance with the harvest-rate limit-reference points shown in Table 1. These harvest-rate limit-reference points are varied by the state of the 3+ biomass of the 3Ps cod stock.

Table 1 Limit levels of harvest rate used in simulations of the 3Ps cod stock.

Stock biomass	Limit harvest level (Lim)	Comment
>200 000 t	20%	Recovered
50 000 – 200 000 t	10%	Stressed
<50 000 t	6.666%	Depleted (F as low as possible)

As, with discrete steps in TAC, harvest rates may sometimes exceed the limit levels specified in the table above, an absolute harvest rate limit of 75% was imposed. The four TAC setting rules explored are described in Table 2.

Rule 1 attempts to mimic the current approach, and the others were developed in response to observed deficiencies in this rule after preliminary experience of the results of deterministic simulations. The rules were simulated against a temperature series of 120 years formed by concatenating the existing temperature series three times. Deterministic simulations were first made to show the general behaviour of the rules. These were then further explored by stochastic simulation, where recruitment was subject to log-normal noise whose intensity was equivalent to that observed in the fit of the stock-recruitment relationship. Simulations were made using Appendix equation A6.

Table 2 Definitions of the four rules investigated for changing TAC on 3Ps cod.

Stock biomass	Rule 1	Rule 2	Rule 3	Rule 4
>200 000 t	If H>Lim., TAC 5 000 t down. If H<0.8*Lim., TAC 5 000 t up	As Rule 1, but 10 000 t step down if last TAC>20 000 t	As Rule 1	As Rule 1
50 000 – 200 000 t	If H>Lim., TAC 5 000 t down. If H<0.8*Lim., TAC 5 000 t up	As Rule 1, but 10 000 t step down if last TAC>20 000 t	As Rule 1	As Rule 1
<50 000 t	If H>Lim., TAC 5 000 t down. If H<0.8*Lim., TAC 5 000 t up	As Rule 1	As Rule 1	As Rule 1
Additional rules	None	None	Extra 5 000 t reduction if productivity low in previous two years	Upper limit on TAC of 30 000 t

RESULTS

ESTIMATION OF CONDITION, GROWTH AND STOCK-RECRUIT RELATIONSHIPS

The condition factor showed no clear relationship with temperature. An average value of 8.64E–6 (kg cm^{-3}) was estimated. Research length-at-age data also showed no temperature effect on growth. The simple Ford–Walford relationship

$$L(a+1,y+1) = 0.881L(a,y) + 13.647 \text{ (cm)}$$

describes the data. The fit is statistically significant with an R^2 of 0.91 (d.f. 230, $p<0.001$). These results lead to estimates of K= 0.127 and L_∞ =114 cm.

Neither the commercial data for W(3,y) nor the research data for L(3,y) showed relationships with the temperature series averaged for their first three years of life. Therefore, the average L(3) level of 31.1cm was adopted.

Results of regressing the logarithm of recruitment data on stock size and temperature, as indicated in equation 2, showed a significant relationship with 3+ and 7+ biomass and a possibly significant relationship with temperature when both terms are included simultaneously. The overall R^2 for the fit with 7+ biomass (which better approximates to the spawning stock biomass) was 0.41 and for the fit with 3+ biomass (as required in the current version of the model) was 0.31. In both fits, the Student's t values of all regression coefficients except the constant were significant ($p<0.02$, d.f. 33). However, ANOVAs of the additional sums of squares (see Table 3), explained by the progressive addition of extra terms, did not show an improved fit when the Biomass*exp(temp.) term was introduced alone. Despite this, the further introduction of the temperature term did cause a significant reduction in sum of squares. Reversing the order of fit of these two terms led to a similar conclusion; both terms must be fitted in order to achieve a significantly better overall fit.

Table 3 ANOVA of the fit of the stock- and temperature-recruitment relationship.

Cause	d.f.	Sum of squares	Mean square	F	p
ln(biomass)	1	1.47	1.47	8.76	0.006
Biomass*exp(temp.)	1	0.08	0.08	0.47	0.499
Temperature	1	0.96	0.96	5.73	0.023
Residual	33	5.54	0.17	1.00	0.500

Together the two terms act by making the relationship of recruitment with biomass initially rise more rapidly with higher temperatures but also making it fall more rapidly at greater stock size when the temperature is high. The relationship of recruitment with spawning stock and temperature is shown in Figure 1. It may also be inferred from the estimates of the regression coefficients (see Table 4). Note that fits were made to a continuous temperature range, but that these are split into three ranges for illustrative purposes in Figure 1 (warm >0.08°C, cool <0.08°C to >–0.32°C and cold <–0.32°C).

Fitting the relationship without a constant term leads to a very similar fit, with the coefficient of the Ln(biomass) term very close to 1.0. These results are shown in Table 5. This results in a stock recruitment relationship of the form

$$R(0) = \text{Biomass}(3+)*\exp\{0.79\text{temp.} - 4.0\text{E}{-}6\text{biomass}(3+)*\exp(\text{temp.})\}$$

where R(0) is the recruitment estimated at age zero with standard natural mortality rate.

This is of the form of a Ricker curve with the constants both modified by Exp(temp.). Examination of its residuals in the Ln(recruitment) regression (Observed – Expected) suggested some systematic decrease with time, indicating that some, as yet undetermined, time effect is making recruitment lower in recent years. The residuals showed very little first order autocorrelation.

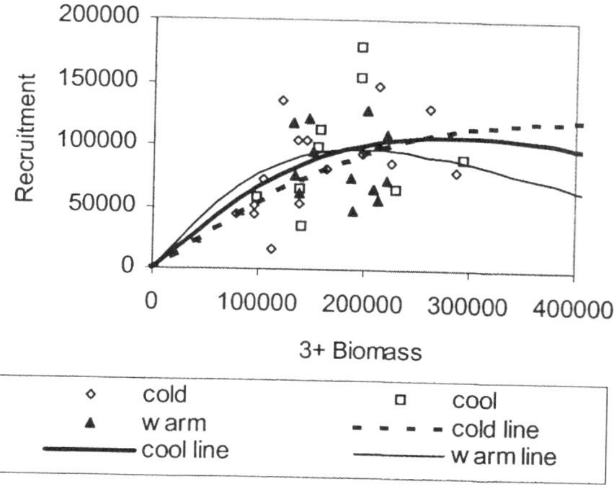

Fig. 1 Relationship between recruitment of the NAFO subdivision 3Ps cod stock and 3+Biomass, and the three-year-centred averages of the temperature of near-surface and 20 m waters over St Pierre Bank. The plot includes estimated fitted lines of the stock-recruitment relationship during warm, cool and cold years.

Table 4 Estimates of the coefficients of the stock- and temperature-recruitment relationship together with estimates of their standard errors (s.e.).

Term	Biomass*exp(temp.)	Ln(biomass)	Temp.	Constant
Estimate	–4.85E–06	1.21	0.916	–2.29
s.e.	1.97E–06	0.337	0.383	3.77

Table 5 Estimates of the coefficients of the stock- and temperature-recruitment relationship fitted with a zero intercept, together with estimates of their standard errors (s.e.).

Term	Biomass*exp(temp.)	Ln(biomass)	Temp.	Constant
Estimate	–4.02E–06	1.007	0.794	0
s.e.	1.41E–06	0.0243	0.322	

Fits of equation 2 to the two half time-series (1959–1976 and 1977–1995) indicated rather similar fits to both halves and gave similar R^2 values. The temperature effects for these half time-series did not attain significant Student's t-values when compared with the estimates of their standard errors. Estimates of the 1977–1995 series were made using the coefficients estimated from the 1959–1976 data series. These estimates and the observed recruitment in the years 1977–1995 show significant positive correlation, but the observed relationship between recruitment and predictor had a curvilinear form. In the statistical fits of equation 2 to both half time-series, the general form of their relationship was similar. Both series showed the same direction of change in recruitment with sea temperature at low and high stock biomass.

Figure 2 shows yield isopleths for 3Ps cod calculated for three different temperature regimes. These were constructed using Appendix equations A7 and A10 together with the growth and recruitment results of the previous section. Yield isopleths are shown for a range of harvest levels. They are also shown for a range of ratios of average weight of fish in the catch to average weight of fish in the sea.

The results of a single stochastic simulation of each of the four specified harvest rules are shown in Figures 3–6. Each figure has three panels showing (a) the catch, (b) the biomass and (c) the harvest rate. Each simulation has recruitment data modified by temperature data that are based upon three concatenated sets of the past 40 years of temperature data, which are further modified by log-normal random numbers.

"Worm" plots showing the results of stochastic simulations were produced to understand something of the range of result to be expected. However, because of the fixed steps in the TAC, these tend to overlie each other and are not shown. As an alternative to worm plots, Table 6 was constructed to show the frequency with which each rule gave a TAC of each possible level. It also shows the mean yield and the standard deviation and standard error from 100 stochastic simulations, taking results from years 41–120 of the simulations (i.e. for the period after the stock had rebuilt).

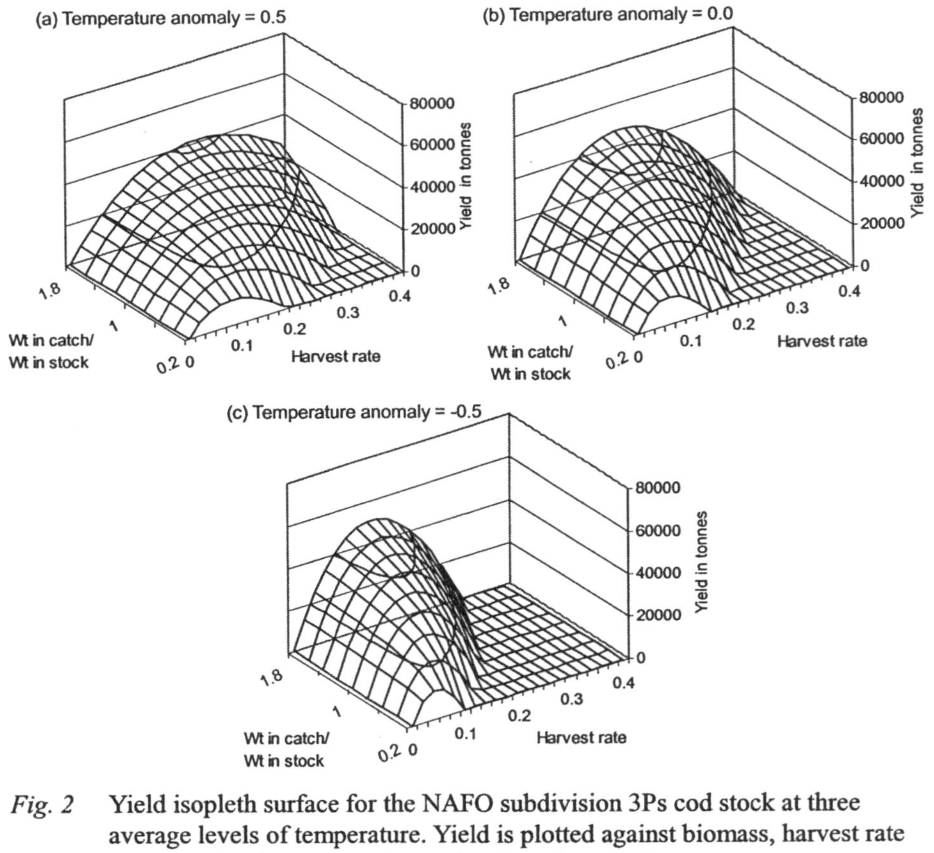

Fig. 2 Yield isopleth surface for the NAFO subdivision 3Ps cod stock at three average levels of temperature. Yield is plotted against biomass, harvest rate and selection (average weight ratio of catch to stock).

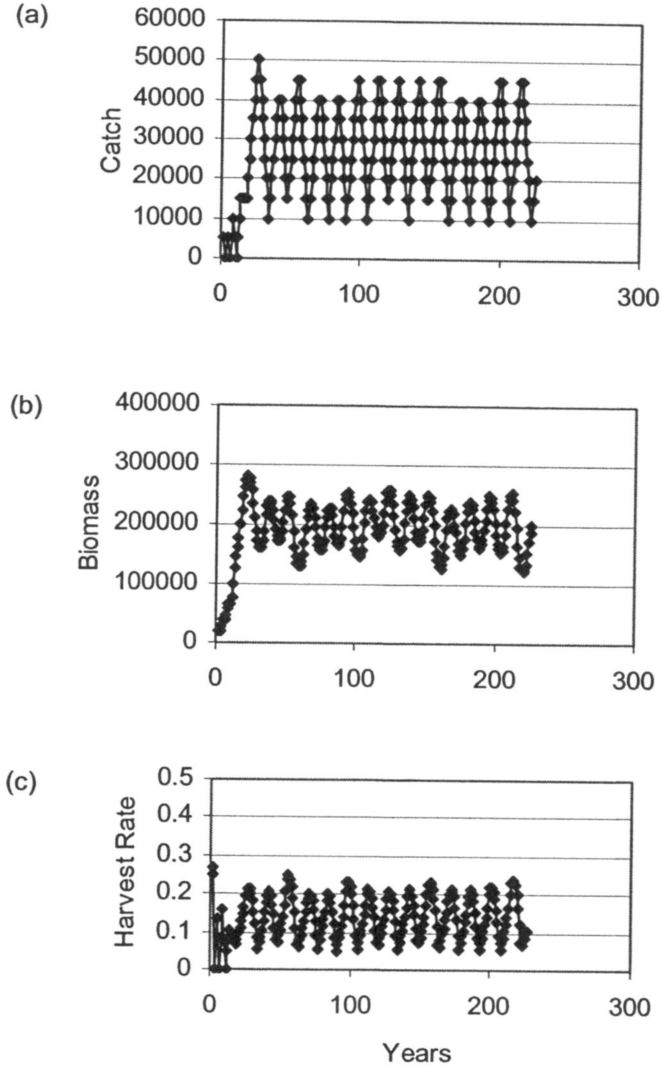

Fig. 3 A single 120-year simulation of (a) catch, (b) biomass and (c) harvest rate of NAFO subdivision 3Ps cod managed according to TAC setting rule 1, which approximates to existing management practices.

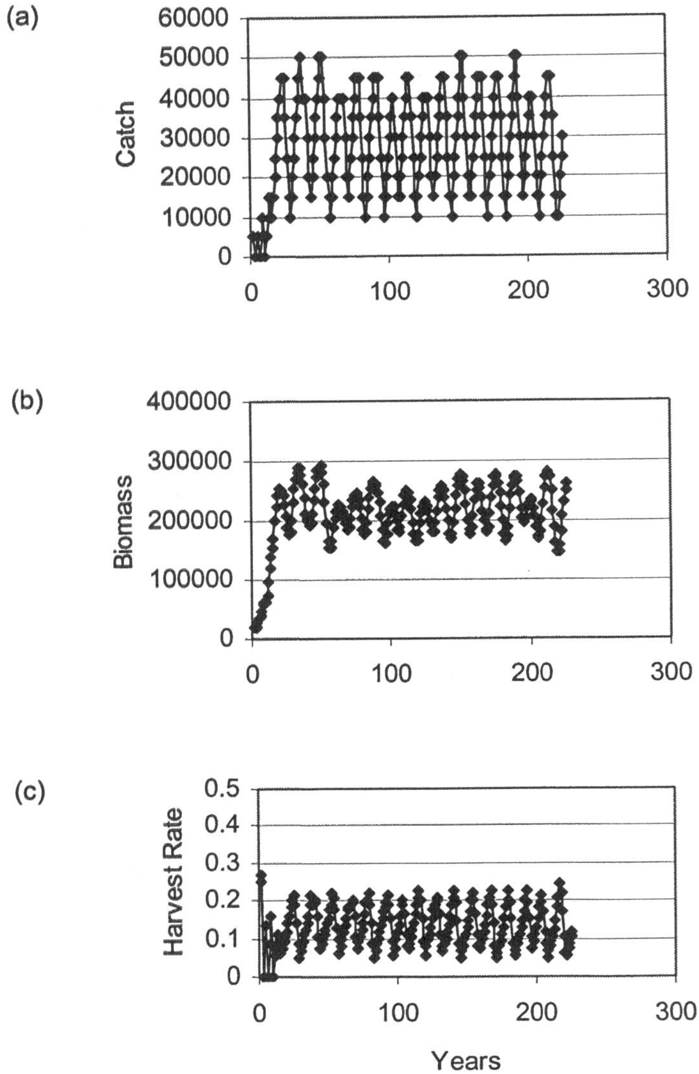

Fig. 4 A single 120-year simulation of (a) catch, (b) biomass and (c) harvest rate of NAFO subdivision 3Ps cod managed according to TAC setting rule 2, which seeks to reduce TAC more rapidly than rule 1 at higher TACs.

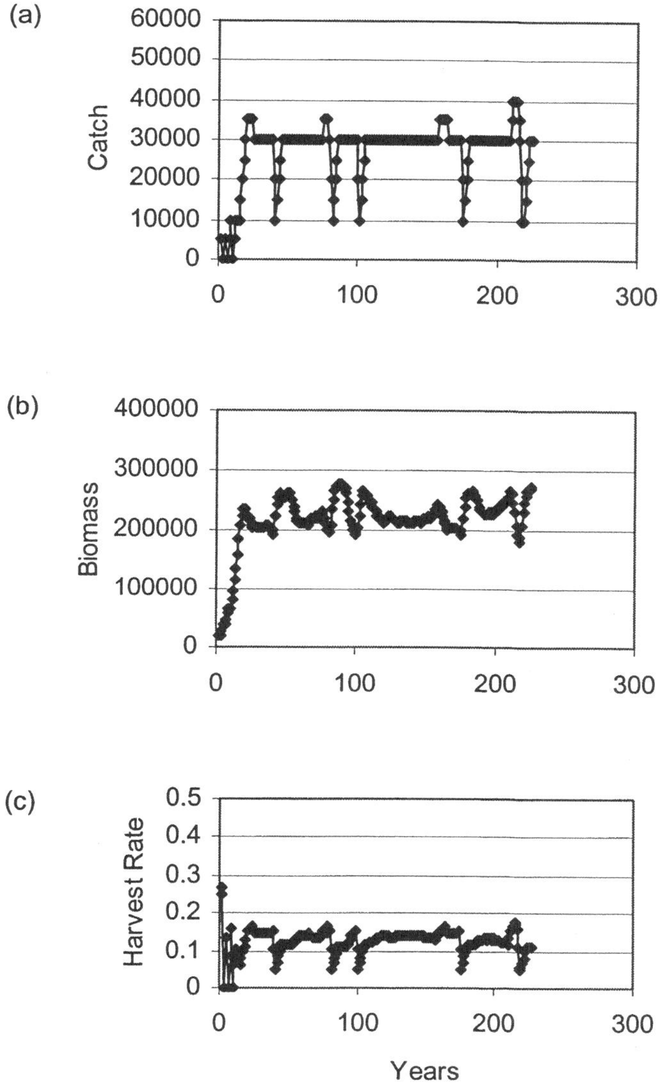

Fig. 5 A single 120-year simulation of (a) catch, (b) biomass and (c) harvest rate of NAFO subdivision 3Ps cod managed according to TAC setting rule 3, which seeks to reduce TACs when productivity is not strong.

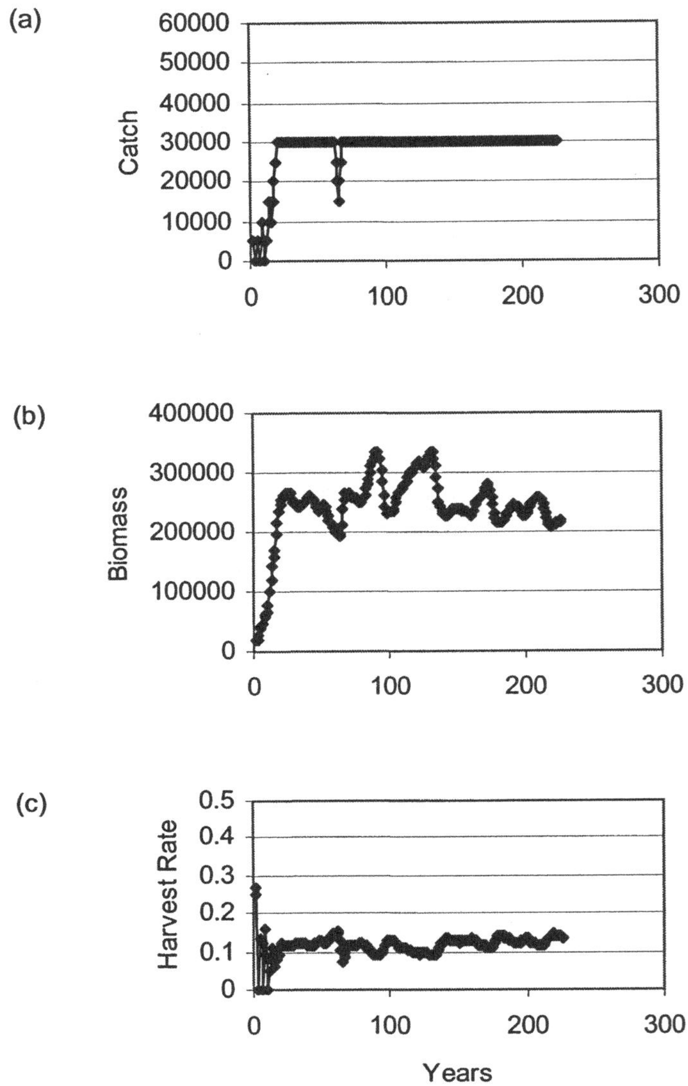

Fig. 6 A single 120-year simulation of (a) catch, (b) biomass and (c) harvest rate of NAFO subdivision 3Ps cod managed according to TAC setting rule 4, which limits the maximum TAC to 30 000 t.

Table 6 Percentage of annual TACs at each level for rules 1–4, together with the average yield and its standard deviation and standard error. These are the results of 100 trials of 80-year simulations.

TAC ('000 t)	Rule 1	Rule 2	Rule 3	Rule 4
0	0.04%	0.00%	0.00%	0.00%
5	0.54%	0.00%	0.06%	0.00%
10	5.53%	5.70%	3.16%	0.00%
15	12.88%	14.46%	2.98%	1.14%
20	12.85%	10.41%	5.91%	2.04%
25	12.35%	11.46%	3.48%	2.44%
30	13.08%	11.25%	56.49%	94.39%
35	13.94%	12.81%	26.46%	0.00%
40	15.83%	14.40%	1.46%	0.00%
45	11.74%	14.81%	0.00%	0.00%
50	1.24%	4.49%	0.00%	0.00%
55	0.01%	0.20%	0.00%	0.00%
60	0.00%	0.00%	0.00%	0.00%
Average yield	29 206	30 326	29 610	29 504
Standard deviation	10 992	11 775	5 869	2 227
Standard error	123	132	66	25

DISCUSSION AND CONCLUSIONS

THEORY

The delay-difference model presented has the virtue of mathematical tractability and mathematical and biological transparency. These are virtues it inherits from the approach of Deriso (1980). It provides analytical solutions of overall long-term yield and provides the basis for simple year-to-year simulations. It has explicit biological parameters (growth, condition factor, initial size, natural mortality, recruitment). A virtue of this explicit biological formulation is that these parameters may be modified by variables subject to climate change (e.g. by temperature) while retaining the mathematical tractability of the formulation. The only apparent constraint on this flexibility is that any such effects must act equally over all sizes in a year, as indeed must the constancy of the parameters themselves. Thus, the model provides a simple yet satisfying way of investigating the first-order effects of climate on yield. The more complex question of how average yield changes in response to annual temperature variations seems likely to be mathematically intense (see Walters & Parma, 1996) and is left for future work.

The formulation has a simple matrix-based structure. This structure facilitates calculations, particularly in a spreadsheet environment. Moreover, it provides a state-vector formulation that could readily be used in a Kalman-filter-like approach to the estimation process. The Kalman filter has already been applied to variants of the Deriso delay-difference model (Kimura *et al.*, 1996), as has a more general Bayesian approach (Meyer & Millar, 1999). Clearly, given the simple matrix-based structures, the model developed here would also fit into such approaches, and these should be further developed.

The length-based approach developed is logically more coherent and practically more flexible than the equivalent weight-based model, but conversely it is less simple. Its growth-at-length equation is asymptotically sensible while the weight-based growth equation may not be (e.g. in the case of 3Ps cod it implies an infinite maximum size). The four population size descriptors B', BA', BL' and N' and the consequent need for four harvest rate terms is an obvious addition in complexity compared with the one or two used in the weight-based approach. Hence, the simplicity of a weight-based approach (e.g that of Kimura *et al.*, 1996) may well have appeal.

In practice, a weight-based model may be adequate for making short- to medium-term simulations at medium to high levels of mortality, where the asymptotic inadequacies of its growth function are not exposed. However, there is a need to question how well it would estimate long-term yield, particularly at low exploitation levels. Some results calculated for 3Ps cod suggest that it may give results considerably higher than the length-based model developed in the Appendix. Certainly the steady state yields of about 70 000 t at average temperatures predicted by a weight-based formulation of long-term yield seem rather high from a stock that barely gave this catch level at the time it was being fished down. It is also worth noting that Kimura *et al.* (1996) found that process error gave upward biases to an analogous weight-based formulation.

The model presented here uses the same growth-in-weight relationship as the well known Beverton & Holt (1957) yield-per-recruit model, and indeed differs from that model only in the way mortality from fishing is formulated. However, the simplification of the catch equation and the formulation of the model as a recurrence relationship allows an explicit solution to be found for long-term yield when recruitment follows a simple stock-recruitment relationship. Therefore, although the parameters of age-based models such as that of Beverton & Holt (1957) might be modified in similar ways, a mathematical solution for overall yield is not possible, and overall yield surfaces have to be calculated numerically (see Sissenwine & Shepherd, 1987). This mathematical tractability, together with the improved description of size selectivity, makes the current approach easier to use than an age-based approach.

The use of different harvest rates for the different population descriptors is a potential strength of the model. The ratio of the average weight of fish in the catch to the average weight of fish in the sea is alone an improvement on the age-of-first capture formulation used by Beverton & Holt (1957), and it has an intuitive appeal. It allows the concept of size selection to be introduced into what effectively are production models. Moreover, the four harvest-rate terms used in the model would allow more subtlety in the depiction of size selection than does the simple ratio of weight in the catch to the weight of fish in the sea. For example, it seems likely that these extra terms could be used to indicate the main differences in action between an asymptotic selection ogive (e.g. from a trawl fishery) and a bell-shaped selection ogive (e.g. from a gillnet fishery). Initial investigation of two such selection curves, acting on a population for which both gave the same Hb and Hn suggests that the gillnet would have higher Ha and Hl than the trawl.

The approximation used in the models, that recruitment be modelled in terms of the biomass of all exploited ages (3+ in the case of 3Ps cod), rather than that of spawning stock, is not entirely satisfactory. In practice for 3Ps cod, this measure has been closely correlated to historic spawning stock biomass (SSB) levels, but this should not be expected to remain the case under all possible exploitation regimes that might be considered. It seems likely that this deficiency could be corrected by including some function of mean weight in the stock-recruitment relationship in the case of the weight-based approach {e.g. SSB ~ B'*f(B'/N'), where f might perhaps be a logistic ogive}. However, attempts to include such factors in the stock-recruitment relationship of 3Ps cod produced little improvement in fit.

Even closer approximations to true spawning biomass should be possible by formulating spawning biomass in terms of linear combinations of numbers, biolength and bioarea, and this possibility deserves further investigation. As the biomass, bioarea and biolength terms when divided by stock numbers provide the first three uncentred moments of the length distribution, it should in principle be possible to use these to estimate approximately how much of the length distribution is of a mature size. In a similar fashion, it should be possible to predict what proportion of the population lies within the selection range of a particular survey or commercial index of abundance. This would be helpful where such indices were obtained from selective gears such as trawls or gillnets that may not adequately sample the smallest or largest sizes of fish.

It is also possible that adequate representations of selection might require the use of some higher moments of the population length distribution. Obvious extensions of Appendix equations A4 and A6 (formed by inserting higher powers of equation A2) could be used to generate these. Indeed, a limited range of higher moments might also be indicated if recruitment is better generated by larger fish. This might indicate a spawning potential measure linked to the 4th rather than the 3rd uncentred moment (see Marshall *et al.*, 1998, and Marteinsdottir & Steinarsson, 1998). This suggests that assessments formulated using the moments of the size distribution rather than the size distribution itself might be an interesting possibility. This is an approach proposed by Fournier & Doonan (1987) that is similar to the model used here.

STOCK-RECRUIT AND OTHER RELATIONSHIPS OF NAFO SUBDIVISION 3PS COD

In the case of 3Ps cod, only the stock-recruitment relationship (Tables 3–5 and Figure 1) shows a reasonably clear relationships with temperature. This relationship is used to provide a rationale for systematic variation in recruitment in short- and long-term simulations of 3Ps cod. Although the current relationship cannot be claimed to have conclusive statistical credentials (see Table 3), it appears consistent through time, and analyses of both first and second time periods indicate the same general form of fit. Moreover, the general form of a lower slope at the origin with cooler temperature and an earlier descending limb at warmer temperature seems not unreasonable.

Growth and condition factor parameters for this cod stock showed little coherent reaction with the temperature series adopted and are therefore estimated at mean levels. One reason for this lack of relationship may be the scope for temperature selection by cod within area 3Ps. As shown by Colbourne & Murphy (2002), bottom temperature in spring can vary widely over a distance of 30 miles, and the temperature at which the highest number of cod occur appears to vary without any relation to the annual mean temperature. Therefore, growth cannot be expected to vary coherently with average temperature series, but needs to be measured with respect to the mean temperature experienced by cod throughout the year. A further problem with respect to growth and condition is the confounding of these factors with survey timing (Brattey *et al.*, 2001).

Lacking a temperature/growth relationship, the 3Ps cod is not a full example of the use of the current model, but was chosen because it is a stock whose response to environmental change is of immediate concern to its management. An alternative to directly fitting the recruitment data to temperature data for a specific stock would be to adopt the results of comparative studies of the effects of temperature on cod growth (e.g. Brander, 1995) and recruitment (e.g. Planque & Fredou, 1999).

YIELD CURVES AND SIMULATIONS

The yield isopleth figures calculated for 3Ps cod show strikingly parabolic relationships with harvest level when sliced along sections corresponding to each level of the average weight ratio of the catch to the population (see Figure 2). The effect of changing temperature (acting through the stock-recruitment relationship alone) is to widen these parabolas at higher temperatures and to narrow and slightly increase their maxima at lower temperatures. Hbcrash, the biomass harvest level that would lead to a collapse of the stock, is thus positively related both to temperature and to the ratio of the average weight in the catch to the average weight in the stock. It follows that a harvest level that appears to be sustainable in a warm period might well turn out to lead to stock collapse in a cold period. This would be a possible explanation of the severe reduction of some cod stocks of the Canadian Atlantic seaboard and the collapse of others.

The fisheries management problem is either to choose a harvest level appropriate to all temperature regimes – inevitably a low level (see Walters & Parma, 1996), or to modulate the harvest level according to the temperature regime. The problem for 3Ps cod is compounded by the fact (if the stock-recruitment curve is to be believed) that cold periods require larger standing stocks than do warm periods. Ideally, this would require the reduction of harvest levels before a cold regime was established. However, this might well be before it could reasonably be predicted. For this reason, the management problem is likely to become one of choosing "a harvest level for all seasons". This might allow population biomass to drift upwards in good periods to anticipate the needs of a less favourable period. It might also allow population biomass to drift downwards in poor periods providing their likely duration can be gauged.

The simulations presented here are designed to inform the development of long-term management plans by clarifying the problems of an idealization of the existing management approach and of indicating some ways in which these might be circumvented to advantage. Long-term management plans for 3Ps cod are under development by an independent body, the Fisheries Resource Conservation Council (FRCC), established by the Canadian Minister of Fisheries to provide him with conservation advice on Atlantic groundfish.

Interim approaches to conservation of 3Ps cod have moved towards an approach of advocating changes of TACs (up or down) by 5 000 t steps, where such modifications appear appropriate in the light of scientific assessments and consultations with the industry and stakeholders. This choice of fixed steps is because 3Ps cod has TACs that are relatively small percentages of the biomass (to European eyes) and because small changes in the TAC for this stock are doubtless within the noise of an inevitably (given the recent moratorium) noisy assessment process. Moreover, small changes in TAC are irritating and incomprehensible to industry and managers. However, there is a counter-argument (D. S. Butterworth pers. comm.) that small regular changes in TAC are to be preferred. The argument is that, where the decision to take a discrete downward step in TAC is a close call, there is a tendency for industry or management to evoke special pleas to put off the painful decision for another year (cf. St Augustine's prayer, "Give me chastity and continence, only not yet"). This danger has to be recognized for management rules using discrete steps in the TAC. It could also suggest that conservative formulations that avoided the need for frequent downward adjustment might be more likely to work in practice than those that were more volatile.

The present management regime can be approximately modelled by a decision rule that moves TACs down by 5 000 t when the harvest rate moves over the level determined as the limit for a particular stock size, and up by 5 000 t when this level is markedly undershot. In practice, the real world approach is a little more conservative than that modelled, because a general principle seems to be emerging that a TAC should not be increased in two successive years.

The simulations presented do differ from reality in several other important respects. First, while all the simulations shown were made against a background of random fluctuations in recruitment about the temperature-mediated mean, they did not include uncertainty in the estimate of stock status or harvest rate. This lack is likely to result in simulations that are more stable than a real management process, because considerable uncertainty clearly exists in the assessments of this "post-moratorium" stock. Second, for reasons of initial simplicity, the simulations were run on the assumption of an equal harvest rate of numbers and biomass. As, in fact, the ratio of average catch weight to average stock weight has historically remained at about 1.4 for 3Ps cod, this aspect of the real stock should give a greater safety margin and hence greater stability than is seen in the simulations. Clearly, simulations should be made over a wide range of such possibilities, but the initial simple set presented gives considerable insight into the problems that the application of such rules might present in the longer term, given the background of decadal climate change.

With the caveats noted above, the simulations shown in Figure 3 are an idealization of the current management approach for 3Ps cod (rule 1). Catch and harvest rate are rather erratic in the early stages of the simulation, but the simulated stock biomass takes off and reaches acceptable levels quite smoothly, though with considerable delay. Subsequently, the biomass varies roughly equally above and below the "recovered" biomass level of 200 000 t. Therefore, while this management rule prevents a collapse, biomass does drop below what is regarded as a recovered level (200 000 t) about 50% of the time. Harvest rates and catches fluctuate in a seesaw fashion between a high of 50 000 t and a low of 10 000 t. This simulation would suggest that this simple rule appears likely to give a bumpy and uncomfortable ride to the fishing industry. It suggests that the management rule does not respond quickly enough to a need to reduce high catches when these cease to be sustainable. Such high catches are unsustainable during the cold, less productive periods of the adopted temperature series.

Figure 4 presents a simulation of a rule (rule 2) designed to reduce large quotas rather more rapidly. This rule adopts a 10 000 t reduction in TAC where this is above 20 000 t. This does result in biomass being more often held above 200 000 t than with the original rule, but catch and harvest rate still seesaw widely.

Some means of damping the fluctuations in catch and harvest rate observed in the first two simulations would seem desirable. Figure 5 shows the results of simulations with a rule (rule 3) designed to reduce fluctuations in catch and harvest rate. The rule is as the first rule but modifies TACs down by 5 000 t when productivity in the preceding two years is less than 5 000 t greater than the average catch in these years. This rule appears to discourage upward fluctuations in catch. Catches only once increased above 35 000 t in this simulation, but during the time-series the catch does need to be reduced sharply from time to time. In general, biomass levels have remained healthy and have been corrected rapidly when they fall below 200 000 t. Harvest rates are mostly stable, but need periodic downward revisions that would seem likely to prove uncomfortable to industry when they occur. Moreover, a worm plot of a number of similar simulations indicated fluctuations to have been wider than those shown in the single simulation.

A further attempt at dampening the fluctuations of catch and harvest rates was attempted by adapting rule 1 by limiting TACs to no higher than 30 000 t. A simulation based upon this rule (rule 4) is seen in Figure 6. In general, the rule results in a constant yield of 30 000 t after recovery is achieved with only one brief drop to 15 000 t. Biomass levels are generally satisfactory and harvest rates remarkably constant. A further simulation (not shown) with an upper TAC limit of 25 000 t resulted in an unchanging catch and higher biomass. From many perspectives the simulation with the 30 000 t upper TAC rule seems to work most satisfactorily. A constant upper limit of TAC seems to allow the stock to slowly build up in times of favourable temperature so that unfavourable times are entered with higher levels of biomass.

The results of 100 of the simulations of each of the 4 rules are shown in Table 6. This shows that, over the 100 runs, rule 4 kept the TAC at 30 000 t for 94% of years and never needed to drop the TAC below 15 000 t in unproductive periods. By contrast, rule 1 spent 6% of the time at TACs of 10 000 t or lower and occasionally the fishery had to be stopped. Overall there was little real difference between the average yield provided by the different rules, the main differences being observed in the standard deviations and the TAC range. On this basis, Rule 4 undoubtedly provides the least variable catch.

In addition to reducing fluctuations, it might be argued that a constant catch would lead to better product value than the fluctuating catch levels seen in the first three simulations. A constant catch would allow development of the forward selling of product, which should help to achieve good prices. The possibility of an upper limit on TAC, a maximum allowable catch (MAC), should therefore be carefully explored with the 3Ps cod fishing industry. More pressing of course is the need to ensure the rapid and safe recovery of the 3Ps cod stock to a 200 000 t biomass level. Currently it is only at about half that level. Currently, maintenance of low exploitation rates on all its components seems the appropriate management approach (FRCC, 2002).

GENERAL

Clearly, climatic factors can have major effects on the yield of fisheries and on their rational management. As demonstrated by the example of the 3Ps cod stock, modest changes in recruitment parameters seem likely to lead to substantial changes in the harvest level that will collapse the stock (Hbcrash). The same might be expected of changes in growth rates. Such changes seem very likely to result from changes in ocean climate. Consequently, consideration of optimum harvest levels and the design of effective harvest rules need to take these possibilities of periodic or permanent regime shift into account. The approaches suggested in this paper provide a simple methodological basis for doing this.

ACKNOWLEDGEMENTS

I am particularly grateful to Eugene Colborne and George Lilly of the Canadian Department of Fisheries and Oceans for providing me with unpublished data. I also wish to thank Prof. D. S. Butterworth and a semi-anonymous referee for their helpful suggestions. The centenary (despite ephemeral name changes) of the Lowestoft "Fish Laboratory" seems an appropriate time to gratefully acknowledge its formative influence and the help I received from friends and colleagues during my time there.

REFERENCES

Beverton, R. J. H. & Holt, S. J. (1957) On the dynamics of exploited fish populations. *Fisheries Investigation Series 2*, **19**. Her Majesty's Stationery Office, London.

Brander, K. M. (1995) The effect of temperature on growth of Atlantic cod (*Gadus morhua* L.). *ICES Journal of Marine Science*, **52**: 1–10.

Brattey, J., Cadigan, N. G., Healey, B. P., Lilly, G. R., Murphy, E. F., Shelton, P. A., Stansbury, D. E., Morgan, M. J. & Mahé, J-C. (2001) An assessment of the cod stock in NAFO Subdivision 3Ps in October 2001. *DFO Canadian Stock Assessment Secretariat Research Document*, **2001/36**. http://www.dfo-mpo.gc.ca/csas/Csas/English/Research_Years/2001/2001_099e.htm

Brattey, J., Cadigan, N. G., Lilly, G. R., Murphy, E. F., Shelton, P. A. & Stansbury, D. E. (1999) An assessment of the cod stock in NAFO Subdivision 3Ps. *DFO Canadian Stock Assessment Secretariat Research Document*, **1999/36**. http://www.dfo-mpo.gc.ca/csas/Csas/English/Research_Years/1999/R_D99_e.htm

Cochrane, K. L., Butterworth, D. S., De Oliveira, J. A. A. & Roel, B. A. (1998) Management procedures in a fishery based on highly variable stocks and with conflicting objectives: experiences in the South African pelagic fishery. *Reviews in Fish Biology and Fisheries*, **8**: 177–214.

Colbourne, E. (2000) Oceanographic conditions in NAFO Subdivisions 3Pn and 3Ps during 1999 with comparisons to the long-term (1961–1990) average. *DFO Canadian Stock Assessment Secretariat Research Document*, **2000/049**.

Colbourne, E. & Murphy, E. F (2002) Recent trends in bottom temperatures and distribution and abundance of cod (*Gadus morhua*) in NAFO Subdivisions 3Pn and 3Ps from the winter/spring multi-species surveys. *DFO Canadian Stock Assessment Secretariat Research Document*, **2002/001**. http://www.dfo-mpo.gc.ca/csas/Csas/English/Research_Years/2002/2002_001e.htm

Colbourne, E., Narayanan, S. & Prinsenberg, S. (1994) Climate changes and environmental conditions in the North Western Atlantic, 1970–1993. *ICES Marine Science Symposia*, **198**: 311–322.

Cushing, D. H. (1982) *Climate and Fisheries*. Academic Press, London.

Deriso, R. B. (1980) Harvesting strategies and parameter estimation for an age-structured model. *Canadian Journal of Fisheries and Aquatic Sciences*, **37**: 268–282.

Dickson, R. R., Briffa, K. R. & Osborn, T. J. (1994) Cod and climate: the spatial and temporal context. *ICES Marine Science Symposia*, **198**: 280–286.

Fournier, D. A. & Doonan, I. J. (1987) A length-based stock assessment method utilising a generalized delay–difference model. *Canadian Journal of Fisheries and Aquatic Sciences*, **39**: 1195–1207.

FRCC. (2002) 2002/2003 Conservation Requirements for Groundfish Stocks on the Scotian Shelf and in the Bay of Fundy (4VWX), in Sub-Areas 0, 2+3 and Redfish Stocks. Report to the Minister of Fisheries and Oceans. FRCC.2002.R.1. Ottawa.

ICES. (1994) Cod and climate change. (Ed. by J. Jakobsson). *ICES Marine Science Symposia*, **198**. 693 pp.

IPCC (Intergovernmental Panel on Climate Change). (1995) *Climate Change 1995: the Science of Global Change*. (Ed. by J. T. Houghton, L. G. Meira Filho, B. A. Callander, N. Harris, A. Kattenberg and K. Maskell). Cambridge University Press. Cambridge, UK. 572 pp.

Kirkwood, G. P. (1992) Background to the development of the Revised Management Procedure. *Report of the International Whaling Commission*, **42**: 236–251.

Kimura, D. K., Balsinger, J. W. & Ito, D. H. (1996) Kalman filtering the delay-difference equation: practical approaches and simulations. *Fishery Bulletin, U.S.*, **94**: 678–691.

Marshall, C. T., Kjesbu, O. S., Yaragina, N. A., Solemdal, P. & Ulltang, Ø. (1998) Is spawner biomass a sensitive measure of the reproductive and recruitment potential of Northeast Arctic cod? *Canadian Journal of Fisheries and Aquatic Sciences*, **55**: 1766–1783.

Marteinsdottir, G. & Steinarsson, A. (1998) Maternal influence on the size and viability of Iceland cod *Gadus morhua* eggs and larvae. *Journal of Fish Biology*, **52**: 1241–1258.

Meyer, R. & Millar, R. B. (1999) Bayesian stock assessment using a state-space implementation of the delay-difference model. *Canadian Journal of Fisheries and Aquatic Sciences*, **56**: 37–52.

Planque, B. & Fredou, T. (1999) Temperature and the recruitment of Atlantic cod (*Gadus morhua*). *Canadian Journal of Fisheries and Aquatic Sciences*, **56**: 2069–2077.

Pope, J. G. (1972) An investigation of the accuracy of virtual population analysis using cohort analysis. *Research Bulletin International Commission for the Northwest Atlantic Fisheries*, **9**: 65–74.

Sissenwine, M. P. & Shepherd, J. G. (1987) An alternative perspective on recruitment overfishing and biological reference points *Canadian Journal of Fisheries and Aquatic Sciences*, **44**: 913–918.

Walters, C. & Parma, A.M., 1996. Fixed exploitation rate strategies for coping with effects of climate change. *Canadian Journal of Fisheries and Aquatic Sciences*, **53**: 148–158.

APPENDIX

Development of a length-based delay-difference model

Weight-based formulation of delay-difference models, e.g. Deriso (1980), Kimura *et al.* (1996), have the virtue of simplicity but the disadvantage of being derived from the assumption that a linear recurrence relationship can be written in the form

$$W(a+1,y+1) = \rho W(a,y) + \varsigma \qquad (A1)$$

where W(a,y) is the average weight at age a in year y and ρ + ς are constants that may be linked to the usual parameters of the von Bertalanffy growth function applied to growth in weight.

In general, Equation A1 is an approximation. In practice, the W(a+1) on W(a) relationship might be expected to be somewhat curvilinear, even if curvature may not be apparent over the ages currently seen in a heavily exploited fishery. Moreover, for the 3Ps stock studied in this paper, equation A1 has a slope estimate of more than 1.0 and hence sets no limit to the maximum size of fish. Therefore, while equations derived from equation A1 may well be adequate for short-term predictions at moderate to heavy harvest levels, they might break down for long-term steady state predictions. This is particularly likely for long-term yield predictions at very low harvest rates, because equation A1 might then predict very large fish in the stock.

These problems of the weight-based model may be avoided by adopting a length-based approach. The Ford–Walford equation for the von Bertalanffy growth in terms of length gives

$$L(a+1,y+1) = L(a,y)*\exp(-K) + L_\infty*\{1-\exp(-K)\} \qquad (A2)$$

When growth is isometric, then

$$W(a,y) = cf*L(a,y)^3 \qquad (A3)$$

where cf is the condition factor.

By considering the cube and square of equation A2, a matrix equation can be developed that provides a recurrence relationship for $L(a,y)^3$, but it has the apparent expense of needing to carry nuisance terms in $L(a,y)^2$ and L(a,y).

$$\begin{vmatrix} L(a+1,y+1)^3 \\ L_\infty*L(a+1,y+1)^2 \\ L_\infty^2*L(a+1,y+1) \\ L_\infty^3 \end{vmatrix} =$$

$$\begin{vmatrix} \exp(-3K) & 3\exp(-2K)*(1-\exp(-K)) & 3*\exp(-K)*(1-\exp(-K))^2 & (1-\exp(-K))^3 \\ 0 & \exp(-2K) & 2\exp(-K)*(1-\exp(-K)) & (1-\exp(-K))^2 \\ 0 & 0 & \exp(-K) & (1-\exp(-K)) \\ 0 & 0 & 0 & 1 \end{vmatrix}$$

$$* \begin{vmatrix} L(a,y)^3 \\ L_\infty*L(a,y)^2 \\ L_\infty^2*L(a,y) \\ L_\infty^3 \end{vmatrix} \qquad (A4)$$

Population numbers are described by the cohort analysis approximation of Pope (1972)

$$N(a+1,y+1) = N(a,y)*\exp(-M) - C(a,y)*\exp(--M/2) \qquad \text{(A5)}$$

C(a,y) is catch in Numbers at age a in year y and M is the natural mortality rate that for present purposes is assumed constant over all ages.

Writing the 4*4 growth matrix in equation A4 as G, then scalar multiplying the left and right hand sides of equation A4 by the equivalent right and left hand sides side of equation A5 and summing over all ages c and older gives

$$\begin{vmatrix} B'(y+1)/cf \\ L_\infty{*}BA'(y+1) \\ L_\infty^{2}{*}BL'(y+1) \\ L_\infty^{3}N'(y+1) \end{vmatrix} =$$

$$G \quad * \quad \begin{vmatrix} \{B'(y)\exp(-M)-CW'(y)*\exp(-M/2)\}/cf \\ L_\infty*\{BA'(y)\exp(-M)-CA'(y)*\exp(-M/2)\} \\ L_\infty^{2}*\{BL'(y)\exp(-M)-CL'(y)*\exp(-M/2)\} \\ L_\infty^{3}*\{N(y)'\exp(-M) - C'(y)*\exp(-M/2)\} \end{vmatrix}$$

$$+ \quad N(c,y+1)* \begin{vmatrix} L(c,y+1)^{3} \\ L_\infty{*}L(c,y+1)^{2} \\ L_\infty^{2}{*}L(c,y+1) \\ L_\infty^{3} \end{vmatrix} \qquad \text{(A6)}$$

In this formulation, c is the age of first capture in the fishery; B'(y), CW(y), N'(y) and C'(y) are the sums of the biomass, catch weight, numbers and catch numbers, and BA' and BL' are respectively the sum of $L(a,y)^2*N(a,y)$ and of $L(a,y)*N(a,y)$ for fish aged c or older. By analogy with biomass, these latter are named the bioarea and the biolength of the stock. Similarly CA and CL are the summations of the products of catch numbers with $L(a,y)^2$ and L(a,y) for all ages c or older. As c is the age of first capture, N(c, y+1) in this formulation may be replaced by the recruitment estimate of the y+1–c year class R(y+1–c) with a suitable correction for natural mortality (Rc(y+1)).

For fixed harvest levels of biomass, bioarea, biolength and numbers (Hb, Ha, Hl, Hn), the steady state column vector of biomass/(cf*Rc), bioarea*L_∞/Rc, biolength*$(L_\infty)^2$/Rc and numbers*$(L_\infty)^3$/Rc designated (Bio/Rc) may be obtained from equation A6 by linear algebra as

$$Bio/Rc = [\text{Inverse}\{I - G*\text{Survivors}\}]*Lc \qquad \text{(A7)}$$

where Survivors is a 4*4 diagonal matrix with diagonal elements of exp(–M) –exp(–M/2)*Hx, where the Hx are Hb, Ha, Hl and Hn respectively and where Lc is the column vector of $L(c)^3$, $L(c)^2(L_\infty)$, $L(c)(L_\infty)^2$,and $(L_\infty)^3$.

Provided B' may be used as a proxy for spawning stock biomass and Rc can be expressed as a function of B', then it may be possible to provide an analytical solution for overall yield. For example, a stock-recruitment relationship of the Ricker functional form

$$Rc = \alpha B'^{*}\exp(-\beta B') \qquad (A8)$$

may be written in biomass-per-recruit terms in the form

$$Rc = \alpha Rc^{*}B'/Rc^{*}\exp(-\beta Rc^{*}B'/Rc) \qquad (A9)$$

Hence, it follows that

$$Rc = \ln\{\alpha B'/Rc\}/(\beta B'/Rc) \qquad (A10)$$

Steady-state biomass and numbers may then be calculated from the scalar product of equation A7 by equation A10. Steady-state yield and catch numbers may then be calculated as

$$CW' = B'^{*}Hb \qquad (A11)$$

$$C' = N'^{*}Hn \qquad (A12)$$

Other formulations of the stock-recruitment relationship [e.g. as a Beverton & Holt formulation $\{Rc = 1/(\alpha/B'+\beta)\}$ or as a power curve formulation $\{Rc = \alpha B'^{\beta}\}$] also allow analytical solutions for B', etc. All of these solutions are also simply formed for a given temperature level when any or all of the parameters K, L_{∞}, M, L(c), α and β of the yield function are functions of aspects of the marine environment. Therefore, the initial objective of developing an overall yield model that can be modified by environmental effects is achieved.

An additional advantage of this approach is that it does not require the assumption of constant fishing mortality rate at age. Note that from equations A11 and A12

$$\{CW'/C'\}/\{B'/N'\} = Hb/Hn \qquad (A13)$$

and therefore

$$Hb/Hn = \text{average catch weight/average population weight} \qquad (A14)$$

which is a charmingly simple description of size selection. Note also that, whereas BA' and BL' appear to be nuisance variables, their inclusion together with Ha and Hl will allow further subtlety in the description of selection curves.

Measuring fish behaviour: the relevance to the managed exploitation of shared stocks

Julian D. Metcalfe and Mike G. Pawson

The Centre for Environment, Fisheries and Aquaculture Science, Lowestoft Laboratory, Pakefield Road, Lowestoft, Suffolk, NR33 0HT, UK

ABSTRACT: From a management perspective, a fish "stock" can be defined as that part of a species' population within which the effects of exploitation on population structure are recognizable. It is a relatively straightforward task to manage the exploitation of fish that remain within the same fishing area throughout their lives, and it is also an advantage if this stock unit has a high degree of biological integrity. However, this is rarely the case. Most fish move between different environments through their lives, and many commercial species exhibit extensive seasonal migrations. Consequently, the identification of individual stocks, as well as knowledge of their spatial distribution and dynamics, are important for any rational management of their fisheries. This applies particularly in situations where a large element of management policy concerns shared fish stocks. Genetic and conventional tagging studies can provide a limited understanding of stock structure and population movements, but they do not provide the resolution needed to understand and quantify the spatial and temporal scales of mixing and movements within populations that is needed to model the interactions and dynamics in relation to providing management advice. Since the late 1960s, CEFAS has developed an approach to overcoming this limitation, using electronic tags to monitor the movements and behaviour of individual free-ranging fish in the open sea for periods from a few days to longer than a year. Such work is advancing our understanding of the stock structure and dynamics of commercially exploited fish species. We illustrate this by reference to plaice stocks around the UK and the influence that stock mixing has on the assessment of stock status. The challenge now is to ensure that this knowledge is built into future assessment and management methodologies.

INTRODUCTION

The objectives of fisheries management vary with one's perspective. At one extreme, a fisher may aim to maximize his financial return for an investment of capital and time, and his goal will be to continue to obtain the highest price for his catches, and for these to be sustained at least in the medium term. To him, the political decisions about who catches what and how much are probably more important than biological sustainability. An environmentalist, on the other hand, may seek to achieve a high level of biodiversity in the aquatic ecosystem, and could be lacking in tolerance of those who seek to exploit its resources. Fisheries biologists lie somewhere in between, recognizing that, whereas biological systems are complex and dynamic, they can support a harvest, provided that human harvesting is well managed. In a wild marine capture fishery, it is only fishing activity that can be managed directly, seldom the production of the resource itself.

The type and quantity of fishing activity influences the dynamics of fish populations, which we have come to regard as composed of "stocks". The biological concept of stocks relies on the premise that fish populations consist of intraspecific groups that have sufficient spatial and temporal integrity to warrant consideration as self-perpetuating units. In the North Atlantic, populations of the more important commercial fish species are distributed across management areas that have been delineated historically (e.g. by international organizations such as ICCAT, NAFO, NEAFC and ICES), but assessment of their abundance and response to exploitation is restricted within these areas. Therefore, "management" stocks are defined principally on a species/area basis.

From a management perspective, a stock may be defined as that part of a population within which the effects of exploitation on population structure are recognizable. If possible, these management (and assessment) stocks should have a high degree of biological integrity. The biological identification of stocks, and a knowledge of their spatial distribution and dynamics, are therefore important aspects of any fisheries management strategy (for a review of stock identity approaches, see Pawson & Jennings, 1996).

If fish stocks remained stationary through time, managing their exploitation would be relatively simple. However, given their life history dynamics and the need to maximize survival from egg to adult (and thus their reproductive success), fish move between those habitats that most suit their various needs. For example, fish may move geographically from nursery areas to adult feeding areas as "recruits", and seasonally as adults between feeding and spawning areas. In this context, the preferred habitat is based on an average state, such that the choice of a particular time and place for spawning may be optimized in relation to spawning success over many years. However, a spawning time and location that suits individuals that are likely to spawn 7–10 times (i.e. when mortality attributable to fishing is low) may be less suitable if mortality increases and stocks become so depleted that most fish only have one or two opportunities to spawn. Furthermore, environmental conditions do not remain constant, and the distributions of fish populations may change over a few generations in response to climate change.

An understanding of the interactions between the environment, fish behaviour and physiology provides us with the ability to predict the likely outcome of such scenarios. We also need to distinguish between the effects of the environment on the population dynamics of a species and those of fishing, if we are to manage the latter effectively. Whereas fishing surveys and observations made by fishers provide some information about which species are where and when, such information only allows broad inferences to be drawn about movements and migrations; it is not sufficient to provide the necessary understanding of the behavioural ecology of fish populations.

There are numerous examples worldwide where a better understanding of these aspects of commercially exploited fish stocks would improve our ability to assess their abundance and structure, and to better manage their exploitation. However, it is not the purpose of this paper to review a range of such cases in detail. Instead, we use the specific example of plaice (*Pleuronectes platessa*) on the European continental shelf to illustrate the potential for using the detailed and quantitative information on movements and behaviour that telemetry can provide to inform management decisions through appropriate spatially and temporally structured models.

THE RELEVANCE OF MIGRATION AND STOCK STRUCTURE TO ASSESSMENT AND MANAGEMENT

Whereas fishers use their knowledge of fish migration to help them exploit fish populations more effectively, fisheries managers need to take account of the effects of fish migration to ensure that exploitation is sustainable. Knowledge of migratory processes is critical to understanding stock structure, and it is therefore essential to a good assessment of stock size (Rijnsdorp & Pastoors, 1995). We need to know, for instance, whether the capture of a particular species in one assessment area at a certain time of the year and in another area at another time, is the result of one stock migrating between the two areas or two separate stocks migrating into each area at different times. Understanding how patterns of movement correlate with spatial and temporal changes in the environment improves understanding of the ecological basis of migratory strategies, i.e. why fish migrate. Such knowledge provides the basis for developing predictive models of how changes in environmental factors, such as food availability or climate, may affect migratory movements and how, in turn, these are likely to affect the structure and distribution of fish stocks.

Where stock assessments are based on data from fishing or acoustic surveys, we cannot just assume that the fish population remains within the survey area, and we know that migration can be a major source of error in acoustic surveys (MacLennan & Simmonds, 1991). An understanding of the spatial and temporal pattern of migrations enables surveys to be targeted appropriately, and the effects of migration to be taken into account in models used to estimate stock abundance. Similarly, migration is a critical element in the effectiveness of management measures that include protected areas (i.e. areas having gear restrictions, being seasonally closed to fishing, or genuinely "no-take"; Horwood *et al.*, 1998; Horwood, 2000), a feature of many developing stock recovery plans. It is essential to know the average duration of a fish's stay within a closed area, as well as its vulnerability to fishing once it moves outside, in order to evaluate the resultant changes in fishing mortality (and yield to the fishery). Clearly, understanding the spatial and temporal patterns of fish movement (and fishing effort) is a prerequisite to constructing biologically realistic models that can be used to assess or predict the potential effects of different management strategies.

STOCK STRUCTURE

Most marine temperate fish species have wide geographical distributions within which they may exist as a number of essentially distinct populations. In some cases, subpopulations may have sufficient integrity in the medium to long term that they respond independently to the effects of exploitation, and such stocks can be managed separately. If there is little or no transfer of fish between these populations, they may also be genetically distinct. However, the continual transfer of only a few fish can lead to loss of genetic heterogeneity, even though exploitation of one population may have no effect on the other. Genetic analysis can therefore be of value when it indicates that a population consists of more than one stock, but difficulties arise when it fails to provide evidence of stock separation.

For example, early genetic studies on Atlantic cod (*Gadus morhua*) by Jamieson & Thompson (1972) suggested that the population in the North Sea was a single stock, and traditionally it has been managed as a single unit. However, microsatellite DNA analysis has recently shown that there may be four genetically distinct groups in the North Sea (off Bergen, in the Moray Firth, off Flamborough Head, and in the Southern Bight; Hutchinson *et al.*, 2001). The lack of obvious physical barriers to mixing between these different groups suggest that some behavioural and/or environmental factors could play a role in maintaining their genetic integrity, and that these should be taken into account in any re-evaluation of the current assessment and management of North Sea cod. Clearly, an understanding of stock structure is fundamental to predictions of the effects on the species' population of changes in the levels and/or patterns of exploitation.

METHODS FOR UNDERSTANDING MOVEMENTS AND BEHAVIOUR

CONVENTIONAL TAGGING

Simple methods of marking (conventional tagging) have been used since the mid-17th century as a means of increasing our understanding of fish biology. Tagging tells us where individual fish are twice in their life (i.e. when caught and tagged, and when recaptured). If sufficient fish are recaptured between a few months and two years following release, analysis of these data can provide information on stock identity, movements and migration (both rates and routes) that can be used to improve assessments and inform management decisions (see reviews by Seber, 1982; Pawson & Jennings, 1996; Begg & Waldman, 1999; Quinn & Deriso, 1999). Plaice around the UK coast provide a useful example for which we have substantial datasets gathered over many years (Wimpenny, 1953; de Veen, 1978; Rijnsdorp & Pastoors, 1995; Bolle *et al.*, 2001; Dunn & Pawson, 2002).

UK west coast plaice

Biological studies have revealed similarities between the growth and age at maturation of plaice in the north-eastern and south-eastern Irish Sea and the Bristol Channel, and differences in the same parameters between plaice populations in these three regions and those in the western Irish Sea (Horwood, 1990, 1993; Nash *et al.*, 2001). While these differences indicate a lack of mixing, similarities of growth and maturation cannot be taken as proof of stock unity, because they may simply be a characteristic response to similar environmental conditions in each area.

Dunn & Pawson (2002) used data derived from CEFAS tagging studies of plaice in the Irish Sea and Bristol Channel (ICES Divisions VIIa and VIIf/VIIg respectively) to identify stock structure and evaluate mixing within and between each fishery area, taking account of the pattern and intensity of fishing likely to catch plaice as well as any dispersal or migratory behaviour specific to sex and size (and inferred maturity). They found evidence of distinct stocks in the north-eastern Irish Sea and the western Irish Sea, and of a population that includes a migratory contingent moving between the VIIa and VIIf,g "stocks" (Figure 1). There is also a dispersing adolescent phase (largely pre-fishery recruits) that results in low levels of mixing between all these stock components. From this evidence, Dunn & Pawson (2002) developed a hypothesis about plaice stock structure and movements with which to evaluate the suitability of the ICES stock assessment areas for plaice on the west coast of England and Wales.

From a fisheries management perspective, the most important feature of this work is the movement of plaice between the two management areas. A proportion of the larger,

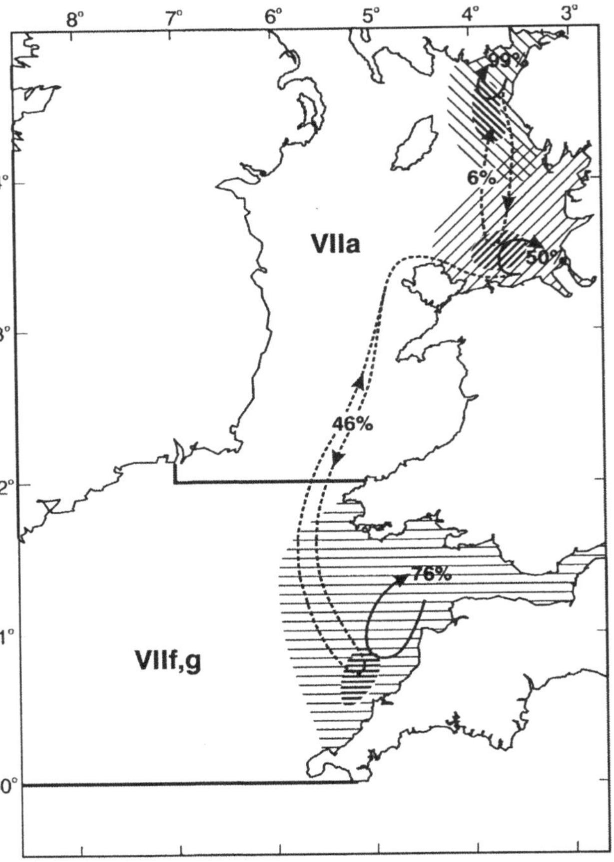

Fig. 1 A schematic diagram showing principal "stock" areas and movements of plaice tagged by CEFAS on the west coast of England and Wales, in relation to main feeding areas (light shading), spawning areas (dark shading) and ICES assessment/management areas (after Dunn and Pawson, 2002). The solid arrows and percentage values represent the proportion of recruits that remain in the three stock areas, and the broken arrows and percentage values represent the return migrations between these stocks.

maturing or mature plaice that were tagged and released in the south-eastern Irish Sea (VIIa) during summer and autumn were recaptured during the spawning season in the Bristol Channel (VIIf,g). Some of these plaice appeared to recruit to this area permanently, but the majority returned in the intervening periods to feed in the Irish Sea. It is estimated that around one-quarter to one-half of the adult plaice in the south-eastern Irish Sea may belong to the migratory contingent. A similar phenomenon has been observed for bass tagged in the same area (Pawson *et al.*, 1987).

Irrespective of the origin of the adult fish moving between spawning areas in VIIf,g and summer areas in VIIa, this "stock" mixing complicates assessments based on fisheries and survey catch data (Dunn & Pawson, 2002). In essence, it leads to an underestimate of total stock biomass and an overestimate of the size of the spawning stock in VIIa; the bias is reversed in VIIf,g. The magnitude of such errors depends upon the timing and the scale of the respective fisheries, and the weight that catch data from such fisheries achieve in the VPA assessment. This is especially relevant when VPA estimates of spawning stock biomass are compared with those obtained from egg production methods (Armstrong *et al.*, 2001). It also implies that the inferred relationship between spawning stock biomass and recruitment in each of these two assessment areas is unlikely to be biologically sound, and this is a major concern for medium-term predictions and the setting of precautionary biomass reference points for the stocks (Frank & Brickman, 2000).

North Sea plaice

Along with its European colleagues, the Lowestoft Laboratory has been conducting research on the movements and migrations of plaice in the North Sea since its inception in 1902. The general pattern of plaice migration has been established from trawl surveys (Wimpenny, 1953) and conventional tagging experiments in the Southern Bight of the North Sea and the English Channel (see review by Harden Jones 1968; de Veen, 1978; Rijnsdorp & Pastoors, 1995; Bolle *et al.*, 2001). Houghton & Harding (1976) showed that 20–30% of the plaice catch from the eastern Channel in winter contained migratory North Sea fish from a group that entered the Channel in autumn and left rapidly after spawning. Plaice tagged in the Channel during summer were not recaptured outside the Channel, but appeared to be members of two groups that returned to specific (east or west Channel) spawning areas each winter. An analysis combining estimates of spawning stock biomass with fishing intensity at the recapture sites suggested that, of the plaice spawning in the Channel during January and February, 20% spent the summer in the western Channel (VIIe), 24% in the eastern Channel (VIId), and approximately 56% migrated to the North Sea (IVb,c). Few plaice tagged in the southern North Sea during January and February were recaptured in the Channel (De Clerk, 1977).

ELECTRONIC TAGS, TELEMETRY AND FISH BEHAVIOUR

Clearly, analyses of conventional tagging data can provide a semi-quantitative description of population movements. However, the results are inevitably confounded by the integration of both fish behaviour and fishing activity (Rijnsdorp & Pastoors, 1995; Bolle *et al.*, 2002). Though the analysis can accommodate spatial variations in fishing effort, this is often not known, and movements of fish into unfished areas, or changes in fish behaviour which alter their catchability, cannot easily be accounted for. Neither can conventional tagging tell us much about how fish migrate. Therefore, to obtain a better quantitative assessment of these factors, we need a more detailed understanding of fish movements in both space and time.

Since the late 1960s, electronic tags that transmit acoustic signals have been used to track the movements of individual free-ranging fish for limited periods (Arnold & Dewar, 2001). Such work has yielded substantial advances in our understanding of how some species of fish migrate. We have shown, for example, that plaice in the Southern Bight of the North Sea use selective tidal stream transport during their pre- an post-spawning migration (Greer Walker *et al.*, 1978, Metcalfe *et al.*, 1992). Fish exhibiting such behaviour leave the seabed and move into midwater at about the time of slack water, and swim down tide for the major part of the ensuing northgoing or southgoing tide. As the tide turns again, the fish return to the seabed where they remain for the duration of the opposing tide. This behaviour allows fish to move rapidly (25 km per day is not exceptional) between feeding and spawning grounds, while considerably reducing the energy cost of migration in comparison with swimming continuously over the same distance (Metcalfe *et al.*, 1990).

Knowing that plaice in the Southern Bight use the tidal streams for migration has allowed the development of computer simulation models that predict rates and scales of geographical movement by combining patterns of behaviour (vertical movements) with tidal stream vectors (Arnold & Holford, 1995). However, acoustically tagged fish can only be tracked for a few days, and early versions of such models (Arnold & Cook, 1984) were based on relatively small quantities of behavioural data. Consequently, only simple assumptions could be made about how behaviour might change in time and space.

More recently, advances in micro-electonic technology have led to the development of electronic "data storage" or "archival" tags that are small enough to be carried by fish. These devices record and store environmental data such as pressure (depth), temperature (internal and external) and ambient daylight that can be used to derive detailed information about fish behaviour and movements at much finer temporal and spatial scales than is possible with conventional tagging. For example, over the European continental shelf, tidal information (times of high and low water, and tidal range) derived from pressure measurements can be used to geolocate fish whenever they remain stationary on the seabed for a full tidal cycle (Metcalfe and Arnold, 1997; Hunter *et al.*, 2002). In the open sea, records of ambient daylight can be used to derive latitude (from day length) and longitude (from the time of local noon; Hill, 1994; Metcalfe, 2001). Further, depth information can be used to derive behavioural information such as vertical movements and patterns of activity. Because there is no need for human observers to follow the fish (provided it is recaptured), it is now possible to monitor many fish simultaneously over entire migrations (Metcalfe & Arnold, 1997).

Since 1993, CEFAS has released 786 plaice bearing electronic data storage tags (DST) into the North Sea. About 200 of these tags have been returned through the commercial fishery, yielding 24 000 days of data for geolocation (from tidal data), behaviour and environmental temperature. Recent detailed analysis of the data from these tags has revealed that the adult plaice population in the North Sea forms three geographically discreet feeding aggregations during summer, which then disperse over the southern North Sea and eastern English Channel to spawn in winter (Hunter *et al.*, in press). This is different from our previous understanding of the North Sea plaice population based on conventional tagging data (de Veen, 1978), that suggested that there were isolated plaice subpopulations that aggregated during the winter spawning then dispersed during summer over distinct but overlapping feeding grounds.

Data from these DST experiments also reveal that about one-third of plaice released in the Southern Bight visit the eastern English Channel in December and January. In contrast, analysis of the movements of conventionally tagged plaice of a similar size and released at similar times indicates that only 13% of plaice released in the Southern Bight visit the eastern English Channel then (Hunter *et al.*, in press). This difference between DST and conventional tagging experiments is not observed in the central North Sea and German Bight, where the movements of plaice derived from the two approaches are relatively similar. The differences between DST and conventional tagging experiments in the Southern Bight may possibly be due to the fact that the fish migrate to their spawning grounds by selective tidal stream transport. Such behaviour results in migrating plaice spending significantly more time off the seabed, when they would be less available to commercial beam-trawlers. In contrast, plaice in the central North Sea and German Bight do not appear to migrate by selective tidal stream transport (Hunter *et al.*, in press), and therefore these fish would be equally available to commercial beam-trawlers whether or not they are migrating.

In addition to our work on plaice, electronic data storage tags are being used to study the movements of species such as cod (Righton *et al.*, 2001), salmon (Walker *et al.*, 2000), thornback rays (Buckley & Metcalfe, 2002) and tuna (Gunn *et al.*, 1994; Block *et al.*, 1998, 2001). The new technology provides detailed information about fish movement, and about how behaviour changes in time and space over extended periods (in excess of one year). For example, Righton *et al.* (2001) showed how activity (inferred from vertical movements) of cod changes seasonally in the North Sea, but differs from cod activity patterns in the Irish Sea over similar periods. Although we have yet to understand the basis for such regional differences in behaviour, it is very likely to be important to proper understanding of the behavioural ecology of this threatened species.

Despite such technical advances, the use of data storage tags with many species remains limited because the prospect of the fish being caught and the tags returned is low. However, a major innovation, the "pop-up" tag, has increased the probability of data recovery. These tags are attached externally and have a release mechanism that causes the tag to detach from the fish at a predetermined time and rise to the sea surface, where the data can be recovered by airborne radio or satellite (Nelson, 1978; Hunter *et al.*, 1986). Such devices are now available commercially, and are being deployed on large pelagic species such as tuna (Block *et al.*, 1998, 2001; Lutcavage *et al.*, 1999) and basking shark (Sims *et al.*, 2003). Recent studies on Atlantic bluefin tuna indicate that fish tagged in the western Atlantic make transatlantic migrations and utilize spawning grounds in both the Gulf of Mexico and the eastern Mediterranean (Block *et al.*, 2001). This suggests that the current assumption by ICCAT that stock mixing between western and eastern Atlantic is low (Punt & Butterworth, 1994) may not be correct.

Although data transmission capabilities from "pop-up" tags are currently rather limited, further developments in this field give the prospect of much improved data- recovery rates, while further miniaturization will allow the technology to be applied to small species.

FROM INDIVIDUAL DATA TO POPULATION MODELS

There are several approaches to predicting the spatial distributions of fish populations (Giske *et al.*, 1998). In behavioural ecology, such models may involve assumptions about population dynamics such as optimal foraging theory, life history theory, ideal free distribution, and game theory. Such models are undoubtedly intellectually appealing, and they may help in developing ideas about why fish move, but input variables are often hard to parameterize and the model predictions are difficult to test, particularly in the marine environment. To quote Walters & Maguire (1996), referring to the collapse and subsequent closure of the northern cod fishery off Newfoundland in the 1990s, "Such models are useful to compare various management approaches and to try to understand nature, but their specific results should not be mistaken for reality." A more pragmatic approach, underpinned by biological understanding, is needed if we are to describe and explain the spatial distributions of fish populations. Conventional tagging, combined with radio or acoustic tracking and DST experiments, can provide useful information about fish behaviour which, in turn, can be used to construct biologically realistic simulation models of population dynamics.

We are now using our extensive dataset for plaice as the basis for a behaviour-based simulation model of plaice movement for the North Sea and eastern English Channel. The strength of such an approach is that it allows use of data-based methods to calculate movement, as well as changes in the rate, distance and direction of movement that depend on observed changes in behaviour through the annual migratory cycle. This is different from the deterministic and rule-based modelling approaches more commonly applied in fishery models that may "embrace clever but naïve abstractions of reality" (Rose, 1997). Although biologically realistic, behaviour-based movement models are not intended to provide an alternative to traditional stock assessments, they can be used to test the performance of simpler models used in the assessment and management of fish stocks. For example, examining the effects of mixing between plaice populations in the English Channel and the Southern Bight of the North Sea reveals that effective management of the eastern English Channel plaice depends largely on the management of North Sea plaice (Kell *et al.*, submitted). Management measures imposed on the North Sea stock also cause changes in the level of bias between the real and the perceived states of the eastern English Channel stock. However, the North Sea stock is barely affected by measures imposed on the Channel stock. Furthermore, under some conditions, fishing mortality on the Channel stock could rise to very high levels without being detected in North Sea populations.

In addition, large and detailed behavioural datasets can be used to derive other management-related outputs. For example, the depth data derived from electronic tagging studies with plaice can provide a detailed understanding of how their vertical distribution varies in space and time (both seasonally, and through the day/night cycle). This can be used to generate temporally and spatially explicit indices of availability to fishing or acoustic survey gears and so can be used to optimize survey strategies, or to tune survey data. This ability to improve survey data may become increasingly important in situations where stock assessments come to rely much more on surveys from research vessels, because the availability of commercial catch data shrinks concomitant with drastic reductions in catches (e.g. in the North Sea in 2002) or even complete closure of entire fisheries (e.g. Newfoundland cod in the 1990s) for the purpose of conservation.

CONCLUSIONS

As world fisheries continue to be heavily exploited, with drastic reductions in catches or even closures of entire fisheries being necessary to conserve stocks, there is a need for management based on science that takes more account of the fundamental biology of the fish. This applies not only to traditionally exploited species such as cod and tuna, but also to newly developed commercial fisheries, such as those for deep-water species such as orange roughy (*Hoplostethus atlanticus*) and round-nosed grenadier (*Coryphaenoides rupestris*). For most of these species we know little of their behaviour, or how environmental factors influence their population dynamics.

We have used the example of plaice in UK waters to show how tagging experiments using conventional tags, and various types of electronic tag, can provide information about the behaviour and geographical movements of exploited fish populations relevant to assessment and management. The challenge now is for behavioural ecologists and assessment scientists to ensure that this new knowledge is built into future assessment and management methodologies, and that the outputs are taken through into management advice (Schnute & Richards, 2001).

ACKNOWLEDGEMENTS

The authors are grateful to their colleagues, particularly Ewan Hunter, and to two anonymous referees for their constructive comments.

REFERENCES

Armstrong, M. J., Connolly, P., Nash, R. D. M., Pawson, M. G., Alesworth, E., Coulahan, P. J., Dickey-Collas, M., Milligan, S. P., O'Neill, M., Witthames, P. R. & Woolner, L. (2001) An application of the annual egg production method to estimate the spawning biomass of cod (*Gadus morhua* L.), plaice (*Pleuronectes platessa* L.) and sole (*Solea solea* L.) in the Irish Sea. *ICES Journal of Marine Science*, **58**, 183–203.

Arnold, G. P. & Cook P. H. (1984) Fish migration by selective tidal stream transport: first results with a computer simulation model for the European continental shelf. In: *Mechanisms of Migration in Fishes*. (Ed. by J. D. McCleave, G. P. Arnold, J. J. Dodson and W. H. Neill,) pp. 227–261. Plenum Press, London.

Arnold, G. P. & Dewar, H. (2001) Electronic tags in marine fisheries research. In: *Electronic Tagging and Tracking In Marine Fisheries*. (Ed. by J. Sibert and J. Nielsen), pp. 7–64. Kluwer, Dordrecht, The Netherlands (Reviews: Methods and Technologies in Fish Biology and Fisheries 1).

Arnold, G. P. & Holford, B. H. (1995) A computer simulation model for predicting rates and scales of movement of demersal fish on the European continental shelf. *ICES Journal of Marine Science*, **52**, 981–990.

Begg, G. A. & Waldman, J. R. (1999) An holistic approach to stock identification. *Fisheries Research*, **43**, 35–44.

Block, B. A., Dewar, H., Blackwell, S. S., Williams, T. D., Prince, E. D., Farwell, C. J., Boustany, A., Teo, S. L. H., Seitz, A., Walli, A. & Fudge, D. (2001) Migratory movements, depth preferences, and thermal biology of Atlantic bluefin tuna. *Science*, **293**, 1310–1314.

Block, B. A., Dewar, H., Williams, T., Prince, E. D., Farwell, C. & Fudge, D. (1998) Archival tagging of Atlantic bluefin tuna (*Thunnus thynnus thynnus*) Marine Technology Society Journal, **32**, 37–46.

Bolle, L. J., Hunter, E., Rijnsdorp, A. D., Pastoors, M. A., Metcalfe, J. D. & Reynolds, J. D. (2001) Do tagging experiments tell the truth? Using electronic tags to evaluate conventional tagging data. *ICES Document*, **CM 2001/O:2**, 15 pp.

Buckley, A. A. & Metcalfe, J. D. (2002) The movements and behaviour of thornback ray, *Raja clavata*, in the Thames Estuary. Proceedings of the 4th European Elasmobranch Association Meeting, Livorno (Italy). (Ed. by M. Vacchi, G. La Mesa, F. Serena and B. Séret). ICRAM, ARPAT and SFI: 191–199.

De Clerk, R. (1977) The migration of plaice on the spawning grounds "Noord-Hinder". *ICES Document*, **CM1977/F:40**, 9 pp. de Veen, J. F. (1978) On selective tidal transport in the migration of North Sea plaice (*Pleuronectes platessa*) and other flatfish species. *Netherlands Journal of Sea Research*, **12**, 115–147.

Dunn, M. R. & Pawson, M. G (2002) The stock structure and migrations of plaice populations on the west coast of England and Wales. *Journal of Fish Biology*, **61**, 360–393.

Frank, K. T. & Brickman, D. (2000) Allee effects and compensatory population dynamics within a stock complex. *Canadian Journal of Fisheries and Aquatic Sciences*, **57**, 513–517.

Giske, K., Huse, G. & Fiksen, Ø. (1998) Modelling spatial dynamics of fish. *Reviews in Fish Bioliology and Fisheries*, **8**, 57–91.

Greer Walker, M., Harden Jones, F. R. & Arnold, G. P.(1978) The movements of plaice (Pleuronectes platessa L.) tracked in the open sea. *Journal du Conseil, Conseil International pour l'Exploration de la Mer*, **38**, 58–86.

Gunn, J., Polacheck, T., Davis, T., Sherlock, M. and Betlehem, A. (1994) The development and use of archival tags for studying the migration, behaviour and physiology of southern bluefin tuna, with an assessment of the potential for transfer of the technology to groundfish research. *ICES Document*, **CM 1994/Mini: 2.1**, 23 pp.

Harden Jones, F. R. (1968) *Fish Migration*. Arnold, London, 325 pp.

Hill, R. D. (1994) Theory of geolocation by light levels. In: *Elephant Seals: Population Ecology , Behavior, and Physiology* (Ed. by J. B. Le Boeuf and R. M. Laws), pp. 227–236. University of California Press, Berkeley.

Horwood, J. W. (1990) Fecundity and maturity of plaice (*Pleuronectes platessa*) from Cardigan Bay. *Journal of the Marine Biological Association of the United Kingdom*, **70**, 515–529.

Horwood, J. W. (1993) Fecundity and biomass of plaice (*Pleuronectes platessa* L.) in the northern Celtic Sea. *ICES Journal of Marine Science*, **50**, 315–323.

Horwood, J. W. (2000) No-take zones: a management context. In: *The Effects of Fishing on Non-Target Species and Habitats: Biological Conservation and Socioeconomic Issues*, (Ed. by M. J. Kaiser and S. J. de Groot). Blackwell Science, Oxford, 302–311.

Horwood, J. W., Nichols, J. H. & Milligan, S. (1998) Evaluation of closed areas for fish stock conservation. *Journal of Applied Ecology*, **35**, 893–903.

Houghton, R. G., & Harding, D. (1976) The plaice of the English Channel: spawning and migration. *Journal du Conseil, Conseil International pour l'Exploration de la Mer*, **36**, 229–239.

Hunter, E., Aldridge, J. N., Metcalfe, J. D. & Arnold, G. P. (2002) Geolocation of free-ranging fish on the European continental shelf as determined from environmental variables. 1. Tidal location method. *Marine Biology*, **142**, 601–609.

Hunter, E., Metcalfe, J. D., Reynolds J. D. & Arnold, G. P. (in press) Subdivision of North Sea plaice population: evidence from electronic tags. *Journal of Animal. Ecology*

Hunter, J. R., Argue, A. W., Bayliff, W. H., Dizon, A. E., Fonteneau, A., Goodman, D. & Seckel, G. R. (1986). The dynamics of tuna movements: an evaluation of past and future research. *FAO Fisheries Technical Paper*, **277**, 1–78.

Hutchinson, W. F., Carvalho, G. R. & Rogers, S. I. (2001). Marked genetic structuring in localised spawning populations of cod *Gadus morhua* in the North Sea and adjoining waters, as revealed by microsatellites. *Marine Ecology Progress Series*, **223**, 251–260.

Jamieson, A. & Thompson, D. (1972) Blood proteins in North Sea cod (*Gadus morhua* L.) In: *Proceedings XII European Conference on Animal Blood Groups and Biochemical polymorphism*. (Ed. by G. Kovacs and M. Papp). W. Junk, The Hague, pp. 585–591.

Kell, L. T., Scott, R. & Hunter, E. (submitted) Implications for current management advice for North Sea plaice. 1. Migration between the North Sea and English Channel. *Journal of Sea Research*.

Lutcavage, M. E., Brill, R. W., Skomal, G. B., Chase, B. C. & Howey, P. W. (1999) Results of pop-up satellite tagging of spawning size class fish in the Gulf of Maine: do North Atlantic bluefin tuna spawn in the mid-Atlantic? *Canadian Journal of Fisheries and Aquatic Sciences*, **56**, pp. 173–177.

MacLennan, D. N. & Simmonds, E. J. (1991) *Fisheries Acoustics*. Chapman and Hall, London. Metcalfe, J. D. (2001) Summary report of a workshop on daylight measurements for geolocation in animal telemetry. In: *Electronic Tagging and Tracking In Marine Fisheries*. (Ed. by J. Sibert and J. Nielsen), pp. 331–342. Kluwer, Dordrecht, The Netherlands (Reviews: Methods and Technologies in Fish Biology and Fisheries **1**).

Metcalfe, J. D. & Arnold, G. P. (1997) Tracking fish with electronic tags. *Nature*, **387**, 665–666.

Metcalfe, J. D., Arnold, G. P. & Webb, P. W. (1990) The energetics of migration by selective tidal stream transport: an analysis for plaice tracked in the southern North Sea. *Journal of the Marine Biological Association of the United Kingdom*, **70**, 149–162.

Metcalfe, J. D., Fulcher, M. & Storeton-West, T. J. (1992). Progress and developments in telemetry for monitoring the migratory behaviour of plaice in the North Sea. In: *Wildlife Telemetry: Remote Monitoring and Tracking of Animals*. (Ed. by I. G. Priede and S. M. Swift), pp. 359–366. Ellis Horwood, London.

Nash, R. D. M., Witthames, P. R., Pawson, M. G. & Alesworth, E. (2001) Regional variability in the dynamics of reproduction and growth of Irish Sea plaice, *Pleuronectes platessa* L. *Journal of Sea Research,* **44**, 55–64.

Nelson, D. R. (1978). Telemetering techniques for the study of free-ranging sharks. In: *Sensory Biology of Sharks, Skates, and Rays*. (Ed. by E. S. Hodgson and R. F. Mathewson), pp. 419–482. Office of Naval Research, Department of the Navy, Arlington, Virginia.

Pawson, M. G. & Jennings, S. (1996) A critique of methods for stock identification in marine capture fisheries. *Fisheries Research*, **25**, 203–217.

Pawson, M. G., Kelley, D. F. & Pickett, G. D. (1987). The distribution and migrations of bass *Dicentrarchus labrax* L. in waters around England and Wales as shown by tagging. *Journal of the Marine Biological Association of the United Kingdom*, **67**, 183–217.

Punt, A. E. & Butterworth, D. S. (1994) Use of tagging data within a VPA formalism to estimate migration rates of bluefin tuna across the North Atlantic. *Collective Volume of Scientific Papers. International Commission for the Conservation of Atlantic Tunas*, **44**, 166–182

Quinn, T. J. & Deriso, R. B. (1999) *Quantitative Fish Dynamics*. Oxford University Press, New York.

Righton, D. R., Metcalfe, J. D. & Connolly, P. (2001) Electronic tags reveal behavioural differences between North Sea and Irish Sea cod. *Nature*, **411**, p. 156.

Rijnsdorp, A. D. & Pastoors, M. A. (1995) Modelling the spatial dynamics of fisheries of North Sea plaice (*Pleuronectes platessa* L.) based on tagging data. *ICES Journal of Marine Science*, **52**, 963–980.

Rose, G. A. (1997). The trouble with fisheries science! *Reviews in Fish Biology and Fisheries*, **7**, 365–370.

Schnute, J. T. & Richards, L. J. (2001). Use and abuse of fisheries models. *Canadian Journal of Fisheries and Aquatic Sciences*, **58**, 10–17.

Seber, G. A. F. (1982) *The Estimation of Animal Abundance*, 2nd edn. Griffin, London.
Sims, D. W., Emily J. Southall, E. J., Richardson, A. J., Reid, P. C. & Metcalfe J. D. (2003) Foraging and migratory behaviour of basking sharks over seasonal scales: no evidence for winter hibernation. *Marine Ecology Progress Series*, **248**, 187–196
Walker, R. V., Myers, K. W., Davis, N. D., Aydin, K. Y., Friedland, F. D., Carlson, H. R., Boehlert, G. W., Urawa, S., Ueno, Y. & Anma, G. (2000) Diurnal variation in the thermal environment experienced by salmonids in the north Pacific as indicated by data storage tags. *Fisheries Oceanography*, **9**, 171–186.
Walters, C. & Maguire, J-J. (1996) Lessons for stock assessment from the northern cod collapse. *Reviews in Fish Biology and Fisheries*, **6**, 125–137.
Wimpenny, R. S. (1953) *The Plaice*. Arnold, London. 145 pp.

The rise and fall of cod (*Gadus morhua, L.*) in the North Sea

R. Colin A. Bannister
Centre for Environment, Fisheries and Aquaculture Science, Pakefield Road, Lowestoft, Suffolk NR33 0HT, UK

ABSTRACT: The recent history and current state of the North Sea cod (*Gadus morhua*) stock is described using the 2000 stock assessment carried out by the International Council for the Exploration of the Sea (ICES), which covers the period since 1963. This shows that the stock is heavily exploited, seriously depleted, and suffering from reduced recruitment. Under the precautionary approach framework, and as required by an EU–Norway management agreement, the ICES Advisory Committee on Fishery Management (ACFM) has advised a severe reduction in fishing mortality and the implementation of a plan to rebuild spawning stock safely and rapidly to the precautionary level. Stakeholders are sceptical about the justification for this advice and the severity of the measures proposed, and the paper discusses the issues arising from this. ACFM has consistently advised strong management, and in many years managers agreed the advised Total Allowable Catch (TAC) or even a lower one, but these have so far failed to reduce fishing mortality. Contributory factors are discussed, e.g. the problem of the mixed gadoid fishery, the weakness of TAC management, and evidence that in several years the assessment overestimated spawning biomass and hence the TAC. Current stock trends are compared with provisional literature estimates for the period back to 1920, showing that the marked increase in landings and fishing mortality in the 1960s and 1970s coincided with a major increase in cod recruitment (part of the so-called "gadoid outburst"), the possible causes of which are enumerated from the literature. Since 1987 the decline in spawning biomass to an all-time low coincides with a downturn in recruitment to the historical level, and the persistence of a high fishing mortality that is now unsustainable. To avoid collapse, the North Sea cod stock requires effective implementation of a recovery plan to increase spawning biomass and to maintain or increase recruitment. Recovery may take 10 years or more. Interim measures have been agreed in the form of a low TAC and an increase in the minimum mesh size for the gadoid fishery, but it is too soon to see the benefits of these measures, even assuming that they are complied with.

INTRODUCTION

There are nine major commercial fisheries for cod (*Gadus morhua*) distributed across the Northeast Atlantic from the Celtic Sea and Irish Sea to the North Sea, Faroes, the Northeast Arctic, Iceland, and east Greenland, and a further 10 fisheries distributed from west Greenland to Labrador, Newfoundland and Georges Bank (see, for example, Figure 1 in Garrod, 1977). In the North Sea, cod are caught principally by otter trawlers in mixed fisheries for cod, haddock (*Melanogrammus aeglefinus*) and whiting (*Merlangius merlangus*) in the central and northern North Sea, but also by coastal otter trawl fisheries in the Skagerrak and Eastern Channel, and by gillnet fisheries off the Danish east coast and in the Southern Bight. Cod are also part of the bycatch in the *Nephrops* fisheries north of the Dogger Bank and along the east coast of the UK, as well as in the beam trawl fishery for flatfish in the southern North Sea. The North Sea gadoid fishery became particularly important to UK fishers in the 1970s and 1980s because it expanded at a time when access to distant water fisheries ceased following the establishment of 200 mile exclusive economic zones in the late 1970s (Macer & Easey, 1988).

The North Sea cod stock is assessed by the International Council for the Exploration of the Sea (ICES), the principal provider of independent scientific advice for commercially exploited fish stocks in the Northeast Atlantic. The assessment covers cod caught in the Eastern Channel (ICES Division VIIe), the North Sea (Divisions IVa, b, c) and the Skagerrak (Division IIIa). Cod spawn in the northern North Sea, around the Dogger Bank and off the Dutch coast (Daan, 1978), but the contemporary distribution of spawning is not well defined owing to a lack of up-to-date systematic ichthyoplankton surveys of the North Sea, and because cod and haddock eggs are difficult to distinguish in the early planktonic stages. Brander (1994) shows the late pelagic stages as occurring mainly along the southern edge of the Norwegian Trench in the northeastern North Sea, but trawl survey data summarized by Heessen (1993) show one- and two-year-old cod mainly close to the coasts of Germany, Denmark and northeast England. Cod aged 3 and older are found in the Southern Bight and off northeast England, but the main concentrations are in the northern North Sea from Scotland to Norway, and into the Skagerrak. Cod therefore occur widely, but their distribution changes markedly between the planktonic, juvenile and adolescent stages. The Norwegian Trench forms the northeastern boundary of the stock, whereas the northwestern boundary lies in the vicinity of the Orkney and Shetland Islands. Despite recent studies (Hutchinson *et al.*, 2001), knowledge of the genetic status of the stock is still incomplete and North Sea cod are assessed and managed as one stock.

The ICES North Sea cod assessment covers the period since 1963, during which recorded cod landings declined from a peak of 354 000 t in 1972 to 50 000 t in 2001. The stock is now severely depleted, and in October 2000 the ICES Advisory Committee on Fishery Management (ACFM) advised that cod fishing should be severely curtailed to the lowest possible level because of the possibility of stock collapse (ICES, 2001a). The European Union (EU) and Norway, joint managers of this shared stock, subsequently reduced the Total Allowable Catch (TAC) for 2001 by 50%, and the EU Commission commenced negotiations to develop a recovery programme. ACFM advice was similarly severe in 2001 (ICES, 2001c), and in 2002 ICES advised that the North Sea cod fishery should be closed in 2003 (ICES, 2002a). Similar advice was given for the Irish Sea and West of Scotland cod stocks, which are also seriously depleted.

The severity of this advice and its potential consequences caused UK catchers to question whether the scientific advice was justified, or whether it arose from a false perception about the state of the stock, perhaps because of the influence of factors other than fishing. Stakeholders have since continued to challenge the severity of the management actions proposed. Such reactions echo events in the mid-1970s, when managers were first faced with the recruitment failure of the North Sea stock of herring (*Clupea harengus*). This followed severe overfishing throughout the 1960s on immature and mature herring of all three North Sea herring substocks, but there were significant delays in taking management action because scientific views differed as to the role of overfishing and environmental change (Burd 1974, 1978; Nichols 2001).

The difference between the herring and the cod story in the North Sea, however, is that the herring stock collapsed because the fishery was previously unregulated, whereas the cod fishery has nominally been subject to ICES advice and to EU management for a number of years. The apparent failure to conserve the cod stock, coupled with the criticism by stakeholders, led the UK Management Group of Directors to call for an informal appraisal of North Sea cod science, which was undertaken in 2001 by the author and Robin M. Cook of Fisheries Research Services, Aberdeen. The present paper has been developed from that unpublished review. It first describes the current assessment results for North Sea cod, and the justification for the present advice, then reviews the consistency of advice and the effectiveness of management over the last decade or so, discussing associated compliance and assessment issues. It then compares the recent trends in recruitment, fishing mortality and spawning biomass with those described in the literature by a historical analysis for the period since 1920. That comparison indicates that ecological and environmental factors must also be in play, but that fact strengthens rather than weakens the case for taking strong management action to safeguard the stock. The paper concludes by describing current proposals and time-scales for rebuilding the stock, and the likely difficulty of achieving agreement and recovery in a shared stock. Although the 2002 centenary celebrations of ICES and of CEFAS demonstrate that the Northeast Atlantic has been at the heart of the development of fisheries science and management for the past 100 years, this paper emphasizes just how difficult it is to manage shared stocks effectively.

THE ICES ASSESSMENT AND ADVISORY PROCESS

Scientific advice on the management of North Sea cod is based on annual stock assessments and forecasts produced by the ICES Working Group on the Assessment of Demersal Stocks in the North Sea and Skagerrak (WGNSSK). This working group carries out a full analytical assessment using extended survivors' analysis (XSA; Shepherd, 1992), a development of the virtual population analysis (VPA) method (Gulland, 1965; Pope, 1972), as described by Darby & Flatman (1994). The assessment uses estimates of the annual number at age in the recorded international landings from 1963 up to the last full year of catch data just prior to the assessment year, and is calibrated using landings per effort for the principal commercial fleets, plus indices of abundance from several national and international research vessel surveys. The analysis estimates the matrix of annual fishing mortality (F) and population number (Pn) at age, and thence the number of one-year-old recruits (R) and spawning stock biomass (SSB, calculated from the vectors of population number and weight at age for ages above the mean age of first maturity). The most recent estimates of F, R and SSB in the unconverged part of the assessment can change between assessment years, depending on the most recent landings and their age structure, and whether there are changes to the configuration and calibration of the assessment, as detailed in the reports of WGNSSK each year. The results that feature in this paper are mainly those obtained by WGNSSK in 2000 (ICES, 2001b), supplemented by brief references to assessments in other years.

The principal forecast is short term, in which the population number at age surviving at the end of the last full year of catch data (i.e at the start of the assessment year) is projected to the start of the following TAC year in order to predict the catch and residual stock corresponding to a range of fishing mortality options for the TAC year. The results are presented in a fishing mortality–catch options table that forms the basis for the advice. The forecast depends on the estimates of new recruitment that enter the stock in the assessment year and at the start of the TAC year, and upon whether or not the working group constrains fishing mortality by the TAC already operating in the current assessment year. The new recruitment is either predicted using survey data, or is estimated as the average of some appropriate historical period. For North Sea cod, WGNSSK routinely investigates whether to estimate incoming year-class strength by weighting various survey estimates, but in the 2000 assessment the working group opted to use the geometric mean of recruitment for the years 1988–1997, when recruitment was low (ICES, 2001b). WGNSSK and some other ICES assessment working groups also use a medium-term forecast programme incorporating a stock–recruit relationship in order to estimate the probability of SSB being above a specified precautionary value after 10 years.

The formal conduit through which ICES gives advice to client Commissions and the EU on the management of fish stocks in the ICES area is the Advisory Committee on Fisheries Management (ACFM), which consists of an elected chair plus a nominated expert from each ICES member country. ACFM reviews the various working group assessments and forecasts, and decides what advice to recommend in the ACFM report alongside an assessment summary and the catch options table. Since 1999, the TAC advised by ICES has been selected from those catch options that are acceptable within the framework of the precautionary approach, whose concepts and reference point methodology are described in ICES (1997a) and ICES (1998). In this framework, ICES aims to ensure the sustainability of a stock by keeping it above B_{lim}, a minimum SSB defined as that below which the probability of reduced recruitment increases, or the recruitment dynamics are unknown. Owing to assessment uncertainty, however, SSB may only be above B_{lim} with high probability if it is above a higher level, B_{pa} (pa here stands for precautionary approach), which takes uncertainty into account. The corresponding fishing mortality values are F_{lim} and F_{pa}. ICES uses B_{pa} and F_{pa} to define whether stocks are inside or outside safe biological limits, and hence whether management action should be taken to reduce F and increase SSB above B_{pa}. ACFM therefore determines whether the current F and SSB are inside biological safe limits (below F_{pa}, and above B_{pa}), and then identifies the catch options in the upcoming TAC year that correspond to values of F and SSB predicted to be below F_{pa} and above B_{pa}. ACFM will also show the probability of SSB being above or below B_{pa} in the medium term when such a forecast is available. If stocks are already below B_{lim} or above F_{lim}, ACFM will advise that a rebuilding plan is required to increase stocks rapidly and safely to B_{pa}.

For stocks with full analytical assessments, B_{lim} was generally derived using the values of recruitment and SSB estimated by the 1998 assessment (ICES, 1998), but in some cases ICES continued to use values of MBAL, the minimum biologically acceptable level of SSB, a concept introduced several years before precautionary reference points were calculated (Serchuk & Grainger, 1992). MBAL was defined as either the SSB below which the probability of poor recruitment increased (which is synonymous with B_{lim}), or the SSB where management action should be taken because of concerns about poor recruitment in the future (which is operationally analogous to B_{pa}, but may be different computationally). North Sea cod is an example where ICES implemented the previous MBAL of 150 000 t as B_{pa}, from which was derived the B_{lim} of 70 000 t. The corresponding fishing mortality reference points are $F_{pa} = 0.65$ and $F_{lim} = 0.86$.

THE STATE OF THE NORTH SEA COD STOCK IN 2000

The principal results of the ICES North Sea cod assessment in 2000 are summarized below and in Figure 1 and Table 1, based on the summary in ICES (2001a):

- Fishing mortality (averaged for ages 2–8) doubled from 0.47 in 1963 to 0.91 by 1983, then over the next two decades remained high in the range 0.72–0.95. Recent fishing mortality ($F_{status\ quo} = F_{'97-'99}$) was 0.90, which is above F_{lim}.
- SSB (the weight of mature cod aged 4 and older) declined fourfold from a peak of 277 000 t in 1971 to an historical low of only 67 000 t by 2000, below both B_{pa} and B_{lim}.
- Recruitment of one-year-old cod averaged 453 million (CV 52%) between 1963 and 1987, a period with frequent large year classes, but only 202 million (CV 45%) in the period 1988–1997, when there were no large year classes.
- Recorded landings from 1966 to 1987 varied in the range 200 000–354 000 t, but then fell dramatically to 96 000 t by 1999, and later to 50 000 t in 2001.

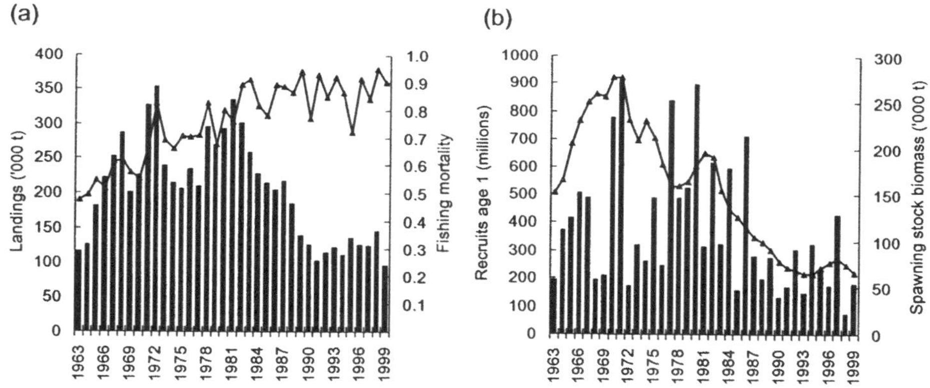

Fig. 1 Trends in (a) landings (bars) and fishing mortality (line), and (b) recruitment (bars) and spawning stock biomass (line) of North Sea cod, 1963–1999, from the ICES assessment made in 2000 (ICES, 2001a).

Table 1. North Sea cod SSB and F from the 2000 assessment (ICES, 2001a), and the ICES reference points.

Parameter	2000 assessment	Limit reference point	Precautionary reference point
Spawning stock biomass	SSB_{2000} = 66 700 t	B_{lim} = 70 000 t	B_{pa} = 150 000 t
Fishing mortality	$F_{'97-'99}$ = 0.90	F_{lim} = 0.86	F_{pa} = 0.65

In addition to the high fishing mortality, the pattern of exploitation on North Sea cod is poor (Table 2). Cod are partially selected by the fishing gear at ages 1 and 2, and the full fishing mortality operates from age 3, so in combination with the high rate of natural mortality (M) adopted for the younger ages, each year class is depleted rapidly. On average, only 4% of the cod recruiting at age 1 survive to the mean age of first maturity at age 4, so there is a very narrow age range in the stock and very few North Sea cod survive to breed. Immature cod average <2 kg in weight compared with a weight range of 4–13 kg for mature fish. The low cod SSB is therefore caused by a combination of poor recruitment, poor survival and a dominance of fish of low average weight.

The high proportion of immature cod in the catch arises because cod in the central and northern North Sea are caught in a mixed fishery with haddock and whiting, which attain a smaller maximum size than cod. The trawl mesh size permitted in the North Sea mixed fishery has hitherto been 100 mm or less, to allow sufficient quantities of haddock and whiting to be retained in this mixed fishery, whereas a mesh size of at least 140–160 mm is required to allow immature cod to escape (STECF, 2001).

Table 2. Age-specific vectors of catch number (Cn), fishing mortality (F), natural mortality (M), population number (Pn), weight and maturity, from the 2000 North Sea cod assessment (after ICES, 2001b).

Age	Cn 1999 ('000)	F (1997–1999)	M	Pn end of 1999 ('000)	Weight (kg)	Proportion mature	Average Pn 1963–1997 ('000)
1	5 941	0.06	0.8	202 000	0.67	0.01	391 744
2	9 731	0.63	0.35	76 384	1.03	0.05	156 857
3	32 225	1.07	0.25	14 217	2.13	0.23	49 642
4	4 034	1.00	0.2	14 913	4.0	0.62	16 254
5	1 446	0.91	0.2	2 172	6.3	0.86	6 655
6	626	0.83	0.2	800	8.2	1.0	2 874
7	223	0.95	0.2	426	9.7	1.0	1 212
8	91	0.90	0.2	183	10.9	1.0	541
9	14	0.85	0.2	45	12.3	1.0	234
10	10	0.94	0.2	8	13.3	1.0	96

Based on this assessment, the ICES advice is justified on the grounds of fisheries biology, the precautionary approach, and the terms of a shared-stock management agreement between the EU and Norway, implemented in 1999. Biologically, the combination of a high mortality rate, poor exploitation pattern, very small stock size and reduced recruitment gives obvious cause for concern about the future of North Sea cod if the fishing rate is not severely curtailed. In the stock–recruitment diagram in Figure 2 the data points for the past decade occur in the lower left quadrant well to the left of B_{pa} (150 000 t), a strong symptom of recruitment–overfishing. Even without considering the effect of reduced recruitment, $F_{status\ quo}$ (0.8 –0.9) is four and six times F_{max} (0.23) and $F_{0.1}$ (0.14) respectively, values that can be used as targets for optimal harvesting. Table 3 shows the corresponding reduction in yield and SSB per recruit at the current level of fishing mortality (ICES, 2001c).

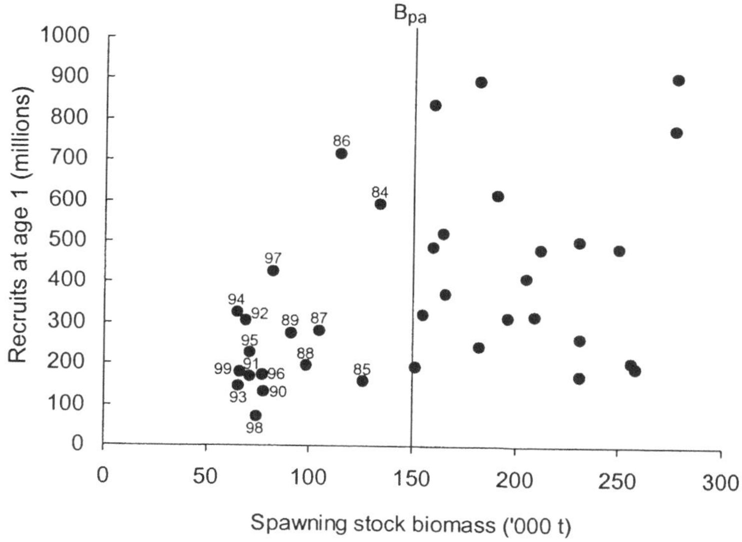

Fig. 2 The relationship between one-year-old recruits and spawning stock biomass of North Sea cod, from the ICES assessment made in 2000 (ICES, 2001a).

Table 3. Values of North Sea cod yield per recruit (Y/R) and SSB per recruit (SSB/R) at $F_{'98-'00}$, F_{max} and $F_{0.1}$ (after ICES, 2001c).

Parameter	$F_{'98-'00} = 0.83$	$F_{max} = 0.23$	$F_{0.1} = 0.14$
Y/R (kg)	0.524	0.718	2.838
SSB/R (kg)	0.381	2.833	4.560

From a precautionary viewpoint, a stock where SSB is below B_{lim} and F is above F_{lim} is well outside safe biological limits. This constrains ICES to recommend a substantial reduction in F and the development of a rebuilding plan, as advised in October 2000. ACFM recommended the lowest possible catch level because, in the forecast for 2001, not one TAC option would bring SSB_{2002} above B_{pa}. The objective for recovery is described in the 1999 shared-stock management agreement between the EU and Norway (ICES, 2001a), which can be paraphrased as:

- "every effort shall be made to maintain a minimum SSB greater than 70 000 t (B_{lim});
- fishing will be restricted by a TAC consistent with a value of F no higher than 0.65 (F_{pa});
- if SSB falls below the reference point of 150 000 t (B_{pa}), F will be adapted to ensure a safe and rapid recovery of SSB to more than 150 000 t;
- the exploitation pattern will be improved to reduce discarding and enhance SSB"

The proposed EU recovery plan and its implications are described later in this paper.

CONSISTENCY OF ADVICE AND EFFECTIVENESS OF MANAGEMENT

The sequence of scientific advice and management for the North Sea cod fishery since 1987, the year when SSB and recruitment began to approach their current low levels, is shown in Table 4. This table summarizes the underlying trends in SSB and F, the year-on-year advice, the agreed management targets and the resulting fishery landings, based on the sources cited in the table legend. The table addresses the claim by some stakeholders that the severity of the ICES advice in 2000 reflects either inadequate previous advice, or previously inadequate action by managers.

Table 4. North Sea cod advice, TAC and landing statistics, 1987–2001 (after ICES, 2001c) and the realized SSB and F (after ICES, 2002a).

Year	ACFM advice	SSB ('000 t)	F_{2-8}	Advised TAC ('000 t)	Agreed TAC ('000 t)	Official landings ('000 t)	ACFM landings ('000 t)
1987	Recover SSB	105	0.88	100/125	175	167	182
1988	0.7 of F_{1986}	99	0.86	148	160	142	157
1989	Stabilize SSB	91	0.94	124	124	110	116
1990	0.8 of F_{1988}	78	0.77	113	105	99	105
1991	0.7 of 1989 gadoid effort	71	0.93	–	100	87	89
1992	0.7 of 1989 gadoid effort	69	0.85	–	100	98	97
1993	Zero cod effort, or 0.7 of 1989 gadoid effort	65	0.92	–	101	94	105
1994	Zero cod effort, or 0.7 of 1989 gadoid effort	65	0.86	–	102	87	95
1995	0.7 of recent gadoid effort	71	0.72	–	120	112	120
1996	0.8 of F_{1994}	76	0.92	141	130	104	107
1997	0.8 of $F_{1995} = 0.65$	80	0.86	135	115	100	102
1998	F_{1998} less than F_{1996}	72	0.99	153	140	114	122
1999	F = 0.6, to rebuild *SSB*	61	1.06	125	132	80	78
2000	F less than 0.55	54	0.83	<79	81	62	59
2001	Lowest possible catch			0	48.6		

THE ADVICE

ACFM advice is generally formulated in relation to fishing mortality, although from 1991 to 1995 the advice was given in terms of fishing effort. Table 4 shows that, since 1987. ACFM sought to stabilize SSB (1989) or to rebuild it (1987, 1999); to cut fishing mortality or fishing effort by 20 or 30% (1988, 1990, 1991, 1992, 1995, 1996, 1997); or to reduce F below a specified value (1998, 1999, 2000). In 1992, however, ACFM advised that "... in isolation fishing effort on cod should be reduced to zero in order to increase the stock towards its 'lowest desirable level' (150 000 t SSB) at the fastest possible rate. Recovery of the cod stock would require, at minimum, a marked and sustained reduction of effort or even closure of the fishery. Considering that effort on cod is to a large extent directed through mixed demersal fisheries, ACFM recommends that in 1993 fishing effort in the directed fisheries on North Sea roundfish stocks except saithe should be limited to 70% of the 1989 fishing effort."

Similar advice was given in 1993, and in 1996 ACFM also pointed out the possibility of stock collapse: "Recent analyses which examine the stock–recruitment relationship suggest that the stock may collapse under fishing mortality rates above 0.75. Present fishing mortality is above this level."

ICES advice has therefore consistently aimed to be restrictive, and as early as 1992 it included advice to close the cod fishery in 1993. Severe advice on cod was also accompanied by weaker advice for the mixed fishery, but until recently it is the mixed fishery advice that prevailed.

THE EFFECTIVENESS OF MANAGEMENT

In practice, North Sea cod has been managed through TACs, irrespective of whether ICES advised a cut in fishing mortality or in fishing effort. In 1987 and 1988, the TAC agreed by EU managers was substantially above that corresponding to the advice (Table 4), but for 1989 and 1990, and from 1996 to 1999, the agreed TAC was generally the same or below that proposed, although this was on the basis of the advice for the mixed fishery rather than harsh advice for the cod stock in isolation. The TAC changed little between 1990 and 1994, when ACFM advised a reduction in effort. In 2001 the TAC was well above the zero catch implied by the advice, but was nevertheless a 50% cut from that of the previous year. Since 1989, therefore, TACs have been in line with, or below, those advised for the mixed fishery, but the most severe TACs aimed at the cod stock alone were not agreed until very recently.

Table 4 also shows how the agreed TAC compares with the recorded landings of North Sea cod. In most years, landings were either close to or less than the agreed TAC, with the largest undershoot in the period 1996–1999. The two columns of landings in Table 4 are those reported officially by national agencies, and landings adjusted by ACFM to take account of discrepancies identified during working group meetings.

Given that ACFM consistently advised at least some reduction in fishing mortality or effort for the mixed fishery, that managers agreed TACs that were usually equal to or more conservative than the mixed fishery advice, and that recorded landings were usually equal to or less than the agreed TAC, we should expect to have observed a corresponding reduction in realized fishing mortality and some stabilization or recovery in SSB. Regrettably, Table 4 shows that this did not happen. Fishing mortality on cod has continued to fluctuate at a high level, and SSB has continued to decline. We must therefore conclude that management of the North Sea cod stock has not been effective. The most likely reasons for this are, *inter alia*, problems resulting from the use of TACs especially for a mixed fishery, bias in the assessments, or the influence of non-fishery (e.g. environmental or ecological) factors, although the latter will tend to affect SSB only, and not F.

The EU system of TACs and quotas allows EU resources to be partitioned between countries using historical allocation keys, but a TAC does not necessarily restrict fishing activity unless other measures are taken. When a TAC is restrictive, fishers should reduce their catch (i.e. the number of fish killed), but unless fishing activity is also curtailed they may continue to catch fish as before, but adjust to the required landings by discarding, high-grading, misreporting or misallocation, so leaving true fishing mortality unchanged. This is particularly likely in a mixed fishery, where a prospective cut on one species may be undermined by continued fishing opportunity on others. Over the past few years it has frequently been rumoured that recorded landings in the UK are underestimated because fish are landed over and above the TAC, to the maximum permitted tolerances in the reporting system, or by deliberate evasion. In the nature of things, the degree of non-compliance cannot be established, but in years when recorded landings of cod have been markedly below the TAC, this must be due to either non-compliance or to overestimation of the TAC, or both.

ASSESSMENT BIAS AND UNCERTAINTY

Landings could be markedly less than a TAC if the forecast overestimates the predicted catch because the assessment has overestimated SSB and underestimated F. There is evidence of such assessment bias in the case of North Sea cod. Van Beek (2000) compared the landings predicted at $F_{status\ quo}$ for the TAC year against the observed landings. He found that predicted landings exceeded observed landings in 14 out of 17 years since 1983, with an average difference of +32% (range –17 to +64%). For the assessments of 1983–1997 assessments, Van Beek & Pastoors (1999) applied the observed landings to the stock predicted by the assessment for the TAC year in order to calculate a "predicted" or assessment fishing mortality. This was in the range 0.5–0.7, whereas the realized fishing mortality subsequently estimated in the converged part of a later assessment ranged from 0.8 to 1.0. Realized F was therefore higher than the assessment F, a result that was also obtained for North Sea plaice (*Pleuronectes platessa*). Darby (pers. comm.) has also pointed out that North Sea cod weight-at-age has been declining sharply over the past decade, which could cause an overestimate of SSB and TAC when current weight-at-age is used in a forecast. Finally, retrospective plots generated by the XSA assessments for North Sea cod (Figure 3.4.7 in ICES, 2001a) show that, since 1992 and especially since 1996, the terminal unconverged portion of previous assessments overestimated SSB and underestimated F compared with the values estimated later by the 2000 assessment run. Recruitment was less affected. Such observations provide evidence of assessment bias, and imply that in some years the TAC for North Sea cod may not have been restrictive, especially since 1996. Consequently, even when landings were equal to or less than the TAC, fishing mortality would not necessarily be reduced.

Assessments can produce different perceptions about the state of a stock from year to year because of trends in the data or because of changes to the configuration of the assessment model. In Figure 3 the historic trend in cod fishing mortality from the 2000 assessment is compared with the results from assessments in each odd year back to 1989. In the converged part of each assessment, fishing mortality increased to a peak value above 0.9 by 1989, but the terminal years of the 1997 and 1999 assessments depict a subsequent decline in F to below F_{pa} (0.65), whereas in the 2000 assessment F remains at 0.9. There is an associated difference in the trend in SSB (not shown). In the 1999 assessment SSB reaches a minimum of 65 000 t in 1993, but it recovers to 128 000 t in 1999 as F declines (ICES, 2000), whereas in the 2000 assessment, SSB rises to 82 000 t in 1997, but falls back to 66 000 t in 1999 (ICES, 2001b). Consequently, it would be perceived in 1999 that management was having an effect, whereas in 2000 it was concluded that management was not working.

The difference between the results of the 1999 and 2000 assessments can be attributed to the trend in recorded landings, and a change in the configuration of the assessment. Cod landings increased from 1997 to 1998, but fell sharply in 1999. Estimated SSB was therefore boosted by a rise in landings in the terminal year of the 1999 assessment, but reduced by a sharp fall in landings in the terminal year of the 2000 assessment. In addition, the 1999 assessment was tuned using all available commercial catch per unit effort and survey data (ICES, 2000), whereas the 2000 assessment eliminated commercial data that

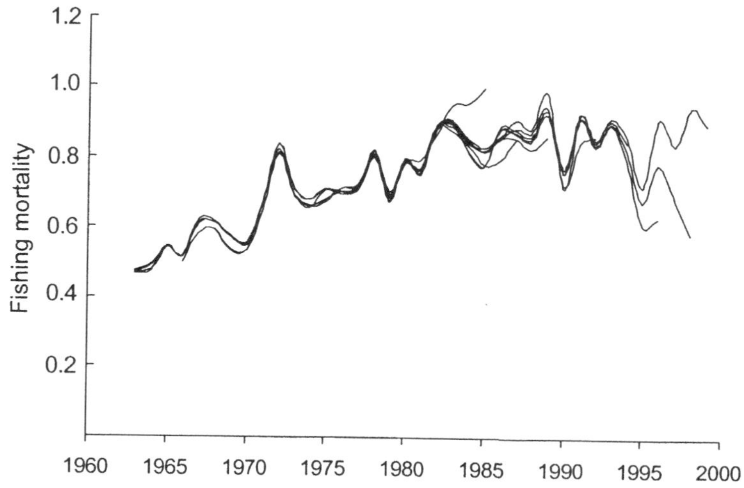

Fig. 3 Historical trends in the fishing mortality of North Sea cod estimated by ICES assessments in 2000, and successive odd years from 1999 back to 1989 (reports of the ICES Working Group on the Assessment of Demersal Stocks in the North Sea and Skagerrak).

were either too erratic (English seine), or were biased by a change in the way recent effort data were recorded (Scottish data; ICES, 2001b). The 2000 assessment also used different tuning windows and weightings following a pre-assessment sensitivity analysis (Cook, 2000; C. D. Darby & M. T. Smith, pers. comm.).

Nevertheless, the impact of assessment uncertainty needs to be kept in perspective. The 1999 and 2000 assessments give different perceptions about the most recent values of F and SSB in the vicinity of F_{lim} and B_{lim}, but all assessments describe a major decline in stock during the1980s, and show that the stock has been outside safe limits (above F_{pa} and below B_{pa}) throughout the 1990s.

THE HISTORICAL AND ECOLOGICAL PERSPECTIVE

ACFM advice for North Sea cod is based on assessment data commencing in 1963, but it is helpful to put the current evolution of the stock in perspective by considering a much longer time-scale, using data compiled by Heessen (1993), Daan *et al.* (1994) and Pope & Macer (1996). Based on data and calculations in ICES (1969) and ICES (1992), Daan *et al.* (1994) reconstructed long-term trends in F back to 1914. They estimated total effort from landings and catch per unit effort data, and then converted this to F by scaling from VPA results for the period 1960–1966, when the two analyses overlapped. SSB was then estimated from the relationship B = Y/F, and recruitment from yield-per-recruit values at each F. Pope & Macer (1996) subsequently used English age, size composition and weight-at-age data to construct total catch at age for the period 1920–1962, in order to estimate historical trends in F, Pn, R and SSB using VPA. Pope & Macer (1996) considered their VPA estimates to be tentative because they had raised and extrapolated from incomplete data, but these are the only results available for describing the long-term dynamics of cod in the North Sea.

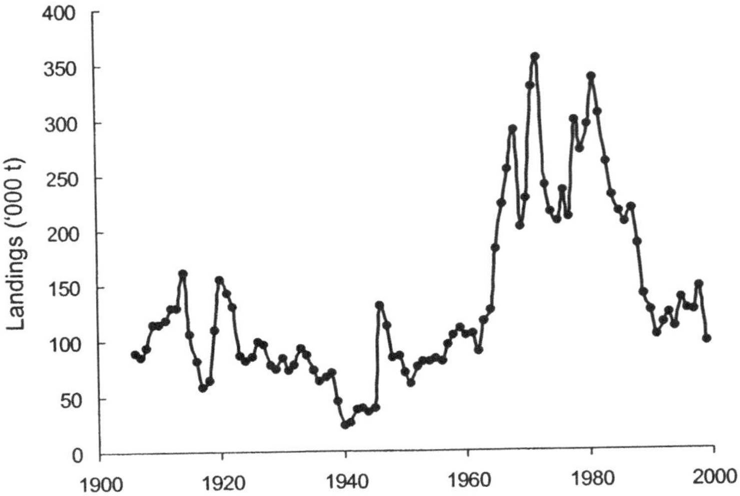

Fig. 4 Landings of North Sea cod from 1903 to 1999 (1903–1962 from Daan *et al.*, 1994; 1963–1999 from ICES, 2001a).

Figure 4 shows the trend in landings since 1903 using data tabulated in the ICES 2000 assessment, and in Daan *et al.* (1994). As first noted by Heessen (1993), the rapid increase in landings from 1963 (116 000 t) to 1972 (354 000 t) followed a long period from 1903 when landings were much lower (30 000–160 000 t). Since 1980 landings have fallen back rapidly to below 100 000 t. Figure 5 shows the underlying long-term trends in fishing mortality, recruitment and spawning biomass obtained by plotting together the results from the ICES assessment and the historical data kindly made available by Pope & Macer. Fishing mortality from 1920 up to 1963 was mainly in the range 0.3 to 0.6, with occasional values of 0.8 (Figure 5a). After 1963 it increased rapidly from 0.4 towards 0.9 by 1980, and has since remained high. Recruitment of one-year-old cod fluctuated in the range 100 to 300 million up to 1963, then increased rapidly following the frequent recruitment of good year classes ranging from 200 to 900 million cod, before returning to the previous range after 1987 (Figure 5b). SSB varied in the range 100 000–275 000 t up to 1970 (Figure 5c), then declined within the previous range until 1985, but subsequently declined beyond that range. Assuming that the historical VPA is reasonably reliable, the period of the ICES North Sea cod assessment incorporates an unprecedented rise in recruitment, fishing mortality and landings, but whereas recruitment and landings have returned to former levels, fishing mortality has remained very high, and SSB has fallen to an unprecedented low level. These events raise questions as to the cause of the initial increase in recruitment, whether the subsequent decline in SSB and recruitment are solely the result of a high fishing rate or include a reversal of that cause, and whether the current high F is sustainable.

The period of increased recruitment has been called the gadoid outburst (Cushing, 1980), because the landings, recruitment and SSB of haddock and whiting also increased (Pope & Macer, 1996), although their recruitment pulses were earlier than that of cod and were of shorter duration (Hislop, 1996). There has been much speculation about the cause of the gadoid outburst, but few definitive conclusions. Postulated causes include

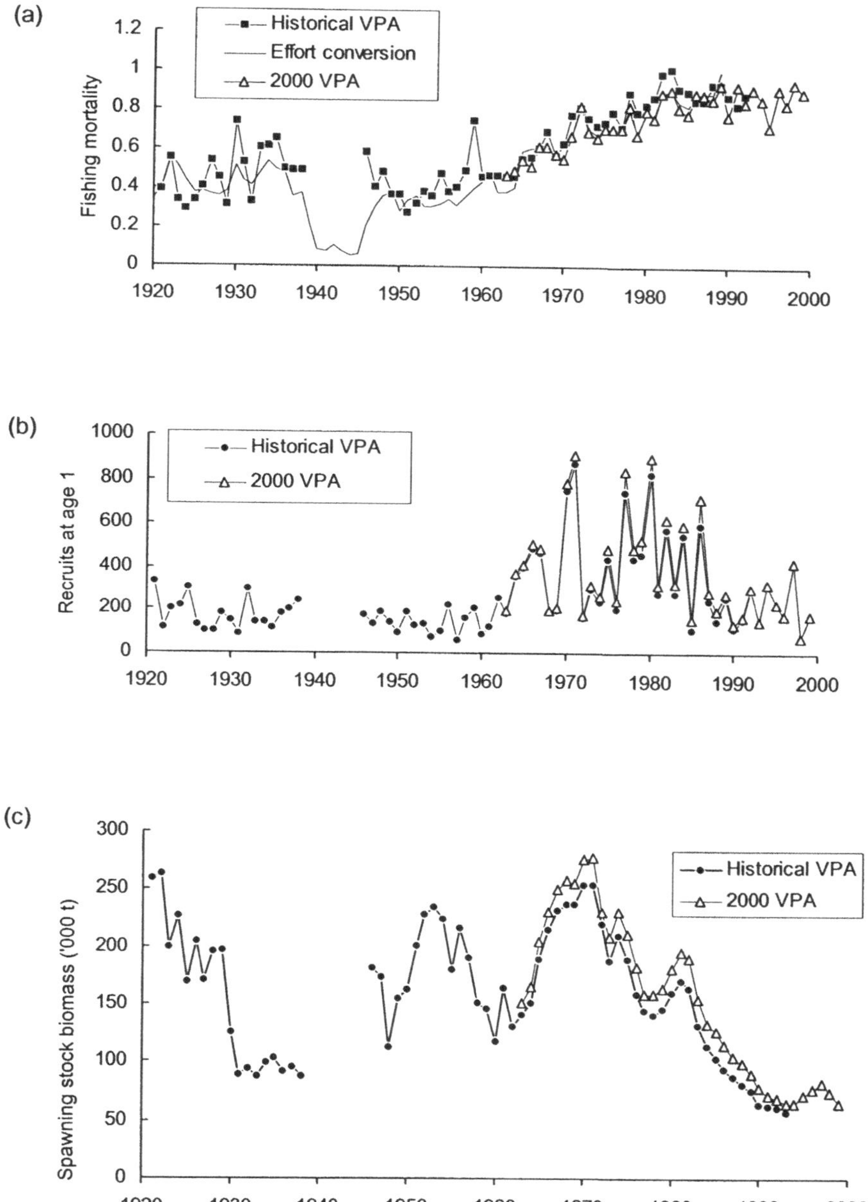

Fig. 5 Long-term trends in (a) fishing mortality, (b) recruitment and (c) spawning stock biomass of North Sea cod. Historical VPA results from Pope & Macer (1996), effort conversion data from Daan *et al.* (1994), and 2000 VPA results from the 2000 ICES XSA assessment (ICES, 2001a).

changes in North Sea pelagic biomass, changes in the timing of production, and changes in temperature as a result of the North Atlantic Oscillation (NAO). Cushing (1980) noted that the gadoid outburst began when the North Sea herring stock was collapsing from 1964 to 1976, following the prolonged overfishing of the 1950s (Burd, 1974, 1978). The pelagic biomass in the North Sea was also reduced by the collapse of North Sea mackerel in the 1970s. Daan *et al.* (1994) and Hislop (1996) pointed out that, because herring prey on gadoid eggs and juveniles (see also Pálsson, 1994) and there is overlap between the diet of gadoid larvae and post-larvae, and adult herring, the collapse of herring could have benefited the survival of cod. On the other hand, Hislop (1996) also noted that the southern and central North Sea herring stocks were much reduced as early as the 1950s, yet cod recruitment did not increase until a decade later. The decline in cod SSB after 1980, however, does coincide with the recovery of the herring stock following the closure of the herring fishery. The idea of an oscillation between pelagic and demersal biomass needs further investigation.

The possibility of productivity changes was raised by Cushing (1984). He suggested that, after 1962, the seasonal peak of *Calanus* abundance in the North Sea was later, perhaps increasing the survival prospects of cod by matching better with the production of cod larvae. There was no evidence for a relationship between the timing of *Calanus* production and haddock recruitment, however. Brander (1992) further examined data on the timing of *Calanus* production using Continuous Plankton Recorder data, but he could not corroborate the link between the copepod production cycle and the recruitment of cod. To account for the subsequent decline in cod recruitment it would have to be postulated that the timing of *Calanus* production has reverted in recent years.

The Fourth ICES/GLOBEC Backward-Facing Workshop undertook a comprehensive review of the long-term changes in North Sea gadoid stocks, plankton, temperature and water transport data (ICES, 2001d). It showed that the extreme recruitment in gadoid stocks in the 1960s occurred when the North Atlantic environment was also extreme, with minimum southwesterly winds, reduced storminess and very weak inflows of Atlantic water into the northern North Sea, all linked to a strong negative phase of the NAO. Conversely, during the 1990s when gadoid recruitment decreased, the NAO became progressively more positive, with increased southwesterly winds, stronger inflows of oceanic water and a rise in sea temperature, approximately 0.75°C in the case of the mean annual surface temperature of the North Sea. The workshop could not specify a causal process because of our poor understanding of how the North Sea ecosystem functions, but the work of Planque & Frédou (1999) suggests that temperature is either a candidate or a proxy for the process. They investigated the relation between interannual variation in recruitment and sea surface temperature in nine cod stocks. The relationship was positive for Arctic stocks, negative for the North Sea, Irish Sea and West of Scotland, and neutral elsewhere. The North Sea, Irish Sea and West of Scotland cod stocks all suffer from a combination of very high fishing mortality and stock depletion (ICES, 2001a), but the most marked decrease in recruitment of all three stocks was at about the same time, in 1987 (Figure 6). Conceivably, the effect of fishing has been compounded by the negative impact on recruitment of increased sea temperature in the last decade or so, when the NAO entered its most positive phase, but the biological mechanism has not been identified.

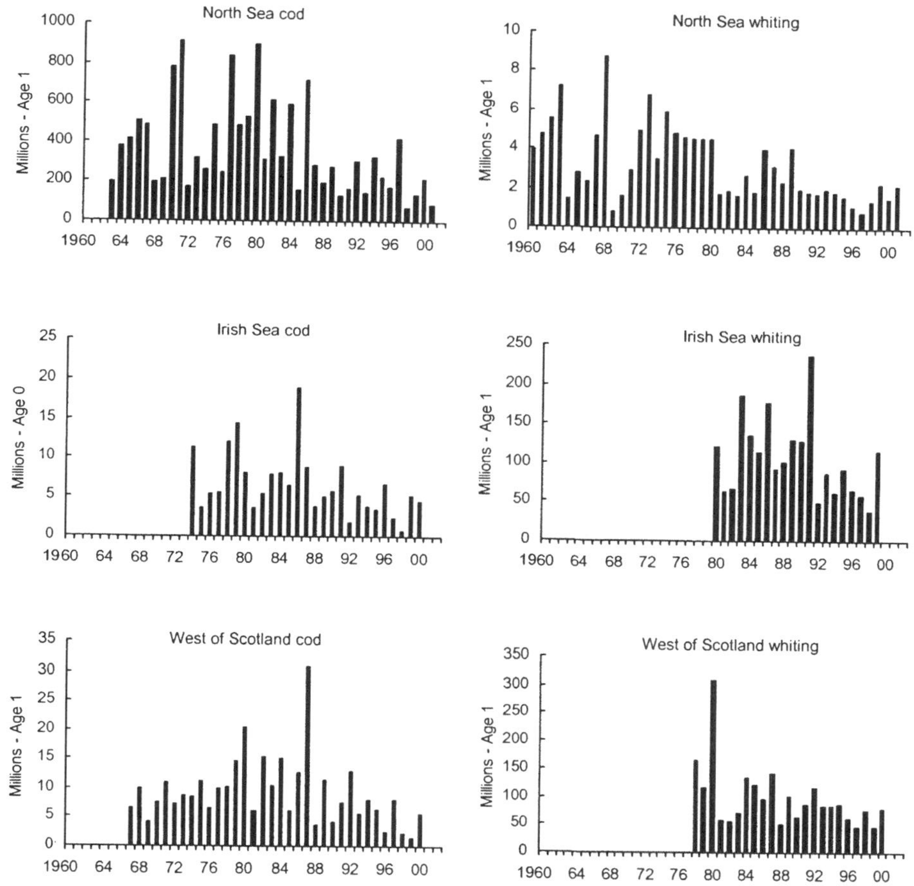

Fig. 6 Contemporary trends in recruitment to the stocks of cod and whiting in the North Sea, Irish Sea and West of Scotland (after ICES, 2001a).

More than a decade ago, Holden (1991) suggested that the decline in the landings and recruitment of cod in the North Sea could be interpreted as a reversion to historical levels following the gadoid outburst. Some catchers currently argue the same point in order to imply that stock decline is not attributable to fishing. The critical difference between history and now is that, even if recruitment has reverted to historical levels, fishing mortality remains at its highest level of 0.9, twice that observed before the gadoid outburst, whereas SSB has fallen below historical levels. The fear of stock collapse and recruit failure is therefore very real (Cook *et al.*, 1997). A survival/replacement analysis carried out by Pope & Macer (1996) suggests that, even if a fishing mortality of 0.8 were to allow the stock to replace itself when average recruitment is high, such a fishing mortality is unsustainable when average recruitment is low. They postulated that F must fall to at least 0.3 to ensure replacement at the current levels of recruitment and SSB. Therefore, whether the recent stock decline results from overfishing alone, or is also attributable to a reversal of the gadoid outburst and or climate change, the present fishing mortality is unsustainable and needs to be reduced substantially.

The prospects for cod recovery not only depend on whether an appropriate plan can be agreed, but on the likely degree of compliance, and whether the stock will respond biologically.

The EU–Norway management agreement requires that fishing mortality on North Sea cod should be adapted to ensure a safe and rapid recovery in SSB to >150 000 t (B_{pa}), and the EU has been negotiating a formal recovery plan for the cod and northern hake stocks since 2001. Proposals for cod (COM (2001) 724 final) consist of a target (to reach the B_{pa} of 150 00 t inside 5–10 years), a recovery trajectory (a 30% annual increase in SSB in the case of cod), and a control rule (a TAC that increases SSB by 33%, subject to a maximum F of F_{pa}, and a maximum permissible yearly change in the TAC of 50%). The SSB target is notionally an increment large enough to be detectable taking into account assessment uncertainty. These proposals are not yet agreed owing to prolonged discussions between the different sectors (country, gear type, directed fisheries, mixed fisheries, and bycatch fisheries), coupled with the underlying objections of many stakeholders to the scientific justification, scale and duration of the proposed plan. The EU and Norway have therefore implemented interim measures. The North Sea cod TAC was reduced from 81 000 t in 2000 to 48 600 t in 2001 (a 50% reduction in F), while an area comprising all or part of 45 statistical rectangles in the eastern and northern North Sea was closed from mid-February to the end of April 2001 in order to curtail fishing for spawning cod (Commission Regulation (EC) 259/2001 of 7 February 2001). Because this closure displaced fishing into more sensitive areas of the North Sea (Rijnsdorp *et al.*, 2001), it was not repeated in 2002, but the 2002 TAC was maintained at 49 300 t. In addition, otter trawlers qualifying as participants in the northern North Sea mixed fishery were required to increase mesh size from 100 to 110 mm for 2002, followed by an increase to 120 mm from 2003 (Council Regulation (EC) 2056/2001 of 19 October 2001). Smaller mesh sizes were retained in the fisheries for saithe, *Nephrops,* and the cod and flatfish fisheries south of 56°N, subject to restrictions on the level of cod bycatch.

The possible route to stock recovery has been investigated several times by modelling. In 1993, stock rehabilitation options were examined by a scientific North Sea Task Force established by the EU following the 1992 ACFM advice to close the fishery. The Task Force simulated the expected effect of spawning season closures, closed areas, increased mesh size and reduced fishing mortality. Results suggested that closing spawning areas in the first quarter of the year, or even closing large areas to cod fishing for the whole year, would not produce significant benefits if the displaced effort were to be redistributed. Taking discarding into account, mesh sizes of 120–160 mm could increase cod SSB in the long term by factors of 1.42–2.79, but anything above 120 mm would severely reduce the retention of haddock and especially whiting. It was estimated that a sustained 25% reduction in fishing mortality could increase cod SSB by a factor of 2.35 based on stock–recruit data, but would take 10 years to achieve (Anon., 1993). In 2001, comparable results were obtained by a new EU expert group, which modelled effort reduction and mesh size scenarios using assessment data for 1999, in this case uncorrected for discarding (STECF, 2001). It was shown that recovery to B_{pa} necessitates a doubling of the cod survival rate from recruitment to spawning age. To achieve this requires a minimum decrease in fishing mortality of 30%, or an increase in the mesh size of the dominant 100 mm otter trawl fleet by 30–40 mm, but as before the latter will significantly reduce the retention of haddock and whiting.

In 2002, STECF experts simulated recovery scenarios based on the SSB and TAC constraints proposed in the EU recovery plan, using a long-term prediction model incorporating the cod stock–recruit relationship, and assumptions about assessment uncertainty (Anon., 2002, STECF, 2002). Predicted recovery times varied from six to nine years, and in some cases longer. Recovery will obviously take even longer if compliance is poor or assessment uncertainty higher than assumed, or if SSB or recruitment deteriorate further in the meantime. Unfortunately, the draft 2002 North Sea cod assessment (ICES, 2003) indicates that, despite the nominal restrictions in 2001, assessed fishing mortality remained high in 2001, and cod SSB_{2002} has deteriorated further to 40 000 t. Recovery time could therefore extend beyond 10 years.

It must be stressed that if and when recovery is achieved, scientists will continue to advise restrictions in order to maintain fishing mortality below F_{pa}. Recovery will not be sustainable if fishing mortality simply returns to current levels.

DISCUSSION

As already noted there has been mounting criticism of various aspects of North Sea cod science by stakeholders, at least in the UK. A primary concern is that the assessment is inaccurate because it works in arrears, and that adolescent and adult cod on the fishing grounds are more abundant than the assessment results derived from catch-at-age and research vessel survey data. Regarding the latter, the evidence from surveys is that, as abundance has declined, the distribution of cod has become increasingly parsimonious, thus increasing the disparity between the mean density and the density in local aggregations (C. D. Darby, pers. comm.), from which catchers tend to extrapolate to the entire stock. Catchers also claim that the decline in abundance in the central North Sea has been overestimated because fishing has shifted farther north to the edge of the Norwegian deep water, citing this as a behavioural response to rising water temperature. A previous investigation of the effect of temperature on the distribution of North Sea cod (Heessen & Daan, 1996) does not provide evidence for such an effect, however. It is more likely that, because of the heavy exploitation and the decline in recruitment, there has been a greater decrease in the abundance of cod in the southern North Sea nursery grounds than on the adolescent and adult grounds in the central and northern North Sea (Heessen, 1993).

A second concern is that the assessments do not account for all factors that affect the cod stock, including climate change, the effect of the 16–30 mm mesh industrial fisheries, seismic surveys and coastal dredging, and consumption by marine mammals. The issue of climate change has already been dealt with, but if temperature does reduce the survival of juvenile cod, it is even more necessary to encourage a higher spawning stock size in order to increase cod egg production and compensate for the combination of overfishing and environmental change.

The concern about the industrial fisheries for sandeel (*Ammodytes* spp.) around the Dogger Bank and for Norway pout (*Trisopterus esmarkii*) in the northern North Sea is that they exploit species that are an important part of the marine food chain, in areas where a large bycatch of juvenile whitefish may also take place. There are already some area–season restrictions on fishing for Norway pout and sandeel. Pout fishing is prohibited in a box off the northeast coast of Scotland, where large whitefish bycatches were formerly taken. In Shetland waters, sandeel fishing is prohibited in spring during the breeding season for seabirds, and an experimental closure of the sandeel fishery in the Wee Bankie area off the Scottish east coast has been in operation since 2000 (Council Regulation (EC) 850/98 of 30 March 1998) in order to assess whether this improves breeding success in local seabird colonies. From a fisheries viewpoint, the estimated catch and species composition of the North Sea small-meshed fisheries is reviewed annually by ICES scientists. The total catch and bycatch of the small-meshed fisheries has declined from 1.86 million t in 1974 to 1.15 million t in 2001, and the estimated gadoid bycatches in 2001 were 6000 t of haddock, 7000 t of whiting and 192 t of cod (ICES, 2003). The haddock and whiting bycatches are incorporated in the haddock and whiting assessments, but fishers question whether bycatch estimates based on sampling a compressed and deteriorating catch at the landing place are valid. Regarding the North Sea sandeel stock itself, the ICES 2000 assessment shows that fishing mortality and SSB have fluctuated without trend over the past 20 years, and that the sandeel fishery and stock have been sustained by regular good year classes (ICES, 2001a). This assessment aggregates data for six or seven different subareas of the North Sea, however, and ICES has pointed out that there is concern about the possibility of localized depletion, but unless the aggregated assessment is seriously in error there is no evidence that the sandeel stocks are in sufficient danger to affect the fish that depend on them for food. ICES collected more than 300 000 fish stomachs from the North Sea in 1981 and 1991 (ICES, 1989, 1997b). These show that, of the total amount of sandeels consumed by predators, only 2% is consumed by cod (as compared to 40% by mackerel), and that this represents <10% of the overall diet of cod (E. D. Bell, pers. comm.), implying that predator-prey interactions between cod and sandeels may not be very strong. The sandeel fishery operating northeast of the Dogger Bank does so in areas where juvenile cod could be settling in summer (Bromley & Kell, 1999), so this requires further investigation.

UK catchers have strong views on the issue of stock recovery. They are sceptical about the possibility of recovery if recruitment has been depressed by climate change, and they have indicated their opposition to implementing additional management measures until benefits have been demonstrated from the current interim measures, including the effects of technical measures and possible vessel decommissioning schemes. An ability to demonstrate whether stocks are actually recovering will depend partly on the severity of the recovery measures and hence on its magnitude and time-scale, partly on the degree of compliance and the accuracy of the landings data, and partly on the problem of assessment uncertainty in the unconverged part of the assessment. In the short term, much may depend on the utility of survey results. Regarding climate change, O'Brien *et al.* (2000) examined the effect of temperature on cod recruitment at low, medium and high levels of SSB, and showed that the temperature effect is weaker at low SSB, implying best survival during the early phases of a recovery programme. Even if future recruitment remains unchanged at the present level, however, the equilibrium yield- and SSB-per-recruit values shown earlier in Table 3 suggest that a change in F from the current value (0.83) to F_{max} (0.23) could increase expected long-term yield by a factor of 1.4, and expected SSB by a factor of 7, a substantial benefit. Such benefits will obviously be reduced if there is a further effect of temperature on recruitment in the future, however.

SUMMARY AND CONCLUSIONS

The combination of recent and historical data shows that a unique feature of the North Sea cod fishery has been a major expansion based on the high recruitment levels generated at the time of the gadoid outburst in the 1970s, later followed by an equally marked decline in SSB and a return in recruitment to previous levels owing to the absence of large year classes. The expansion was unprecedented, but fishing mortality has since remained high, and the present combination of high F, low R and very low SSB is also unprecedented and unsustainable. The gadoid outburst followed the collapse of the North Sea herring stock and coincided with a markedly negative phase of the NAO. The decline in recruitment since 1987 coincides with a severely reduced SSB, recovery of the herring stock, and a switch from a negative to a strongly positive phase of the North Atlantic Oscillation (NAO).

Despite restrictive scientific advice since 1987, including advice in 1992 and 1993 to close the cod fishery, management measures have failed to achieve either a significant reduction in fishing mortality, which is still above F_{lim}, or an improved exploitation pattern. The mixed fishery issue is partly to blame for this because the North Sea mesh size has remained too low to prevent large numbers of cod being caught at ages 2 and 3, so fewer than 5% of one-year-old cod survive to maturity at age 4 or 5. There is also evidence that ICES assessments overestimated stock size and underestimated fishing mortality in the 1990s, so although managers generally adopted TACs intended to be restrictive, they may not have been so in practice. It is also rumoured that non-compliance with the TAC has been a factor in recent years, but this is difficult to pin down.

Despite stakeholder criticisms, there is little doubt that the North Sea cod stock has declined dramatically since the 1980s, whichever assessment is used, and whatever the assessment shortcomings. The principal arguments are about the detailed trend in the last few years. The recently introduced reference points, and the newly established EU–Norway agreement, have inevitably led to stronger scientific advice, resulting in severely reduced TACs since 2001, and tough negotiations about recovery plans, but so far there is no halt to the decline in SSB. The stock therefore remains in danger of collapse and, on the basis of the successful rehabilitation of the North Sea herring stock once the fishery was closed in 1976, it is arguable whether cod recovery can really take place in the absence of a fishery closure.

The differences of view between the fishing industry and scientists on the cod stock and its management are hardly a surprise. As discussed by Payne & Bannister (2003), who compared the management of North Sea cod and herring with that of comparable species in southern Africa, they can be interpreted as a symptom of the exploitation of a shared common property resource, where there is no incentive to accept economic hardship for the common good. It is emerging that a contributory factor may also be that modern-day catchers feel too far outside the assessment and advisory process to understand and identify with its conclusions and proposed solutions. It is arguable that compliance and conservation could be better served if fishers were to embrace the issues by being more closely involved through, for example, science–fisheries partnerships. If nothing else, therefore, the cod crisis has highlighted the issue of stakeholder participation, which is set to achieve greater attention within ICES and the new Common Fisheries Policy. The crisis with stakeholders has also identified a number of scientific challenges that must be addressed by the scientific community. These concern the accuracy and timeliness of the assessments, their sensitivity to the most recent changes in the fishery, the effect on stocks of non-fishery factors, and how well we will be able to monitor stock recovery. It is conceivable that the dialogue between scientists, fishers and managers would be enhanced by the use of simpler, more robust methods that are less demanding of accurate fisheries data.

ACKNOWLEDGEMENTS

Thanks are due to the many colleagues who have assessed the North Sea cod stock over the years, to Keith Brander, Niels Daan, Henk Heessen, Bob Dickson and John Casey, for various insightful views on cod and the ocean, to Robin Cook who worked with me on the UK Directors review, and to the helpful comments of two referees. I thank ex-colleagues John Pope and Tim Macer for use of the data table from their valuable historical analysis. I am indebted to Joe Horwood for his support during my role as an adviser, to members of the UK fishing industry for their ongoing questions about the state of the stocks, and to Andy Payne for his patient encouragement during completion of this manuscript. On the occasion of the CEFAS Symposium, and my impending retirement from fisheries science, I acknowledge H A Cole, Ray Beverton, John Gulland, David Cushing, David Garrod, Brian Jones, Arthur Lee, Tony Burd and Eric Edwards, who all helped and guided my early career, and also the many other national, international and ICES colleagues with whom I have worked over the years. I dedicate the paper to the memory of two very special colleagues, Joop de Veen and Mike Nicholson.

REFERENCES

Anon. (1993) Unpublished Report of the North Sea Task Force, October and November, 1993 (mimeo), submitted to the European Commission, Brussels.

Anon. (2002) Report of a two-day Meeting of Scientists from Norway and the Community on the Evaluation of Harvest Control Rules for North Sea Cod, 18-19 March 2002, Brussels. Appendix to STECF, 2002. 45 pp.

Brander, K. M. (1992) A re-examination of the relationship between cod recruitment and *Calanus finmarchicus* in the North Sea. *ICES Marine Science Symposia*, **195**: 393–401.

Brander, K. M. (1994) Spawning and life history information for North Atlantic cod stocks. Report No. 205, International Council for the Exploration of the Sea, Copenhagen.

Bromley, P. J. & Kell, L. T. (1999). Vertical migration and spatial distribution of pelagic 0-group gadoids (cod, haddock, whiting and Norway pout) prior to and during settlement in the North Sea. *Acta Adriatica*, **40**: 7–17.

Burd, A. C. (1974) The North-East Atlantic herring and the failure of an industry. In: *Sea Fisheries Research*. (Ed. by F. R. Harden Jones), pp. 167–192. Paul Elek (Scientific Books), London.

Burd, A. C. (1978) Long term changes in North Sea herring stocks. *Rapports et Procès-verbaux des Réunion Conseil International pour l'Exploration de la Mer*, **172**: 137–153.

Cook, R. M. (2000) Provisional Assessment of North Sea Cod. Unpublished working paper presented at the North Sea Commission Fisheries Partnership, Peterhead, July 2000.

Cook, R. M., Sinclair, A. & Steffánsson, G. (1997) Potential collapse of North Sea cod stocks. Nature, **385**, 521–522.

Cushing, D. H. (1980) The decline of herring stocks in the North Sea and the gadoid outburst. *Journal du Conseil International pour l'Exploration de la Mer*, **39**: 70–81.

Cushing, D. H. (1984) The gadoid outburst in the North Sea. *Journal du Conseil International pour l'Exploration de la Mer*, **41**: 159–166.

Daan, N. (1978) Changes in cod stocks and cod fisheries in the North Sea. *Rapports et Procès-verbaux des Réunion Conseil International pour l'Exploration de la Mer*, **172**: 39–57.
Daan, N., Heessen, H. J. L. & Pope, J. G. (1994) Changes in the North Sea cod stock during the twentieth century. *ICES Marine Science Symposia*, **198**: 228–243.
Darby, C. D. & Flatman, S. (1994) Virtual Population Analysis: version 3.1. (Windows/DOS) user guide. *Information Technology Series, MAFF Directorate of Fisheries Research, Lowestoft*, **1**. 85 pp.
Garrod, D. J. (1977). The North Atlantic cod. In: *Fish Population Dynamics* (Ed. by J. A. Gulland), pp. 216-239. Wiley, London.
Gulland, J. A. (1965) Estimation of mortality rates. Annex to the Report of the Arctic Fisheries Working Group. *ICES Document*, **C.M. No. 3:1**.
Heessen, H. J. L. (1993) The distribution of cod (*Gadus morhua*) in the North Sea. *NAFO Scientific Council Studies*, **18**: 59–65.
Heessen, H. J. L. & Daan, N. (1996). Cod distribution and temperature in the North Sea. *ICES Marine Science Symposia*, **198**: 244–253.
Hislop, J. R. G. (1996) Changes in North Sea gadoid stocks *ICES Journal of Marine Science*, **53**: 1146–1156.
Holden, M. J. (1991) North Sea cod and haddock stocks in collapse or the end of the "gadoid outburst"? *ICES Document*, **C.M. 1991/G: 41**.
Hutchinson, W. F., Carvalho, G. R. & Rogers, S. I. (2001) Marked genetic structuring in localised spawning populations of cod *Gadus morhua* in the North Sea and adjoining waters, as revealed by microsatellites. *Marine Ecology Progress Series*, **223**: 251–260.
ICES. (1969) Report of the Working Group on Assessment of Demersal Species in the North Sea. *ICES Cooperative Research Report, Series A*, **9**.
ICES. (1989) Database Report of the Stomach Sampling Project 1981. (Ed. by N. Daan). *ICES Cooperative Research Report*, **164**.
ICES. (1992) Report of the Study Group on Ecosystem Effects of Fishing Activities, Copenhagen, April 1992. *ICES Document*, **C.M. 1992/G: 11**.
ICES. (1997a) Report of the Study Group on the Precautionary Approach to Fisheries Management. February 1997. *ICES Document*, **C.M. 1997/Assess: 7**.
ICES. (1997b) Database Report of the Stomach Sampling Project 1991. *ICES Cooperative Research Report*, **219**.
ICES. (1998) Report of the Study Group on the Precautionary Approach to Fisheries Management. February 1998. *ICES Document*, **C.M. 1998/ACFM: 10**.
ICES. (2000) Report of the Working Group on the Assessment of Demersal Stocks in the North Sea and Skagerrak. ICES Headquarters, 11–20 October 1999. *ICES Document*, **C.M. 2000/ACFM: 07**.
ICES. (2001a) Report of the ICES Advisory Committee on Fishery Management, 2000. *ICES Cooperative Research Report*, **242**. 911 pp.
ICES. (2001b) Report of the Working Group on the Assessment of Demersal Stocks in the North Sea and Skagerrak, October, 2000. *ICES Document*, **C.M. 2001/ACFM: 07**.
ICES. (2001c) Report of the ICES Advisory Committee on Fishery Management, 2001. *ICES Cooperative Research Report*, **246**. 557 pp.

ICES. (2001d) Workshop on Gadoid Stocks in the North Sea during the 1960s and 1970s. The Fourth ICES/GLOBEC Backward-Facing Workshop. *ICES Cooperative Research Report*, **244**. 55 pp.
ICES. (2002a) Report of the Working Group on the Assessment of Demersal Stocks in the North Sea and Skagerrak. Hamburg, 19–28 October 2001. *ICES Document*, **C.M. 2002/ACFM: 01**.
ICES. (2003) Report of the Working Group on the Assessment of Demersal Stocks in the North Sea and Skagerrak. ICES Headquarters, 11–20 June, and RIVO Ijmuiden 7–8 October 2002 *ICES Document*, **CM 2003/ ACFM: 02.**
Macer, C. T. & Easey, M. W. (1988) The North Sea cod and the English fishery. *Laboratory Leaflet Ministry of Agriculture, Fisheries and Food, Directorate of Fisheries Research, Lowestoft, UK*, **61**. 22 pp.
Nichols, J. H. (2001) Management of North Sea herring and prospects for the new millenium. In *Herring Expectations for a New Millenium* (Ed. by F. Funk, J. Blackburn, D. Hay, A. J. Paul, R. Stephenson, R. Toresen & D. Witherell), pp. 645–665. Fairbanks; University of Alaska Sea Grant, **AK–SG–01–04**.
O'Brien, C. M., Fox, C. J., Planque, B. & Casey, J. (2000) Climate variability and North Sea cod. *Nature*, **404**: p. 142.
Pálsson, O. K. (1994) A review of the trophic interactions of cod stocks in the North Atlantic. *ICES Marine Science Symppsia*, **198**: 553–575.
Payne, A. I. L. & Bannister, R. C. A. (2003) Science and fisheries management in southern Africa and Europe. *African Journal of Marine Science*, **25**: 1-23.
Planque, B & Frédou, T. (1999) Temperature and the recruitment of Atlantic cod (*Gadus morhua*). *Canadian Journal of Fisheries and Aquatic Sciences*, **56**: 2069–2077.
Pope, J. G. (1972) An investigation of the accuracy of virtual population analysis using cohort analysis. *Research Bulletin International Commission for the Northwest Atlantic Fisheries*, **9**: 65–74.
Pope, J. G. & Macer, C. T. (1996) An evaluation of the stock structure of North Sea cod, haddock, and whiting since 1920, together with a consideration of the impacts of fisheries and predation effects on their biomass and recruitment. *ICES Journal of Marine Science*, **53**: 1157–1169.
Rijnsdorp, A. D., Piet, G. J. & Poos, J. J. (2001) Effort allocation of the Dutch beam trawl fleet in response to a temporarily closed area in the North Sea. *ICES Document*, **C.M. 2001/N: 01**.
Serchuk, F. M. & Grainger, R. J. R. (1992) Development of the basis and form of ICES fisheries management advice: historical background (1976–1990) and the new form of ACFM advice (1991-??). *ICES Document*, **C.M. 1992/Assess: 20**.
Shepherd, J. G, (1992) Extended survivors' analysis: an improved method for the analysis of catch at age data and catch-per-unit-effort data. Working paper 11, ICES Multispecies Assessment Working Group, June 1992, Copenhagen, Denmark: 22 pp.
STECF. (2001) North Sea cod recovery plan. Report of the Scientific Meeting on Improvement of the Selectivity of Fishing Gears, Brussels, 5–9 March 2001 (mimeo).
STECF. (2002) Report of subgroup on review of stocks (SGRST). Scientific, Technical and Economic Committee for Fisheries of the Commission of the European Communities (STECF). Evaluation of Cod Recovery Plans. Brussels 20–22 March 2002. Commission Staff Working Paper. 151 pp (mimeo).

Van Beek, F.A. (2000) A note on the Working Group performance of Short Term Predictions. Working Paper 2 in *ICES Document*, **C.M. 2001/ACFM: 07**. Report of the Working Group on the Assessment of Demersal Stocks in the North Sea and Skagerrak, October 2000.

Van Beek, F. A. & Pastoors, M. A. (1999) Evaluating ICES catch forecasts: the relationships between implied and realised fishing mortality. *ICES Document*, **C.M. 1999/R: 04**.

Managing Arabian Gulf sailfish - issues of transboundary migration

John Hoolihan
Environmental Research and Wildlife Development Agency, P.O. Box 45553, Abu Dhabi, United Arab Emirates.

ABSTRACT: The Arabian Gulf coastline covers eight political entities. Relatively small in geographical dimension, this semi-enclosed basin reduces exclusive economic zones into constrictive juxtaposed boundary areas. These zones are substantially smaller than the familiar "200 mile limit" EEZs characteristic of coasts facing open oceans, exhibiting distance limits approximating 60 nautical miles in the northern Gulf to 15 miles near the Strait of Hormuz. This substantially increases the incidences of straddling and transboundary fish movements across the territorial waters of Gulf nations. The sailfish *Istiophorus platypterus* is illustrated as a case example. Tagging data have shown transboundary migratory patterns of sailfish within the Gulf. The importance of sailfish to the United Arab Emirates (UAE) results primarily from revenue generation in the recreational fishing (and tourism) sector, the need to maintain biodiversity levels for a healthy marine ecosystem and the species' inherent intrinsic value. Rates of tag recapture attributed to transboundary migration into Iranian waters indicate significant incidental sailfish by-catch in drifting gillnets, raising concerns that multiple users may exceed sustainable limits. Federal Law 23 (1999) requires the UAE to manage its fisheries optimally within sustainable limits. Development of any sound management plans will require understanding of migration patterns, habitat preferences and other life history parameters, as well as the participatory input from affected multinational stakeholders. This paper purports to describe early indicators contributing to the possible overexploitation of this Gulf fishery, thus suggesting that in-depth stock assessment and management considerations are warranted.

INTRODUCTION

Overexploitation of fisheries and associated issues has developed into a worldwide concern. The region known both as the Arabian Gulf and the Persian Gulf (hereafter referred to as the Gulf) is no exception. The area, which in the media is most often associated with petroleum production and political conflict, is not immune to the negative impacts of overfishing. Traditionally, and particularly before the advent of petroleum exploration, fishing and pearl collection played a major role in the economic activities of Gulf residents. In terms of size and capacity the Gulf is comparatively small. However, its coastline covers eight political entities. The eastern coastline is made up entirely by

Iran, but the western one includes Iraq in the north, followed in a southward direction by Kuwait, Saudi Arabia, Bahrain, Qatar, the United Arab Emirates (UAE) and Oman. Together, the eight countries share and exploit the Gulf for reasons of transport, fishing and petroleum extraction.

The Gulf is a semi-enclosed basin that opens into the Gulf of Oman via the narrow Strait of Hormuz, the single inflow/outflow exchange point of saline water with the adjoining Arabian Sea. The only other water sources feeding the Gulf originate from freshwater river inflow in the north (e.g. Shatt Al Arab) and limited rainfall. Physically, the dimensions of the Gulf approximate 540 nautical miles length and 182 miles maximum width, narrowing to 30 miles in the Strait of Hormuz. Mean depth is 36 m (Reynolds, 1993). Its confined size constricts the Gulf's nations' exclusive economic zones (EEZs) into juxtaposed boundary areas. The familiar "200 mile limit" of coasts facing open oceans is not possible. Instead, EEZs are limited to approximately 60 miles in the north, but just 15 miles in the Strait of Hormuz. Consequently, many Gulf fish species constitute transboundary or straddling stocks.

Gulf fisheries include a variety of demersal and pelagic species that are important to both commercial and recreational fishing sectors. Included are the highly migratory tuna and tuna-like species of the family Scombridae, of which the five considered most important locally are listed in Table 1. In terms of regional consumption and preference, the narrow-barred Spanish mackerel is arguably the most important of the five species, for all stakeholder nations.

Table 1 Important large pelagic species in the Gulf.

Species	Common name
Scomberomorus commerson	Narrow-barred Spanish mackerel
Scomberomorus guttatus	Indo-Pacific king mackerel
Euthynnus affinis	Kawakawa
Thunnus tonggol	Longtail tuna
Istiophorus platypterus	Sailfish

All these highly migratory species move across political boundaries, i.e. they exhibit transboundary migration. Regional development over the past two decades has increased environmental degradation caused by factors such as increased vessel traffic, coastal development, petroleum extraction and the pollution associated with each. Additionally, an expansion in population size and affluence has resulted in increased fishing effort in terms of manpower, vessels and gear.

At present, a rigorous time-series of catch and effort data for the major species is not available. However, the results of tagging studies and limited landing surveys of Gulf sailfish provide early warning indicators of possible overexploitation from regional fishing practices, which could lead to the exceeding of sustainable harvest limits. Further, these indicators reveal an emerging problem that warrants the careful attention of the management authorities. The purpose of this paper, therefore, is to describe the problem and to suggest possible recourses to alleviate the negative impacts of overexploitation, with particular consideration given to developing cooperative regional management measures.

GULF SAILFISH RESEARCH

Research conducted on sailfish life history has provided some understanding of the significance of Gulf fishing practices, and it is presented here as a case example. Within the UAE, sailfish *Istiophorus platypterus* are an important resource for recreational fishers, the only member of the billfish group residing in the Gulf. Historically, they have never been a preferred fish for human consumption, at least not by locals in the UAE. As a result, the commercial catch is insignificant and the species is less important than some others. However, rapid growth in tourism and coastal development has created an escalating demand for recreational fishing. This statement is particularly true for sailfish, for which most private anglers and charter operations practice catch and release. Sailfish are seasonal residents in UAE waters, being present from around mid-October to mid-April. Known aggregation sites of the species afford world-class sportfishing opportunities, confirmed by the growing number of foreign anglers travelling to the UAE specifically to pursue them.

Sailfish tagging in the UAE began in the early 1980s, initially voluntarily by expatriate recreational fishers. Tags from several different foreign agencies were deployed, but there were few returns. Details of earlier tagging efforts and the associated data were forwarded to the foreign agencies, and therefore the information remained unknown and unavailable to local scientists. In 1998, the Environmental Research and Wildlife Development Agency (ERWDA) of Abu Dhabi, UAE, established a cooperative tagging programme to further study the migratory movements and biology of sailfish. One objective of the programme was to consolidate the tagging and data collection efforts so that information would be retained for local scientists and managers. Tags are distributed *gratis* and a reward scheme is in place to encourage disclosure of recaptures. Subsequent to the onset of ERWDA's cooperative tagging programme, a definitive spring migration was noted. Sailfish disperse from UAE aggregation areas around mid-April and tag-recapture data (Hoolihan, 2001) reveal migration in a NNW direction, i.e. further into the Gulf (Figure 1). Recaptures of the tagged fish are made and reported by Iranian commercial fishers.

Fig. 1 Sailfish tag deployment and recapture areas.

Initial rates of tag recovery indicated something unusual. Of 1871 tags from various agencies deployed in the Gulf, there were 92 recaptures, or 4.91% (Hoolihan, in press). However, the current recapture rate specifically from tags deployed in ERWDA's cooperative tagging programme is nearer 6.9% (94 of 1364 tags). Both these figures are higher than other established sailfish tagging programmes, in which recapture rates have been generally well below 2% (Jones & Prince, 1996; Peel *et al.*, 1996; Bullen *et al.*, 2000; Holts & Prescott, 2001). Additionally, 10 pop-up satellite tags were deployed in UAE waters during spring 2002 to ascertain the summer location of the sailfish population. Iranian fishers caught three (30%) of these fish before the programmed tag release date. Such a high rate of recovery suggests that fishing mortality on sailfish is high.

IRANIAN FISHERIES

During June of 2001 and 2002, I was afforded opportunity to visit Iran with the objective of investigating sailfish biology and fishing practice. During landing-site surveys in 2002, the presence of ripe gonads gave clear indication that the sailfish being caught were about to spawn. The movement away from UAE waters in late spring may well therefore be to spawn, although more conclusive data are needed to confirm this statement. Most of the catch and the associated tag recoveries originate from the Bushehr Province of Iran, extending from approximately 26°30'N to 29°00'N. Bushehr has 786 licensed fishing dhows, traditional wooden vessels some 15 m long (Anon., 1996). The number of dhows has remained relatively constant since 1995, although fishing effort has escalated as a result of greater engine horsepower and gear improvement. From winter until around mid-July these dhows deploy drifting gillnets targeting narrow-barred Spanish mackerel, longtail tuna and kawakawa. Each net is some 2.3 miles long and 28 m deep.

Some sailfish are caught, and although they are considered to be an incidental by-catch, they are not discarded but are rather sold in local markets. Comprehensive recording of landing data only began in 1996, so N. Niamaimandi (Iran Shrimp Research Centre, Bushehr, Iran, pers. comm.) could only provide total landings for the period 1996–2001 for the five main species (Table 2). Iranian sailfish landings initially increased, peaked in 1999, then dropped sharply again, especially in 2001. Such a decline is supported by verbal accounts of UAE recreational anglers, who reported a decrease in sailfish abundance over the same period. For illustrative purposes, reported landings of sailfish are also listed by number in Table 2, using an average weight of 25 kg per fish, the mean weight of sailfish landed in the UAE prior to 2002.

Table 2 Total annual landings by species (metric tons) for Bushehr Province, Iran, 1996–2001.

Species	1996	1997	1998	1999	2000	2001
Narrow-barred Spanish mackerel	1278	1663	1454	1262	4019	2216
Indo-Pacific king mackerel	1359	723	897	608	1835	464
Kawakawa	2876	4761	5578	8119	10216	6555
Longtail tuna	2110	2834	3189	4735	12099	6571
Sailfish	373	328	318	819	401	224
(Numbers)	(14 920)	(13 220)	(12 720)	(32 760)	(16 040)	(8 960)

DISCUSSION

UAE recreational fishers do not normally record or report the total numbers of sailfish observed, so assertions of apparent declines in sailfish numbers cannot be verified. However, when many fishers independently make similar claims, there does appear to be cause for concern and a need for further investigation. Additionally there is the evidence of declining Iranian landings, high rates of tag recovery and burgeoning driftnet fishing effort, so there is reasonable justification for saying that sailfish are being overfished in the Gulf.

It must be stressed that the Iranian figures are for the Bushehr Province only. However, all tag recaptures with the exception of a few from the original tagging locations are from that Province. This seems to suggest that much of the resource moves to that area at that time of year and that catch data from the Bushehr Province are fairly representative of abundance of the resource as a whole. It should be noted in this context that Iranian waters in the southern Gulf leading out of the Strait fall within the Bandar Abbas Province. Similar fishing practices to those described occur in that area, but fish tagged in that province were recaptured in or near the Bushehr Province. Conventional tag recaptures have been limited to the period ending in mid-July, at which time Iranian fishers cease driftnetting and begin to trawl the lucrative shrimp resources. It has therefore not been possible to ascertain whether sailfish remain in the identified capture area throughout summer. ERWDA's pop-up satellite tagging efforts are intended to clarify the situation. Also, to date there has not been a single tag recovery from outside the Gulf, although sailfish are landed in the Iranian ports of Jask and Chabahar. Jask is located just outside the Strait of Hormuz in the Gulf of Oman, and Chabahar lies on Iran's border with Pakistan. The facts that no tags have been recaptured outside the Gulf and that sailfish do spawn inside the Gulf suggest that the population may be resident in the Gulf year-round, which would be highly unusual for a billfish species.

Given the above statement, and its derivative that the sailfish in the Gulf may well constitute a local stock, it may well be that the declining catches and the high rates of tag recovery are indicative of overexploitation of the resource. If this is the case, then measures need to be implemented urgently to prevent collapse of the Gulf sailfish population. Threats to biodiversity and ecosystem balance are also of concern if it is a resident population. Such problems are not unique to the Gulf fishery, and concerns are mounting worldwide for billfish populations under threat from excessive driftnet and longline harvesting. For example, the populations of Atlantic blue and white marlin have declined to levels well below maximum sustainable yield as a result of overfishing (Farber & Venizelos, 2001). As a result, the International Commission for the Conservation of Atlantic Tunas (ICCAT) has responded by conducting rigorous stock assessments and implementing resolutions aimed at rebuilding the stocks (ICCAT Secretariat, 2002).

Although driftnets are effective and economical to operate, they have been criticized for their detrimental impact on non-target species (Northridge, 1991), to the extent that the United Nations has implemented an international ban on the use of "large-scale" driftnets. In particular, recreational fishers are increasingly voicing objections to the reduction of billfish stocks caused by the commercial tuna industry (Peel, 1996). In most commercial fisheries, billfish are considered to be an incidental by-catch. The same is true for Gulf sailfish, and therein lies the problem. Although sailfish in the Gulf have a lesser economic value than the primary targeted species, they do have value and are sold and consumed, rather than discarded. In terms of management requirements, is there really a difference between catch and by-catch of sailfish? One consequence is that less valued species such as sailfish are given a status of diminished importance, and may be neglected (from a management perspective) altogether. It is becoming increasingly apparent (Wallace, 1996) that, as recreational fishers, environmentalists and the general public become aware of the negative impacts of non-selective commercial fishing practices on certain by-catch species, demands to minimize or eliminate the by-catch will increase.

At present the general conception of sailfish value centres on its pursuit as a recreational gamefish. Generated revenues (in terms of tourism, supporting infrastructure, etc) from such activities can be high and, in many cases, greater than those that could be derived from commercial harvesting. For example, a survey of U.S. Atlantic tournament billfish anglers (Fisher & Ditton, 1992) documents overall expenditure in pursuit of billfish as US$5576 for each billfish caught and US$42 565 for each billfish landed. These values exhibit an indication of the level of revenue potentially lost should the resource in the Gulf continue to decrease as a consequence of overexploitation. Aside from the ecological and economic value of sailfish, they also have intrinsic value, i.e. the existence value society places on species regardless of whether or not direct economic benefit accrues from them. Fisheries policies have been notably affected by this phenomenon in many areas of the world, particularly in relation to the by-catch of so-called charismatic species such as turtles, dolphins and seabirds (Jones, 2000). The public's general knowledge of environmental issues has grown significantly in recent years, providing a higher level of understanding of why all species are important in maintaining ecosystem balance. This knowledge, and a desire to protect ecosystems, effectively raises the intrinsic value of all species.

Currently, there are no regulations or shared management arrangements to control the harvesting of sailfish in either Iran or the UAE. Development and implementation of a management regime should be based on sound scientific evidence. Further, as noted earlier, there is a lack of any quantified certainty that overexploitation is taking place. Although suggestion has been made that the recent population decline can be linked with decreased abundance (as opposed to decreased effort or availability), evidence of a decline in catch rate is needed for purposes of verification.

A first step in addressing this fisheries management issue has already been mooted, namely to seek to collaborate with other Gulf states in undertaking a comprehensive stock assessment of the species. To achieve this, a cooperative dialogue among stakeholder nations is necessary to recognize and to take ownership of the problem. Until a scientifically defensible assessment is available, other interim measures aimed at averting catastrophic stock decline warrant consideration, in line with the UN's precautionary principle (Garcia, 1994). For example, the establishment of no-take zones or seasonal closures could prove beneficial, especially in guaranteeing spawning success. However, any actions taken in this regard need to take into account any resulting socio-economic impact on fishers using the resource area.

Ultimately, a shift in approach from the present fisheries management strategy to an ecosystem approach of integrated fisheries management (IFM) should be considered. IFM has been described (Symes, 2000) as supporting a sustainable level of human use by combining the principles of fisheries and ecosystem management, while ensuring ecosystem integrity and natural biological diversity. In Iran's case there is a pre-existing precedent for supporting this type of management style, because that country has ratified the Convention on Biological Diversity and is therefore obliged to work towards achieving its objectives. The issues surrounding Gulf fisheries are further complicated by the breadth of socio-economic conditions and the varying levels of political will in the region, so any successful management plan will have to address the needs of stakeholders, both in terms of exploitation and environmental protection.

Iran's economic development has been hindered in the past as a consequence of political upheaval. Following the Islamic Revolution in 1979, its government initiated concerted efforts to develop the economy. Part of this effort was the establishment of the Iranian Fisheries Research and Training Organization (Shilat), whose prime objectives were to develop fisheries to maximize sustainable yields and to provide industry, employment and a source of protein for impoverished citizens. The results of this initiative are evident along the coast of Bushehr Province, where many small villages and towns are totally reliant upon the fishing industry. These communities incorporate fishers as well as vessel and net manufacturers and canneries. Any attempts to regulate or restrict fishing activities further would most likely be met with stiff opposition from those affected most.

In comparison, the UAE economy was substantially strengthened by the advent of petroleum production and is no longer dependent on fishing for sustenance. That is not to say that fishing is unimportant in the UAE. There is still strong demand for fresh fish and a desire to retain fishing traditions and practices. There is evidence of overfishing in the UAE, but the economic and political structures allow such problems to be addressed in a manner different from that in Iran. For example, trawling has been outlawed in UAE waters for more than 15 years, and the use of driftnets was recently banned to protect stocks from overexploitation. It is unlikely that such measures would be accepted by Iranian fishers at present.

Development of the tourism industry is vigorously supported in the UAE. Based on experience elsewhere, it is likely that the promotion of recreational catch-and-release fisheries for species such as sailfish would have a much greater revenue-generating potential than commercial harvesting, with the added value of bringing in foreign currency. Releasing large gamefish promotes conservation and helps maintain predator-prey ratios in a balanced ecosystem. In contrast, it is doubtful whether Iran can develop this type of tourism, because the infrastructure necessary to attract and support it, including charter fishing operations, is not currently available in the Bushehr Province.

The development of an effective management plan for Gulf sailfish will require cooperation between stakeholders in those countries sharing the resource. Such broad cooperation will also have to be attuned to the limitations of bilateral negotiations solely between the UAE and Iran, because there are six other nations that share the Gulf with those two countries. The best overall approach might be to coordinate shared stock management initiatives through an outside governing body such as the Indian Ocean Tuna Commission (IOTC). First steps in this regard have been taken in Abu Dhabi by drafting a strategy of initiatives, proposed to ERWDA, aimed at developing a comprehensive sailfish management plan. It contains the following key elements:

- continued baseline research to understand life history, migrations, habitat preferences, mortality levels and the spatio-temporal requirements for spawning;
- enlisting the cooperation of the Iranian government to review and reduce the total catch of sailfish;
- bestowing the status of recreational "gamefish" to sailfish in UAE waters, so affording them protection by encouraging catch and release and discouraging commercial interests;

- planning, developing and implementing management policies through the international mediation of the IOTC.

These proposals are currently under review and consideration. It is hoped that their acceptance and implementation will signify a first and major step towards acquiring the sound stock assessment and life history information needed to conserve and protect the species regionally.

ACKNOWLEDGEMENTS

I am indebted to the Iranian Fisheries Research and Training Organization, and particularly to Nassir Niamaimandi, Director of the Iran Shrimp Research Centre, for assisting with tag recoveries and field sampling and for sharing landing data. I also thank the UAE's Environmental Research and Wildlife Development Agency (ERWDA) for supporting the project and the study financially and administratively.

REFERENCES

Anon. (1996) Fisheries statistics (Islamic Republic of Iran). Fisheries Company of Iran (Shilat). Administrative and Programming Deputy Planning and Development Department (PDD). Tehran, Iran. 63 pp.

Bullen, E., Mann, B. & Everett, B. (2000) *Tagging News*. Oceanographic Research Institute. Durban, South Africa. Vol. **13**: p. 5.

Farber, M. I. & Venizelos, A. (2001) An evaluation of U.S. billfish landings in 1999 relative to 1996. *Collective Volume of Scientific Papers, ICCAT*, **53**: 188-197.

Fisher, M. R. & Ditton, R. B. (1992) Characteristics of billfish anglers in the U.S. Atlantic Ocean. *Marine Fisheries Review*, **55**(1): 1-6.

Garcia, S. M. (1994) The precautionary principle: its implications in capture fisheries management. *Ocean and Coastal Management*, 22: 99-125.

Holts, D. B. & Prescott, D. W. (2001) The Southwest Fisheries Science Center's 2001 Billfish Newsletter. National Marine Fisheries Service, National Oceanic and Atmospheric Administration, U.S. Department of Commerce: 3-8.

Hoolihan, J. (2001) Sailfish tagging in the Arabian Gulf. In: Proceedings of the First International Conference on Fisheries, Aquaculture and Environment in the Northwest Indian Ocean (Ed. by S. Goddard, H. S. Al-Oufi, J. McIlwain & M. Claereboudt). Department of Marine Science and Fisheries, College of Agriculture and Marine Sciences, Sultan Qaboos University, Sultanate of Oman. Volume **1**: 86–90.

Hoolihan, J. (in press). Arabian Gulf sailfish movements – a summary of tagging efforts. *Marine and Freshwater Research.*

ICCAT Secretariat. (2002) Compendium of the Management Recommendations and Resolutions Adopted by ICCAT for the Conservation of Atlantic Tunas and Tuna-Like Species. ICCAT Document, **SEC/2002/010**: 75–83.

Jones, C. D. & Prince, E. D. (1996). The cooperative tagging center mark-recapture database for Istiophoridae (1954–1995), with an analysis of the west Atlantic ICCAT billfish tagging program. In: Report of the Third ICCAT Billfish Workshop. International Commission for the Conservation of Atlantic Tunas, Madrid. 311-321.

Jones, P. J. S. (2000) Economic and socio-cultural priorities for marine conservation. In: The Effects of Fishing on Non-Target Species and Habitats – Biological, Conservation and Socio-Economic Issues (Ed. by M. J. Kaiser and S. J. de Groot). The European Commission Fisheries, Agriculture and Agro-Industrial Research Programme (FAIR). Blackwell Science, Oxford, UK. 354-365.

Northridge, S. P. (1991) Driftnet fisheries and their impacts on non-target species; a worldwide review. *FAO Fisheries Technical Paper*, **320**. Rome, F.A.O. 115 pp.

Peel, E. (1996) Billfish bycatch means "bye catch" for sport fishermen. *Big Game Fishing Journal*, **Summer 1996**: 70–72.

Peel, E. M., Rice, J., Ortiz, M. A. & Jones, C. D. (1996) A summary of the Billfish Foundation's tagging program (1990-1996). In: Report of the Third ICCAT Billfish Workshop. International Commission for the Conservation of Atlantic Tunas, Madrid. 323-327.

Reynolds, R. M. (1993) Physical oceanography of the Gulf, Strait of Hormuz, and the Gulf of Oman – results from the Mt Mitchel Expedition. *Marine Pollution Bulletin*, **27**: 35-59.

Symes, D. (2000) Economic and sociocultural priorities for marine conservation. In: The Effects of Fishing on Non-Target Species and Habitats – Biological, Conservation and Socio-Economic Issues (Ed. by M. J. Kaiser & S. J. de Groot). The European Commission Fisheries, Agriculture and Agro-Industrial Research Programme (FAIR). Blackwell Science, Oxford, UK. 366-380.

Wallace, R. K. (1996) Catch and bycatch: is there really a difference? In: Proceedings of the Symposium Fisheries Bycatch; Consequences and Management, August 1996, Dearborn, Michigan. Alaska Sea Grant Report, **97–02**: 77–80.

Report of Discussion Group 1
International approaches to management of shared stocks: fisheries, management and external driver issues

Douglas S. Butterworth* and Kevern L. Cochrane† (Co-Chairs)
** Department of Mathematics and Applied Mathematics,*
University of Cape Town, Private Bag, Rondebosch 7700, South Africa.
† F.A.O., Via delle Terme di Caracalla, 00100 Rome, Italy.
Matthew R. Dunn and Clive J. Fox (Rapporteurs)
CEFAS, Pakefield Road, Lowestoft, Suffolk NR33 0HT, UK

INTRODUCTION

The group identified six broad issues under the heading "Fisheries, management and external driver issues" that commonly hindered progress in effective management of shared stocks:

1. conflicting and/or poorly defined objectives;
2. weak or absent incentives for negotiation;
3. a general absence of agreed allocation formulae;
4. poorly responsive and inflexible management systems and regimes;
5. poorly defined ownership and authority;
6. heterogeneity in funding and scientific and negotiation capacity among the participants in fisheries on shared stocks.

Each of these is discussed in more detail later, but numerous other factors that influence management of shared resources, but which may be considered more generic resource management issues, were also identified. These apply to many fisheries, not just to those that utilize shared fish stocks. These more general factors include the following.

- *The need to apply the precautionary approach in a consistent and balanced way.* It was recognized that the precautionary approach has still not been operationalized and is open to a wide range of interpretations, from an extreme preservationist standpoint to effective disregard. A generally applicable, operational approach to the precautionary approach in fisheries could facilitate responsible and sustainable utilization of fishery resources.
- *Poor compliance with management regulations.* This was seen as a significant problem in many fisheries. There are many reasons for this, including: inadequate and inappropriate stakeholder involvement in fisheries management; deficiencies in

the systems of ownership or user rights; allocations which are, or are perceived to be, inequitable by some users; poor enforcement and inadequate penalties for contravening regulations. While possible solutions to these problems are as diverse as the problems themselves, one approach specifically highlighted by the group was to provide rewards for compliance, for example related to future allocation of access rights.

- *Inappropriate expectations regarding entitlements or benefits*. It was recognized that participants in a fishery frequently had such inappropriate expectations. Many participants saw their own rights as superseding those of other participants and tended to blame others for management failures. Problems of expectation will be encountered particularly where there is uncertainty about access and user rights, e.g. where new entrants are negotiating for access to existing fisheries. These problems need to be resolved within the framework of allocation negotiations, a matter dealt with in more detail below.
- *Poor communication between all stakeholder groups*. This frequently hinders effective management. The problem areas include, for example, communication between the management agency and fishing interests, between different fishing interest groups, and between scientists and managers within the management agency. This problem needs to be addressed by the management agency, which should be responsible for ensuring that the structures, legal frameworks and resources for ensuring effective communication are all present. It was agreed that scientists, as the generators and presenters of particularly important information, must also take responsibility for ensuring that they contribute to effective communication.
- *The role of current public interests, the media and Non-Government Organizations*. The group recognized the potential roles, both positive and negative, that these could play in fisheries management. Again, a good flow of information to the public, together with improved participation, transparency and accountability were seen as essential for ensuring that positive impacts were obtained from involvement with any of these.

THE ROLE OF SCIENCE

At least in part because the Symposium was primarily scientific, the role and importance of science in effective management of fisheries was discussed at length. While this was agreed to be an issue that is common to all fisheries management, not just that of shared stocks, it was considered to be sufficiently important to be included as a substantive issue in the report.

Arguments based upon science are widely used to support advocacy in fisheries, as with other controversial issues. This is a legitimate practice, but one that results in increasing scepticism about scientific results among non-scientists and often has the result of devaluing even unbiased (independent, "agenda-free") science. It is essential for decision-makers to be able to distinguish between unbiased science and the science used for advocacy. It is also important for decision-makers to ensure that they do have access to unbiased scientific information. This may require turning to independent scientists or independent scientific bodies (such as ICES, where scientists are freed from their role as national advisors) to provide such advice.

Pervasive and unavoidable uncertainty in the information provided to decision-makers presents substantial problems. It is now widely accepted, although still not always put into practice, that the uncertainty in assessments must be presented to decision-makers, and that the latter must consider it in their deliberations. However, it was agreed that it was frequently very difficult to communicate uncertainty effectively to non-scientists. The importance of open and ongoing interaction between scientists and decision-makers, to encourage questioning and mutual education, was emphasized.

The group agreed that scientific advice was commonly limited to biological issues, and therefore that decision-makers did not have adequate access to similarly important social and economic information. Among other consequences, this meant that the advice presented often overlooked key social and economic factors, leading (for example) to the proposal and adoption of management solutions that were socially or economically inappropriate. Conversely, one of the common causes of failure in fisheries management is the tendency for decision-makers to over-emphasize socio-economic considerations at the expense of biological ones. For these reasons, it was considered essential to increase the scope of scientific advice to include social and economic dimensions.

It was also recognized that there was frequently a mismatch in the time horizons of managers/decision-makers and of (predominantly biological) scientists. The latter would typically be underpinned by the paradigm of sustainability and so give emphasis to longer time horizons related to the longevity and life cycles of the species under consideration. In contrast, managers/decision-makers normally focus on shorter time horizons, typically one or two years. Both these perspectives, in isolation, miss aspects of the reality of fisheries.

Responsibility for broadening the scope of the scientific advice, considering both short and long-term results and their implications (Figure 1), so making management decisions while well aware of all consequences, was seen to be the responsibility of all groups involved in fisheries management.

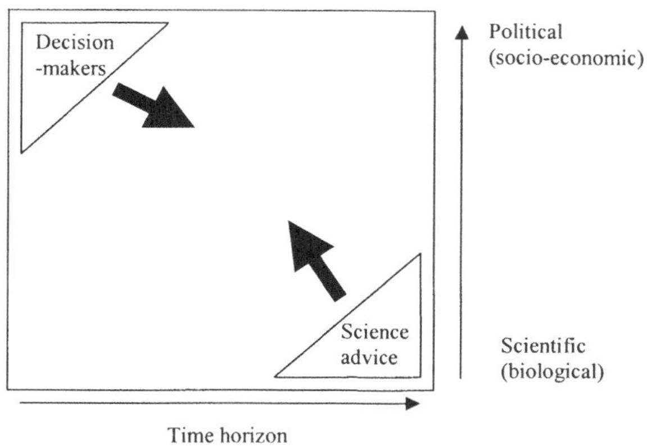

Fig. 1 Relationship between scientific advice, policy and socio-political priorities, and time-scale (note: other stakeholders may fit into this diagram at various locations depending on their interests and perspectives). The arrows show the recommended expansion in perspectives by both functional groups.

Finally, it was considered important to ensure that scientific advice is provided in a timely manner. While scientific knowledge could rarely, if ever, be seen as being final, the principle of ensuring that the best available science is used in decision-making was emphasized.

MAJOR ISSUES RELATED PRIMARILY TO MANAGEMENT OF SHARED STOCKS

These refer to the six broad issues listed above, in the Introduction.

1. CONFLICTING AND/OR POORLY DEFINED OBJECTIVES

This first of the problems considered more specific to fisheries for shared stocks concerned conflicting and/or poorly defined management objectives. While this, again, is pertinent to all fisheries management, the conflicts may be broader and harder to resolve at an international level than when occurring within a national jurisdiction. The problem is exacerbated by the fact that the full underlying objectives for a fishery are rarely specified. Consequently, scientists frequently assume objectives in preparing management options and advice, but their assumptions may be incomplete or even totally wrong. For example, they may mistakenly assume that the objectives for management and negotiation relate only to the value of the resource, whereas in reality the objectives may extend beyond the fisheries themselves. Some examples of underlying objectives presented in the discussions were those of retaining fishing boats as potential minesweepers by the British government following the Second World War, and the generation of interest in fishing to support boat-building in Poland.

Stakeholder and management objectives are frequently conflicting and cannot be achieved optimally and simultaneously. Management should therefore strive to find compromise or otherwise acceptable solutions, and a primary task of science is to present different options to facilitate such solutions. This requires, however, that objectives be explicitly stated.

An additional factor that arises given differing objectives, especially but not exclusively in shared stock fisheries, is that there are frequently disparities in the capacity and power of negotiating parties (see 6. later). This can hinder negotiations and the achievement of genuine compromises. Such inequalities can also influence scientific discussions, and negotiators with access to the greater amount of relevant scientific information are often in a stronger negotiating position.

Transparent and accessible science can assist in the negotiating process. Examples of important scientific information include biological studies to determine stock range, migration paths and mixing rates, all of which could help to clarify resource distribution. Economic studies could indicate the fleet segments that are economically the most viable and/or profitable, assisting negotiations and decision-making concerning, for example, the split of allocations among different fleets. However, it was also recognized that, for many regions and fisheries, such detailed information is not available.

Science can also help to determine the full range and appropriate weights for objectives by demonstrating to managers the potential trade-offs resulting from alternative management decisions. Part of the scientists' role must therefore be to assist in the clarification and definition of objectives, and to evaluate and demonstrate the trade-offs between them. This contrasts with the present situation in many fisheries where scientists are expected to produce a single answer or recommendation. While this relieves decision-makers of responsibility, the likely negative consequences include, *inter alia*, pursuing inappropriate objectives, socio-economic inefficiency and lack of transparency, all of which can lead to management error and failures in compliance.

2. *WEAK OR ABSENT INCENTIVES FOR NEGOTIATION*

The second problem discussed was the lack of motivation for negotiation. If there are no concerns about the status and management, or lack of management, of a fishery, or if the parties believe that they are better off working alone, then there may be no incentive to negotiate or to enforce management measures. Under such circumstances, it may require a crisis of some form before negotiation is taken seriously. However, modern awareness of the non-sustainability of open access and the "tragedy of the commons" have led to a general desire to implement effective management regimes in most fisheries.

With the coming into force of the UN Fish Stocks Agreement on 11 December 2001, States are also under a legal obligation to cooperate in managing shared stocks (Part III, Article 8), possibly through the formation of Regional Fishery Management Organizations.

Where there is a reluctance to cooperate, a role for scientists may be to raise consciousness, and to identify potential problems and encourage negotiation by illustrating that a cooperative solution is better than a non cooperative one. The options will be wider while stocks are still underexploited, which should be an incentive for early negotiation and cooperation in such circumstances. The possible role of game theory in promoting and advising negotiations was emphasized, although it was also recognized that the more standard practice of scenario modelling was also valuable and could prove more flexible.

The present concern about IUU fishing, as reflected in the recent adoption of the F.A.O. International Plan of Action (IPOA) to Prevent, Deter and Eliminate Illegal, Unregulated and Unreported Fishing should also increase motivation for negotiation and cooperation. Paragraph 21 of the IPOA calls for appropriately severe penalties:

"21. States should ensure that sanctions for IUU fishing by vessels and, to the greatest extent possible, nationals under its jurisdiction are of sufficient severity to effectively prevent, deter and eliminate IUU fishing and to deprive offenders of the benefits accruing from such fishing. This may include the adoption of a civil sanction regime based on an administrative penalty scheme. States should ensure the consistent and transparent application of sanctions."

However, it was also suggested that it could be useful to have penalties presented in a positive way to act as incentives, for example by stating that only those who comply will get participatory rights the following year, rather than that those who do not will be penalized by having these rights withheld.

3. *A GENERAL ABSENCE OF AGREED ALLOCATION FORMULAE*

This problem is associated with the absence of agreed objectives, because allocation formulae should reflect the agreed political, social and economic objectives. Without such formulae to guide resource allocation, there are more likely to be conflicts and unsatisfactory solutions, including a lack of equity and a reluctance to compensate those who may sacrifice most in attempting to assist the achievement of greater equitability in allocations.

Article 10, sub-paragraph (b) of the UN Fish Stocks Agreement requires States to " ... *agree, as appropriate, on participatory rights such as allocations of allowable catch or levels of fishing effort;...*". Article 11, dealing with new members or participants, also lists considerations to be taken into account in " ... *determining the nature and extent of participatory right...*". These include:

"*(b) the respective interests, fishing patterns and fishing practices of new and existing members or participants;*
"*(c) the respective contributions of new and existing members or participants to conservation and management of the stocks, to the collection and provision of accurate data and to the conduct of scientific research on the stocks;*
"*(d) the needs of coastal fishing communities which are dependent mainly on fishing for the stocks;*
"*(e) the needs of coastal States whose economies are overwhelmingly dependent on the exploitation of living marine resources; and*
"*(f) the interests of developing States from the subregion or region in whose areas of national jurisdiction the stocks also occur.*"

While qualitative criteria are often employed, the group supported a move towards more precise quantitative criteria, because qualitative criteria are generally not sufficiently precise and are more open to different interpretations. Attention was drawn to the progress being made by ICCAT to develop entry and allocation criteria, and to some of the difficulties that had arisen in attempting to put these into practice.

The potential role of side payments (effectively payment by one party to another for concessions made by the latter) in cases where one or more parties placed a higher value on the resource than did others, was raised. The introduction of side payments in negotiations could increase room for negotiation, facilitating the achievement of an agreed solution.

Where specified allocation formulae were implemented, these should be renegotiated at appropriate intervals. As in the development of objectives, there is a need to identify the real motives of the negotiating parties, so that the achievement of genuine solutions becomes more likely. Where objectives are clearly specified, scientists can contribute to the negotiations by identifying the "envelope" for negotiation. This would include identifying the "areas" where interested parties do not want to be, through establishing a set of constraints. Scientists should also be aware that agreements can trigger other changes, and must try to anticipate unidentified consequences.

Attention was drawn to the F.A.O. Code of Conduct, which states (Article 6, para. 6.1) "*The right to fish carries with it the obligation to do so in a responsible manner...*".

The group recommended a study on the past performance of allocation negotiations, including an analysis of negotiation case histories and an overview of elements in allocation. This study could facilitate a move towards a more formulaic process.

4. POORLY RESPONSIVE AND INFLEXIBLE MANAGEMENT SYSTEMS AND REGIMES

Fish resources and their environment are highly dynamic and can show substantial fluctuations on short time-scales. Management of shared stocks, as with all fish resources, must be able to respond to these changes in, for example, the environment, stock biology, fishing interests, markets, costs of fishing and technology. However, the international decision-making procedures involved in the management of shared stocks are often particularly slow and unresponsive. Scientists must recognize that situations change, and should be able to promote and facilitate the development of arrangements adaptable to change. One of the papers in the current volume (Gavaris and Murawski) entitled "The role and determination of residence proportions for fisheries resources across political boundaries: the Georges Bank example" was cited as a good example of a responsive approach. The paper describes the attempts by Canada and the USA to develop an annual allocation procedure between the two countries for some species caught on the Georges Bank, based on the proportions distributed in each country's EEZ.

5. POORLY DEFINED OWNERSHIP AND AUTHORITY

Poorly defined ownership and allocation rights continue to plague fisheries, and particular problems arise with shared stocks. In accordance with the UN Law of the Sea, the State has jurisdiction over its own EEZ and can allocate ownership or access to resources within this jurisdiction at its own discretion. The UN Fish Stocks Agreement sets out, *inter alia*, to address the problem of managing stocks extending beyond or outside single EEZs, specifically by aiming " *... to ensure the long-term conservation and sustainable use of straddling fish stocks and highly migratory fish stocks through effective implementation of the relevant provisions of the Convention*." (Article 2: Objective. Our underlining. "Convention" refers to the UN Law of the Sea). Article 8 of the UN Fish Stocks Agreement calls on "Coastal States and States fishing on the high seas" to cooperate. However, as discussed above, this Agreement, in its Articles 10 and 11, gives only very general guidelines on participatory rights such as allocations.

The group discussed methods of allocation and referred to the Individual Transferable Quota approach that relies on market forces for allocation and has been implemented in a number of countries (e.g. New Zealand, Iceland). It was recognized that dependence on the market may not be appropriate for some objectives, e.g. equity or ensuring regional balance in fisheries, and that some countries have therefore applied other allocation systems (e.g. Norway, Namibia and South Africa). The same considerations apply on the high seas. Present pressures by new entrants, including coastal states, to participate in already fully exploited fisheries requires urgent resolution of the problems of defining ownership in some cases (e.g. International Commission for the Conservation of Atlantic Tunas, ICCAT, and the Commission for the Conservation of Southern Bluefin Tuna, CCSBT).

6. *HETEROGENEITY IN FUNDING AND SCIENTIFIC AND NEGOTIATION CAPACITY AMONG THE PARTICIPANTS IN FISHERIES ON SHARED STOCKS*

The management of shared stocks is frequently complicated by marked heterogeneity in financial, scientific and management capacities among participants. These heterogeneities may also influence negotiations, affecting the likelihood of achieving genuine solutions. The problem may be most acute where developed and developing countries are participating in the same fisheries.

Part 7 of the UN Fish Stocks Agreement recognizes the special requirements of developing States (Article 24), including their greater vulnerability as well as the particular need to avoid adverse impacts on subsistence, small-scale and artisanal fishers, and on fishworkers and indigenous people. Article 25 calls on States to cooperate in assisting developing States, including through financial assistance, in human resource development, technical assistance, and transfer of technology. However, the group noted that, although the Code of Conduct had a similar requirement (Article 5), the response of donor countries and other relevant organizations and institutions to the requirements of this Article was considered by many developing countries to have been inadequate.

The group recognized that adequate implementation of the provisions of these Articles may be essential for successful management. Scientists can play a part in addressing some of the disparities by drawing attention to them, especially as regards scientific ability, and attempting to resolve the disparity by ensuring the free transfer of data and expertise. This would empower developing countries to be able to participate more effectively in negotiations, to the ultimate benefit of all.

Report of Discussion Group 2
International approaches to management of shared stocks: ecosystem, competition and behavioural issues

Geoffrey P. Kirkwood* and John G. Pope† (Co-chairs)
Renewable Resources Assessment Group, Imperial College, Royal School of Mines, Prince Consort Road, London SW7 2BP, UK
† NRC (Europe) Ltd, The Old Rectory, Staithe Road, Burgh St Peter, Norfolk NR34 0BT, UK
Ewen Bell and John Casey (Rapporteurs)
CEFAS, Pakefield Road, Lowestoft, Suffolk NR33 0HT, UK

INTRODUCTION

The group discussions focused on ecosystem issues in the approach to management of shared stocks, and especially on scientific issues and on how science can contribute to achieving management objectives. While recognizing that issues relating to human competition and behaviour are important in achieving any objectives, the group felt that both were primarily non-scientific issues.

The group noted that man's ability to manage an ecosystem was limited, and that only human activities and their first- and perhaps second-order impacts could be managed, rather than the ecosystem itself. The terminology "ecosystem management" was considered to be misleading, so the more appropriate term "ecosystem approach to management" was adopted for the purpose of the discussion.

Traditionally, fisheries management has set objectives that have tended to be stock-specific and has aimed at managing the population size and fishing mortality rate exerted on each stock. More recently, the scientific problems of setting objectives that consider technical and biological interactions between target fisheries has been addressed (e.g. the ICES Multispecies Working Group has addressed the issues), and some understanding has also been achieved of the impacts of fishing on non-target species and on holistic aspects of the ecosystem. However, progress in achieving objectives that address the inevitable interactions between fishing, its target species and the wider ecosystem has been slow, even in systems under single national jurisdictions. Problems of management are compounded in shared systems where one nation's by-catch is another nation's target. An ecosystem approach to management would inevitably need to have wider agreed objectives. Likely it would be easier to agree on the pessima to be avoided (extirpation of species, destruction of key habitat) than the optima to be achieved (overall system yield or social/economic benefits). Management to avoid pessima would primarily involve managing incidental mortality on other species and mitigating the fishery impacts on non-target species and habitats.

The group recognized that definition of clear management objectives would be critical to the successful development of an ecosystem approach to management. Indeed, it would be certain that each country and each stakeholder would have a different perception of management targets, and that reaching a consensus between the groups would be one of the major challenges to implementation. Under traditional single-species fishery management, the set of main stakeholder groups would be fairly small (i.e. fishers, fisheries managers, scientists). Moreover, an ecosystem approach to management would entail broadening the stakeholder base to include, at least, Non-Government Organizations (NGOs), local community representation, and leisure activity groups (e.g. sport anglers, divers, others using the sea for leisure). In order to achieve cooperation between groups with such diverse backgrounds, programmes of education on the complex structure and function of marine ecosystems would be required. Even after common management objectives had been defined, stakeholders may prefer different, incompatible approaches for meeting the set objectives. An agreed definition of operational objectives, i.e. coherent methodologies for attaining management objectives, would be required first. It seems that these need to have two levels. The first would be concerned with avoiding irreversible harm to elements of the ecosystem. The second and more difficult would be to agree how human benefits could be optimized. Even if this second level could be agreed, it seemed likely that questions of allocation would remain.

AN ECOSYSTEM APPROACH TO MANAGEMENT

Under an ecosystem approach to management, scientists would have a number of responsibilities, as follows:

- to define the relevant ecosystem issues;
- to help clarify ecosystem objectives;
- to maintain and develop systems to monitor the health of ecosystem components and their functions.

Within any discussion of ecosystems, one question that inevitably arises is "what IS an ecosystem and how can it be defined operationally?" Evidently, this question and the provision of a solution will remain fundamental challenges to all stakeholders seeking the adoption of an ecosystem approach to management. To a large extent, scale (spatial and temporal) is the most important factor in defining ecosystems, and the boundaries are likely to prove difficult to determine. For instance, fisheries are generally described as operating on stocks, and the operational definition of a stock is primarily a spatial issue, of particular importance when considering highly migratory or straddling stocks.

Consider an example where juveniles and adults of a species are geographically separated into the coastal waters of two states. Strategies for minimizing the human impact of fisheries on stocks and the ecosystem are likely to favour a fishery on the adult population. This may place one nation at an economic disadvantage, even though it is possible that a greater economic return may accrue to a fishery on the juvenile component if there exists a price differential in favour of juveniles. This is an instance where the setting of operational objectives, such as minimization of economic cost to any one state, may be useful. This illustrates a competition issue that will affect the operational management of the fishery and its resulting impact on the ecosystem.

Once some consensus has been achieved, new metrics for the measurement of human impacts would need to be devised. Currently, impacts of fisheries are largely measured in terms of fishing mortality upon target stocks. In an ecosystem context, such single-issue metrics are unlikely to be sufficient, and new metrics would have to be developed to encompass a much larger ecological scale, possibly including quantification of habitat, biodiversity and trophic level. In addition to these metrics, reference points would need to be set in order to guide the appropriate objectives.

Science would also play a major part in monitoring ecosystems and, within fisheries, this would entail collection of data for a wider range of species and at different spatial and temporal scales than currently undertaken. In the case of larger commercial fishing vessels, this may simply imply an increase in scientific staff time, or the implementation of new technologies (e.g. video monitoring of hauls). It may also be possible to use commercial fishing vessels to increase the spatial and temporal scale of acoustic data for use in fisheries and habitat monitoring. Increased monitoring of smaller vessels, especially in artisanal fisheries, presents a completely different set of problems, in that often there is simply no spare space on board the vessels, and the number of vessels is invariably much greater. Solutions to such problems would need to be found; one possibility would be the education and enablement of fishing crews to monitor their own activities, possibly with financial incentives, such as direct payment of a crew member for data collection. Evidently, monitoring more species on a wider scale would always incur increased financial cost.

Quantification of the indirect impacts of fishing may be assisted by on-board observations and local knowledge of fishers, but there would remain a major role for scientists and dedicated research cruises. Although long time-series of data are essential for ecological studies, the rapid development of human activities prevents data collection on pristine environments. Therefore, new ecological studies may be served best by comparative experimentation through comparing different areas with a range of different exploitation rates, though not at the expense of cutting existing long-term monitoring programmes or establishing new programmes. There is also likely to be an increased need for the classification of habitat sensitivity, mapping and GIS-based approaches. The value of closed areas to fishing (marine reserves or Marine Protected Areas) have the potential to provide reference areas for habitat and biodiversity monitoring, in addition to being a management tool in their own right.

There would be increased scope for a socio-economic study of fishing practices. Whereas fishing in some areas is very much an economics-driven activity, sociological drivers are more important for some, e.g. coastal artisanal fisheries such as in some Pacific islands. In a sociologically driven case, moves to regulate fishery impacts through financial incentives may have little effect.

Application of the precautionary principle to an ecosystem approach to management would require the burden of proof regarding new fishing gears to be more equally shared between science and fishers. In other words, fishers would strive to improve the efficiency of their catching capacity through, *inter alia*, technological development of fishing gears, whereas management would strive to manage the rate of fishing mortality, perhaps through making gear more selective which, from the catcher's perspective, may mean less efficient.

Underlying all issues regarding human impact upon the marine ecosystem are environmental influences. It is important for the credibility of science that scientists are able to distinguish between human and environmental influences. This inevitably means maintaining or increasing monitoring and research, and hence increased costs.

An ecosystem approach to management seems to be developing in special cases that can be sold politically. For example, management of the Alaska pollock fishery is primarily governed by the by-catch of halibut. Furthermore, the Commission for the Conservation of Antarctic Marine Living Resources (CCAMLR) has the concept of ecosystem management enshrined in its convention and management practices.

In summary, it unfortunately remains clear that, to date, progress in developing an ecosystem approach to management has generally been slow. Simply, the issues to be considered are too large and complex. Notwithstanding, the political, scientific and environmental drivers behind development of the approach are already strong and perhaps growing stronger by the day. Therefore, there can be absolutely no doubt that the contribution of science will be fundamental to the successful development of such an approach.

Index

Printed and bound in the UK by
CPI Antony Rowe, Eastbourne

Printed and bound by CPI Group (UK) Ltd, Croydon, CR0 4YY

17/09/2026

14964189-0001